Advanced Green Technology for Environmental Sustainability and Circular Economy

This book elucidates the growing application of greener technology with a circular economic approach and examines the connection among environment, economy, and ecology for an emerging and supportable human society. It focuses on numerous features of environmental sustainability and, more responsibly, labels the technologies and methods essential to overcome growing environmental challenges, including biotechnological methods, cutting-edge research, applications, and procedures.

Features:

- Proposes the latest advances in waste treatment, pollution reduction, and circular economy development based on green technology.
- Considers the relationship between green technological progress and various forms of circular economy.
- Describes resource recycling and recovery.
- Covers advanced technology in bioremediation.
- Includes reports and case studies highlighting the "how-to" on waste-to-energy generation.

This book is aimed at professionals and graduate students in environmental engineering, project management, bioremediation, sustainable development, and waste management.

Maulin P. Shah is Chief Scientist and Head of the Industrial Wastewater Research Lab, Division of Applied and Environmental Microbiology Lab at Enviro Technology Ltd., Ankleshwar, Gujarat, India. His work focuses on the impact of industrial pollution on the microbial diversity of wastewater following cultivation-dependent and cultivation-independent analysis. His major work involves isolation, screening, identification, and genetically engineering high-impact microbes for the degradation of hazardous materials. His research interests include biological wastewater treatment, environmental microbiology, biodegradation, bioremediation, and phytoremediation of environmental pollutants from industrial wastewater. He has published more than 250 research papers in national and international journals of repute on various aspects of microbial biodegradation and bioremediation of environmental pollutants. He is the editor of more than 120 books of international repute and is an active editorial board member in top-rated journals.

Alok Prasad Das serves as Assistant Professor in the Department of Life Sciences, Rama Devi Women's University, Odisha, India. Before that, he was an assistant professor in the Department of Chemical and Polymer Engineering, Tripura Central University, Agartala, India. He has more than 15 years of research experience. His research and innovation work, done in national and international collaborations, has appeared in over 100 publications, and his contributions have been recognized with several awards. His areas of expertise include biological wastewater treatment, bioremediation-based bioleaching of mining pollutants, bioremediation of microplastics and microfiber pollutants, and geomicrobiology of extreme habitats. He is the editor of several books of international repute, and he is an active lead guest editorial board member in top-rated journals.

Greener Technologies for Sustainable Industry and Environment

Series Editors
Pradeep Verma and Maulin P. Shah

Greener technologies, such as microbial-based approaches and sustainable technologies with low net carbon output for energy generation, chemical production, bioremediation, agriculture, and so forth, are preferable to less green alternatives. This series attempts to provide space for scientists, chemical engineers, chemists, academicians, industrialists, and environmentalists to bring out the best of the literature in their area of expertise for developing future sustainable industry and environment.

Algae Refinery
Up- and Down-Stream Process
Edited by Sanjeet Mehariya and Pradeep Verma

Advanced Green Technology for Environmental Sustainability and Circular Economy
Edited by Maulin P. Shah and Alok Prasad Das

For more information about this series, please visit: www.routledge.com/Greener-Technologies-For-Sustainable-Industry-And-Environment/book-series/GTSIE

Advanced Green Technology for Environmental Sustainability and Circular Economy

Edited by
Maulin P. Shah and Alok Prasad Das

CRC Press is an imprint of the
Taylor & Francis Group, an **informa** business

Designed cover image: Shutterstock, Darunrat Wongsuvan

First edition published 2025
by CRC Press
2385 NW Executive Center Drive, Suite 320, Boca Raton FL 33431

and by CRC Press
4 Park Square, Milton Park, Abingdon, Oxon, OX14 4RN

CRC Press is an imprint of Taylor & Francis Group, LLC

Library of Congress Cataloging-in-Publication Data
Names: Shah, Maulin P., editor. | Das, Alok Prasad, editor.
Title: Advanced green technology for environmental sustainability and circular economy / edited by Maulin P. Shah and Alok Prasad Das.
Description: First edition. | Boca Raton, FL : CRC Press, 2025. | Series: Greener technologies for sustainable industry and environment | Includes bibliographical references and index.
Identifiers: LCCN 2024018016 (print) | LCCN 2024018017 (ebook) | ISBN 9781032527925 (hbk) | ISBN 9781032852096 (pbk) | ISBN 9781003517108 (ebk)
Subjects: LCSH: Sanitary engineering. | Environmental engineering. | Sustainable engineering.
Classification: LCC TD7 .A38 2025 (print) | LCC TD7 (ebook) | DDC 628—dc23/eng/20240801
LC record available at https://lccn.loc.gov/2024018016
LC ebook record available at https://lccn.loc.gov/2024018017

ISBN: 978-1-032-52792-5 (hbk)
ISBN: 978-1-032-85209-6 (pbk)
ISBN: 978-1-003-51710-8 (ebk)

DOI: 10.1201/9781003517108

Typeset in Times LT Std
by Apex CoVantage, LLC

Contents

Preface vii
About the Editors ix
List of Contributors xi
Acknowledgments xv

Chapter 1 Comparative Study of Ansys and Aspen Plus Simulation of a Typical Fluidized Bed Column–Advanced Greener Approach for Environmental Sustainability 1

Sourab Baidya, Harjeet Nath, and Abanti Sahoo

Chapter 2 Surfactant and Potable Water Recovery from Gray Water Using Integrated UF and RO Membranes – A Typical Method for Waste Recycling, Reusing, and Recovery of Value-Added Products 18

Harjeet Nath and Sukanta Reang

Chapter 3 *Withania Somnifera*: A Rasayana Herb for Sustainable Management of Human Health and Overall Well-Being 33

Jackson Sugunakara Chary and Anuradha Sharma

Chapter 4 Microbial Nexus for Converting Dairy Wastewater into Value-Added Products 48

Shaon Ray Chaudhuri, Mandakini Gogoi, Ajoy Modak, and Sujan Das

Chapter 5 Novel Microbial Techniques for Pollutant Environment: Their Principles, Advantages, Limitations, and Future Prospects 64

Vikas Kumar, Preeti Pallavi, and Sangeeta Raut

Chapter 6 Microbial Biopolymer from Renewable Feedstock: A Sustainable Approach for the Production of Polyhydroxyalkanoates (PHAs) 89

Shiva Aley Acharjee and Pranjal Bharali

Chapter 7 A Sanitary Engineering Assessment of a Local Community Wastewater Plan for Nitrogen Removal: A Case Study Highlighting Decision-Making Tactics 112

J. T. Tanacredi, R. Reynolds, R. Nuzzi, and R. C. Tollefsen

Chapter 8 Bioremediation by Using Green Nanotechnologies: A Novel Biological Approach for Environmental Cleanup 133

Maheswari Behera, Sunanda Mishra, Prateek Ranjan Behera, Biswajita Pradhan, Debasis Dash, and Lakshmi Singh

Chapter 9 Energy Recovery from Municipal Solid Waste through Pyrolysis Techniques 149

Mukta Mayee Kumbhar, Sunanda Swain, Prajna Sarita Sethy, and Dilpreet Kaur

Chapter 10 Endophytic Microbes in Agarwood Oil Production from *Aquilaria malaccensis* Lam. Engendering Bio-Resources for Socioeconomic Development 168

Bipul Das Chowdhury, Abhijit Bhattacharjee, and Bimal Debnath

Chapter 11 Recent Trends in Conversion of Agro-Waste to Value-Added Green Products 196

Deepika Devadarshini, Swati Samal, Pradip Kumar Jena, and Deviprasad Samantaray

Chapter 12 Green Inventory and Carbon Emissions: A Review 208

Bhabani Shankar Mohanty

Chapter 13 Biological System for Waste Management and Alternative Biofuel Production 223

Oindrila Gupta, Srishti Joshi, Shweta Shukla, and Satarupa Banerjee

Chapter 14 Syngas Production from Lignocellulosic Biomass and Its Applications 242

Diptimayee Padhi, Deepika Devadarshini, Sunanda Mishra, Deviprasad Samantaray, Mahendra Kumar Mohanty, and Pradip Kumar Jena

Index 259

Preface

The Earth's environment is a natural ecosystem that has got its own remedial control. Moreover, the current amount of industrial and man-made activities is extraordinarily gripping the human civilization to search for scientific solutions for the ever-increasing environmental problems for a sustainable and realistic resolution. The present rate of resource exploitation and depletion outlines the destructive ecosystems patterns. The universal environmental complications require a greener solution underlying unity between the global environmental problems irrespective of their intricacy and executing solutions. Integrating waste management through environmental sustainability and economic development is one of the prime milestones in a circular economy. This book aims to publish both original research and to review manuscripts reporting on advanced developed greener technologies and sustainable modification in current technologies for environmental pollution remediation; the management of industrial, mining, and agricultural waste through circular economic approach; and the successful conversion of biomass to biofuels for bioenergy generation. This book delivers a complete understanding of the current developments in greener technologies for endorsing environmental sustainability, edited by experts in the field.

Dr. Maulin P. Shah
Dr. Alok Prasad Das

About the Editors

Maulin P. Shah is the chief scientist and head of the Industrial Waste Water Research Lab, Division of Applied and Environmental Microbiology Lab, at Enviro Technology Ltd., Ankleshwar, Gujarat, India. His work focuses on the impact of industrial pollution on the microbial diversity of wastewater following cultivation-dependent and cultivation-independent analysis. His major work involves isolation, screening, identification, and genetically engineering high-impact microbes for the degradation of hazardous materials. His research interests include biological wastewater treatment, environmental microbiology, biodegradation, bioremediation, and phytoremediation of environmental pollutants from industrial wastewaters. He has published more than 250 research papers in national and international journals of repute on various aspects of microbial biodegradation and bioremediation of environmental pollutants. He is the editor of more than 120 books of international repute (Elsevier, Springer, RSC, and CRC Press). He is an active editorial board member in top-rated journals.

Alok Prasad Das is presently serving as an assistant professor in the Department of Life Sciences, Rama Devi Women's University, Odisha, India. He is an academician; the author of journal articles, several books, and book chapters; and an editorial member of several reputed journals. He has more than 15 years of research experience. His area of expertise includes wastewater treatment, bioremediation, environmental pollution and its sustainable management, microplastic and microfiber pollution and its bioremediation, and geomicrobiology. In addition to this, he also investigates on biosensors for rapid endotoxin detection in fluid systems used for the production of clinically applicable compounds and the development of simple single-step chromogenic methodology for rapid detection of food pathogens and toxins. He is the editor of six books in Elsevier, Springer, and CRC Press. He has published more than 100 research and review articles in international journals of repute. He is an active lead guest editorial board member in top-rated journals. His current Google Scholar Citations is 2,560, and h-index 28.

Contributors

Acharjee, Shiva Aley
Applied Environmental Microbial Biotechnology Laboratory
Department of Environmental Science
Nagaland University
Hq Lumami, Zunheboto-798627
Nagaland, India

Baidya, Sourab
Department of Chemical and Polymer Engineering
Tripura University, Agartala
Tripura – 799022, India

Banerjee, Satarupa
Department of Biotechnology
School of Biosciences and Technology
Vellore Institute of Technology
Vellore-632014, Tamil Nadu, India

Behera, Maheswari
Department of Botany
College of Basic Science and Humanities
Odisha University of Agriculture and Technology
Bhubaneswar-751003

Behera, Prateek Ranjan
Department of Plant Pathology
College of Agriculture
Odisha University of Agriculture and Technology
Bhubaneswar-751003

Bharali, Pranjal
Applied Environmental Microbial Biotechnology Laboratory
Department of Environmental Science
Nagaland University
Hq-Lumami, Zunheboto-798627
Nagaland, India

Bhattacharjee, Abhijit
Department of Forestry and Biodiversity
Tripura University, Suryamaninagar
Tripura, India

Chowdhury, Bipul Das
Department of Forestry and Biodiversity
Tripura University, Suryamaninagar
Tripura, India

Das, Sujan
Microbial Technology Laboratory
Department of Microbiology
Tripura University, Suryamaninagar
Tripura West-799022, India

Dash, Debasis
Department of Botany
College of Basic Science and Humanities
Odisha University of Agriculture and Technology
Bhubaneswar-751003

Debnath, Bimal
Associate Professor
Department of Forestry and Biodiversity
Tripura University, Suryamaninagar
Tripura West-799022, India

Devadarshini, Deepika
Department of Microbiology
CBSH, OUAT
Bhubaneswar-3, Odisha, India

Gogoi, Mandakini
Microbial Technology Laboratory
Department of Microbiology
Tripura University,
Suryamaninagar 799022,
Tripura, India

Gupta, Oindrila
Department of Biotechnology
School of Biosciences and Technology
Vellore Institute of Technology
Vellore-632014, Tamil Nadu, India

Jena, Pradip Kumar
Department of Chemistry
CBSH, OUAT
Bhubaneswar-3, Odisha, India

Joshi, Srishti
Department of Biotechnology
School of Biosciences and Technology
Vellore Institute of Technology
Vellore-632014, Tamil Nadu, India

Kaur, Dilpreet
Department of Zoology
Maharana Pratap Government P. G. College
Hardoi-241001, Uttar Pradesh

Kumar, Vikas
Centre for Biotechnology
School of Pharmaceutical Sciences
Siksha "O" Anusandhan (Deemed-to-Be University) Bhubaneswar-751003,
Odisha, India

Kumbhar, Mukta Mayee
Department of Life Science
Rama Devi Women's University
Bhubaneswar-751022, Odisha

Mishra, Sunanda
Department of Botany
College of Basic Science and Humanities
Odisha University of Agriculture and Technology
Bhubaneswar-751003

Modak, Ajoy
Microbial Technology Laboratory
Department of Microbiology
Tripura University
Suryamaninagar,
Tripura West-799022, India

Bhabani Shankar Mohanty
Department of Statistics and Applied Mathematics
Central University of Tamil Nadu
Thiruvarur, India

Mohanty, Mahendra Kumar
College of Agricultural Engineering and Technology
Odisha University of Agriculture and Technology
Bhubaneswar

Nath, Harjeet
Department of Chemical and Polymer Engineering
Tripura University (A Central University)
Agartala, Tripura-799022, India

Nuzzi, R.
RN Environmental
Hampton, New York

Padhi, Diptimayee
Amity Institute of Biotechnology
Uttar Pradesh, India

Pradhan, Biswajita
School of Biological Sciences
AIPH University
Bhubaneswar-752101

Preeti, Pallavi
Centre for Biotechnology
School of Pharmaceutical Sciences
Siksha "O" Anusandhan (Deemed-to-Be University)
Bhubaneswar-751003, Odisha, India

Raut, Sangeeta
Centre for Biotechnology
School of Pharmaceutical Sciences
Siksha "O" Anusandhan
(Deemed-to-Be University)
Bhubaneswar-751003, Odisha, India

Ray Chaudhuri, Shaon
Microbial Technology Laboratory
Department of Microbiology
Tripura University, Suryamaninagar
Tripura West-799022, India

Reang, Sukanta
Department of Chemical and Polymer Engineering
Tripura University (A Central University)
Agartala, Tripura-799022, India

Reynolds, R.
Chairman
Ad Hoc Group for Clean Water
New York, USA

Sahoo, Abanti
Department of Chemical Engineering
NIT Rourkela
Rourkela, Odisha-769008, India

Samal, Swati
Department of Microbiology
CBSH, OUAT
Bhubaneswar-3, Odisha, India

Samantaray, Deviprasad
Department of Microbiology
CBSH, OUAT
Bhubaneswar-3, Odisha, India

Sethy, Prajna Sarita
Department of Life Science
Rama Devi Women's University
Bhubaneswar-751022, Odisha

Sharma, Anuradha
Department of Molecular Biology and Genetic Engineering
Lovely Professional University
Phagwara-144411

Shukla, Shweta
Department of Biotechnology
School of Biosciences and Technology
Vellore Institute of Technology
Vellore-632014, Tamil Nadu, India

Singh, Lakshmi
Department of Botany
College of Basic Science and Humanities
Odisha University of Agriculture and Technology
Bhubaneswar-751003

Sugunakara Chary, Jackson
Department of Molecular Biology and Genetic Engineering
Lovely Professional University
Phagwara-144411

Swain, Sunanda
Department of Life Science
Rama Devi Women's University
Bhubaneswar-751022, Odisha

Tanacredi, J. T.
Centre for Environmental Research Coastal Oceans Monitoring (CERCOM)
Molloy University
Rockville Centre, New York

Tollefsen, R. C.
Visiting Naturalist
Hampton Bays, New York

Acknowledgments

The editors are thankful to all the authors who have contributed to this book project and made it a success. We are grateful to the authors and reviewers who took part in the evaluation process for sparing their valuable time and giving their efforts for the timely completion of the book. We sincerely appreciate the authors of the chapters, who gave their time and knowledge to this book.

1 Comparative Study of Ansys and Aspen Plus Simulation of a Typical Fluidized Bed Column–Advanced Greener Approach for Environmental Sustainability

Sourab Baidya, Harjeet Nath, and Abanti Sahoo

1.1 INTRODUCTION

Fluidization has been utilized extensively in many industrial applications, such as drying, catalytic decomposition, pyrolysis, coating, etc. *Fluidization* is basically an operation where solid particles are made to behave in a fluid-like state when it contacts with gas or liquid. Despite the extensive application of the fluidized bed reactor, the sophisticated behavior of the solid flow in the reactor bed makes it challenging in designing a flow model. This process involves many variables which are complex to handle, where gas–solid interaction and chemical reaction characterize a complex process (Baron et al. 1990; Debnath, Nath, and Chauhan 2019). It is imperative to understand the hydrodynamic behaviors of solid particles to identify the variations of bubble diameter, solid volume fraction, pressure drop, bubble rise velocity, etc. with bottom zone height and total height. Previously, the influence of fluidization velocity and pressure on hydrodynamics was studied by Olowson et al. Also, the mean bubble rise velocity, the bubble frequency, the mean pierced length, the bubble volume fraction, and the visible bubble flow rate were measured using capacitance probes (Olsson, Wiman, and Almstedt 1995). The hydrodynamic model involves the distribution of solids, gas–solid mixture, velocity, bubble growth, and the correlation

DOI: 10.1201/9781003517108-1

between bubble and emulsion phases. The flux of entrained particles from the column reduces until TDH. Sometimes it is required to recover the entrained solid particle to increase the efficiency and minimize the cost (Baron et al. 1990; Rudra Paul et al. 2019). There are two ways through which the numerical hydrodynamic model can be studied: the Euler–Euler approach or the Euler–Lagrange approach. Due to computational cost, the Eulerian approach is mostly used in the numerical simulation of the fluidized bed (Philippsen, Vilela, and Zen 2015).

The Aspen Plus fluidized bed model is the most up-and-coming tool to get an insight of various parameters, such as interstitial velocity, superficial velocity, etc., which become troublesome to estimate experimentally. Aspen Plus simulation can speculate the valid results when specific experimental data are incorporated which consume less time, and proper optimization is possible for further studies. As per the knowledge of the authors, most of the previous hydrodynamic studies through Aspen Plus have been performed though the equilibrium and kinetic model. The fluidized bed model of Aspen Plus is gaining much popularity because of its accurate prediction of various profiles, height, and velocities. Several parameters, such as particle size and density, terminal velocity, geometry of the vessel, and additional gas supply, have been considered through simulation.

The present study with Aspen Plus aims to provide much detailed analysis of the hydrodynamics behavior in comparison to the data analyzed previously using the Ansys Fluent CFD package and published elsewhere. The simulation is carried out considering silica sand as solid particles having the density of 1,220 kg/m^3, which is in comparison with the biomass having the density of 1,242 kg/m^3 as material density and bed height effects on fluidized bed hydrodynamics.

1.2 EXPERIMENTAL SETUP

The experimental setup consists of a Perspex fluidized bed column of length 0.62 m and 0.10 m width (Figure 1.1). The bed material consists of powdered sawdust material of size 300 μm and density of 1,234.2 kg/m^3. Air is supplied at different velocities through the wind box using air compressor; the flow rate is controlled by a rotameter. Spherical glass beads were packed in the calming section for air to get uniformly distributed (Debnath, Nath, and Chauhan 2019). The instrument details are shown in Table 1.1.

1.3 ASPEN PLUS MODELING

The Aspen Plus fluidized bed model describes an isothermal fluidized bed considering fluid mechanics (one dimensional) and also describes the entrainment of particles. The model considers particle size and density, terminal velocity, geometry of the vessel, option of additional gas supply, impact of heat exchangers on bed temperature, and fluid mechanics. It also considers chemical reactions and their impact on the fluid mechanics and vice versa. The modeling also provides different options/correlations to determine the minimum fluidization velocity, TDH, entrainment of solids from the bed, and distributor pressure drop. The model also considers two

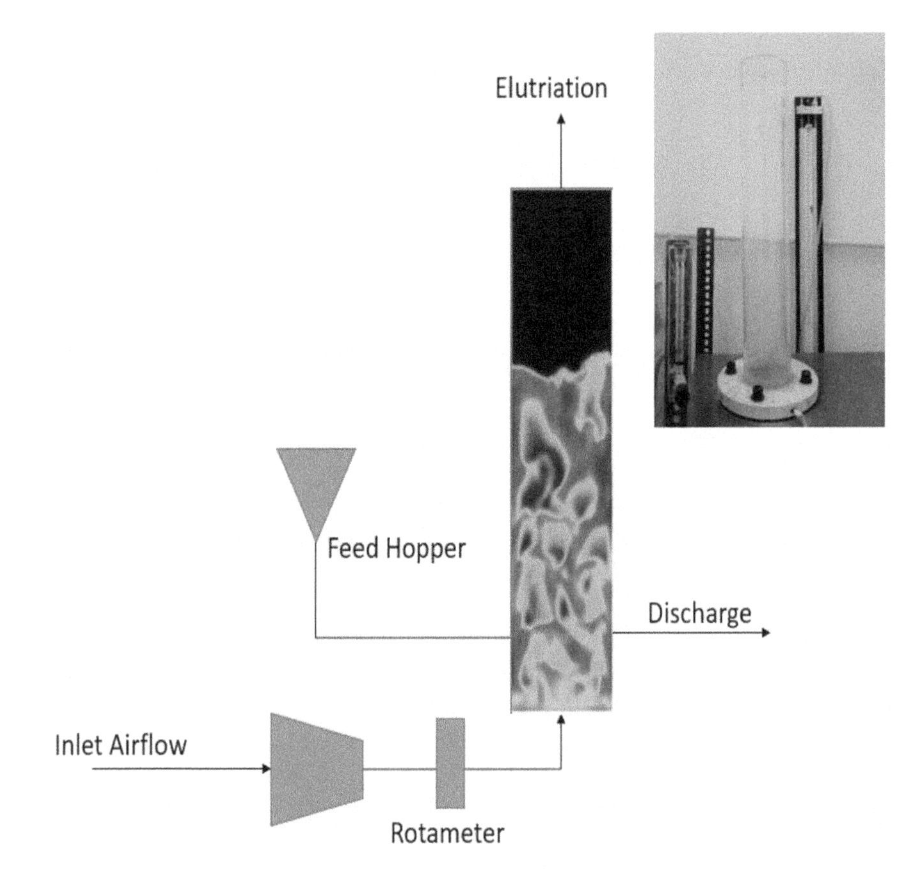

FIGURE 1.1 Fluidized bed column.

TABLE 1.1
Instrument Details of the Experiment

Sl. No.	Instruments	Capacity	Company
1	Air compressor	10 L	Camel
2	Rotameter	0–100 lpm	Indus
3	U-tube manometer	500 ml	Indus
4	Perspex column with feed hopper	–	–

zones, viz., the bottom zone and the freeboard. The model also assumes solids are ideally mixed, and it considers the impact of volume production/reduction on the fluid mechanics as well as change in PSD due to reaction (if available) (Yogendrasasidhar, Srinivas, and Pydi Setty 2018). It is represented in Figure 1.2.

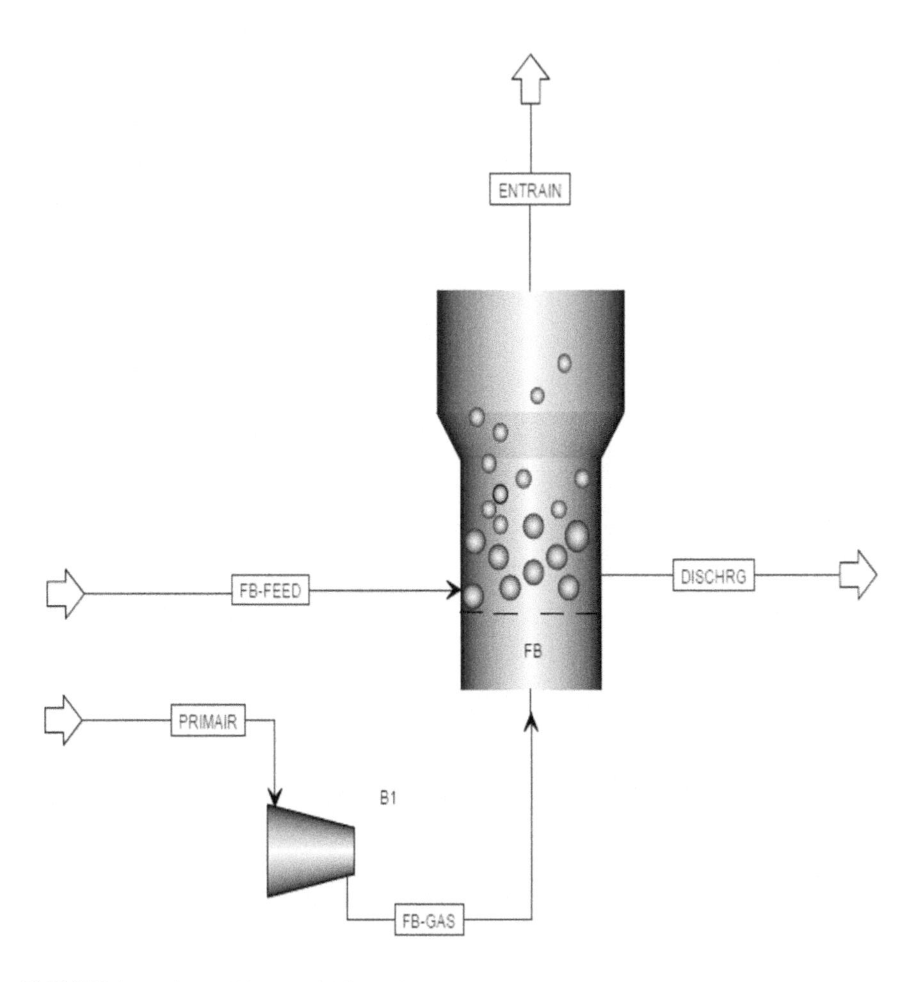

FIGURE 1.2 Aspen Plus model flow sheet.

Ergun equation is suitable for computing the pressure drop across packed-bed reactors in RPlug and packed-bed pipes in Pipe. The minimum fluidization velocity in the model is calculated through Ergun equation:

$$-\frac{dP}{dz} = 150\frac{(1-\varepsilon)^2}{\varepsilon^3}\frac{\mu U}{\phi_s^2 D_y^2} + 1.75\frac{1-\varepsilon}{\varepsilon^3}\frac{\rho U^2}{\phi_s D_p}$$

In the present model of the fluidized bed, the Chan and Knowlton correlation is used to determine TDH.

$$TDH = 0.85 \cdot u^2 \cdot \left(7.33 - 1.2 \cdot \log(u)\right)$$

Nomenclature:

U	=	Superficial velocity
ε	=	Bed voidage
μ	=	Fluid viscosity
D_p	=	Particle diameter
ϕ_s	=	Particle shape factor
ρ	=	Fluid density
u	=	Inlet velocity

1.3.1 Modelling Approach

A continuous fluidized bed model was employed using the Aspen Plus steady-state software. Prior to entering the simulation environment, the component silicon dioxide (SiO_2) as a solid and air was included from the inbuilt database for property estimation. The silica-based solid stream is linked from the left side of the fluidized bed at a flow rate of 1 kg/hr, and the intake gas stream is connected from the bottom, allowing for variable gas velocities. The elutriation of particles from the column is defined by the output stream from the top of the fluidization, whilst the discharged solid particles are defined by the output stream from the right. Under particle size distribution, inlet solid characteristics such as particle size were subsumed as 300 μm (Geldart B). In addition, the solid flow has been kept at room temperature with a pressure of 1 atm. The fluidized bed model has a height of 0.62 m and a width of 0.10 m. The minimum fluidization velocity has been calculated through Ergun equation in Aspen Plus. The transport disengagement height (TDH) was calculated using the Chan and Knowlton model, whereas elutriation was calculated using the Tasirin and Geldart model. Many elements of a fluidized bed model include the design of the distribution plate, orifice diameter, and quantity of orifices, among others. Table 1.2 shows the simulation's operating parameters. These parameters were adopted from the standard literature, and the following assumptions were made in modelling the fluidized bed:

- It is a steady-state process.
- Voidage at minimum fluidization is 0.5.
- The specified bed mass is 3.61 kg, which remains intact with solid and gas flow rate.
- A perforated plate is used with 500 orifices with a diameter of 200 μm and an orifice discharge coefficient of 0.8.

1.4 RESULTS AND DISCUSSION

The simulation study of the fluidized bed was carried out with the parameters obtained from the previous study at different air flow rates using Aspen Plus software. Through simulation, the minimum fluidization velocity, height of the bottom zone, distributor pressure drop, etc. were calculated. The simulation results of

TABLE 1.2
Specifications Used in Simulation

Parameters	Unit	Amount
Solid feed rate	kg/hr	1 kg
Inlet air flow rate	kg/hr	10.26
Geldart powder type	B	–
Bed voidage	–	0.5
Diameter of particle	m	0.0003
Number of orifices	–	500
Orifice diameter	μm	200
Distributor plate type	Perforated plate	–
Height of the bed	m	0.62
Diameter of the column	m	0.10

Aspen Plus were compared with the results obtained from the experimental setup and CFD simulation. The more accurate results were found from Aspen Plus. It can be observed that superficial velocity increases with total bed height, whereas solid volume fraction and interstitial velocity increase initially and start decreasing after reaching a certain point.

1.4.1 Effect of Inlet Air Velocity

Inlet air velocity is the most vital parameter which impacts fluidization largely. For fluidization to take place, the inlet air velocity must be greater than the minimum fluidization velocity. Inlet air velocity reflects pressure drop, bubble volume fraction, bottom zone height, etc. In the experimental setup, the inlet velocity has been given in the range of 0.05 m/s to 0.55 m/s, which has a flow rate of 23.5 lpm to 259.09 lpm. The minimum fluidization velocity has been calculated using the Ergun correlation with Aspen Plus. According to the experimental study with the integration of CFD which used the contours of particles fluctuating and moving upward, minimum fluidization velocity could be predicted between 0.15 m/s and 0.19 m/s, whereas Aspen Plus simulation calculated the minimum fluidization velocity as 0.0612 m/s. At inlet air velocity of 0.05 m/s, superficial gas velocity dropped below the minimum fluidization velocity, due to which fluidization has not taken place. It has been observed that when the inlet air velocity increases gradually to 0.0721 m/s (34 lpm), fluidization takes place smoothly, with no elutriation of particles, and this continues up to 0.0891 m/s (42 lpm). So at 0.0721 m/s, fluidization takes place with minimum elutriation, the results of which have been mentioned in Table 1.3. The particle elutriation rate gradually exceeds 0.0891 m/s, but the freeboard height is less than the calculated TDH based on the solid volume concentration profile and the Chan and Knowlton correlation.

TABLE 1.3
Fluidized Bed Results

Parameter	Amount
Height of bottom zone	0.166010827 m
Height of freeboard	0.453989173 m
TDH calculated from correlation	0.323007345 m
Solids holdup	3.99999995 kg
Number of particles in the bed	127,764,559
Surface area	31.938559 m^2
Distributor pressure drop	0.00492813162 bar
Bottom zone pressure drop	0.0199939955 bar
Freeboard pressure drop	0.0299988196 bar
Fluidized bed pressure drop	0.0499928151 bar
Overall pressure drop	0.0549209467 bar
Minimum fluidization velocity	0.0612929573

1.4.2 Bottom Zone Height

The bottom zone is at the bottom part of the fluidized bed, which has high solid volume concentrations, and the upper dilution zone has lower solid volume concentrations. The bottom zone height, which corresponds to the expanded bed height in the CFD analysis, increases with the increment of inlet air flow rate. Bottom zone height gradually increases from 0.1519 m to 0.4557 m with the augmentation of inlet air flow rate in the previous work. In the present simulation study, the bottom zone height increases from 0.166 m to 0.62 m, which is a good comparison with the previous result.

1.4.3 Fluidized Bed Profiles

1.4.3.1 Superficial Velocity

The gas flow rate is divided by the entire column surface area to get the surface velocity. It has been observed that the superficial velocity of the fluidized bed increases progressively as the height of the fluidized bed rises. The bed experiences local segregation, local mixing, and re-segregation as the gas surface velocity steadily rises. The superficial velocity remains constant after reaching the minimum fluidization velocity and seems not to increase anymore, given by Figure 1.3.

1.4.3.2 Interstitial Velocity

Interstitial velocity is the velocity of the gas between the particles in a dense suspension. It has been observed that as the flow rate increases, the interstitial velocity increases and then begins to decrease exponentially after reaching the bottom zone

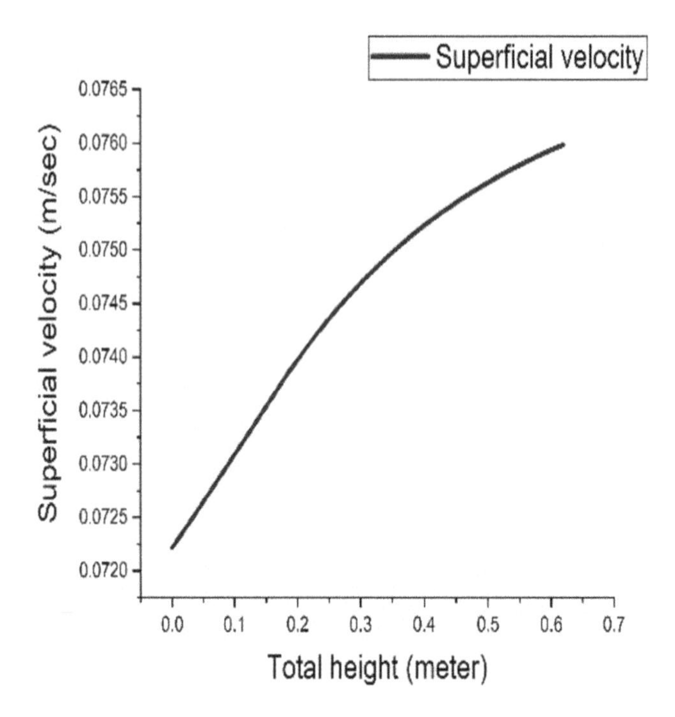

FIGURE 1.3 Total height vs. superficial velocity.

height of 0.16 m. The cause of the decrease might be the increase in porosity as the bed height rises. Figure 1.4 illustrates the behavior of interstitial velocity.

1.4.3.3 Solid Volume Fraction

Aspen Plus analyzed the solid volume fraction, which increases slightly up to the bottom zone height and then decreases gradually. As the bubbles start forming and bubble velocity increases, the void in the dense suspension increases, which, in turn, decreases the solid volume fraction. The profile of solid volume fraction with height can be observed in Figure 1.5.

1.4.3.4 Bubble Diameter

The size of bubbles in a fluidized bed is one of the most important indices since it influences various properties of the bed, including mass and heat transfer rates, as well as bubble rising velocity (Scholarship, Bai, and Zhu 2018). The bubble diameter distribution changes depending on the bed height. It has been observed that bubble diameter increases with the bottom zone and total height due to bubble coalescence and increase in bubble velocity. The profile of bubble diameter can be observed from Figure 1.6 and Figure 1.7

1.4.3.5 Volumetric Bubble Flow

It has been observed that the volumetric bubble flow increases almost linearly with total height and bottom zone height, which can be observed from Figure 1.8 and Figure 1.9

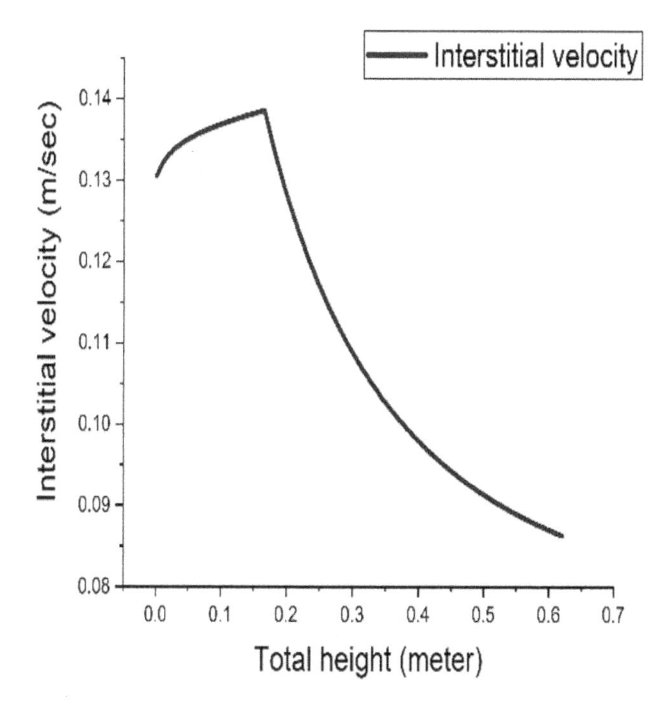

FIGURE 1.4 Total height vs. interstitial velocity.

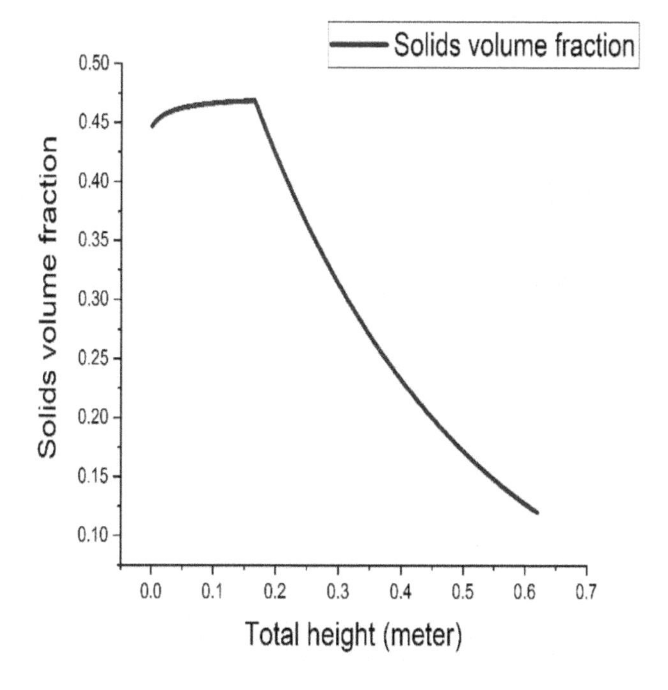

FIGURE 1.5 Total height vs. solid volume fraction.

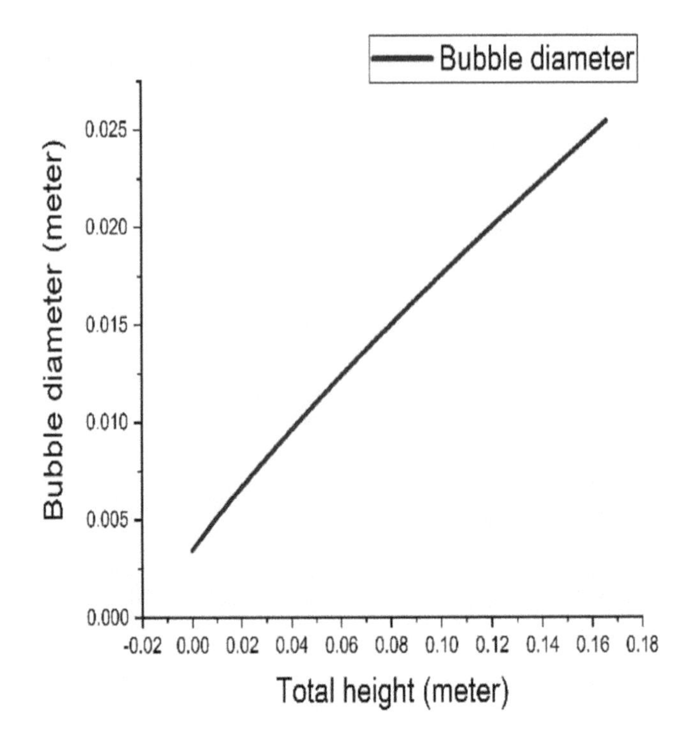

FIGURE 1.6 Total height vs. bubble diameter.

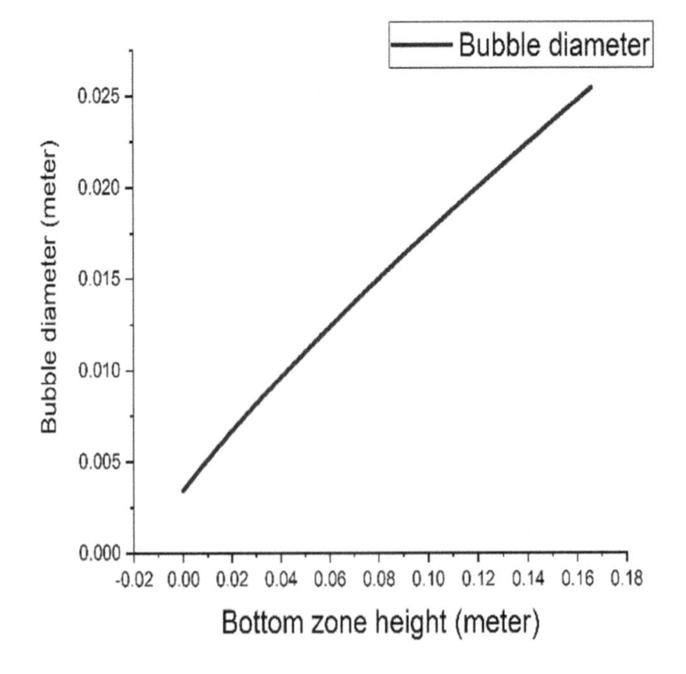

FIGURE 1.7 Bottom zone height vs. bubble diameter.

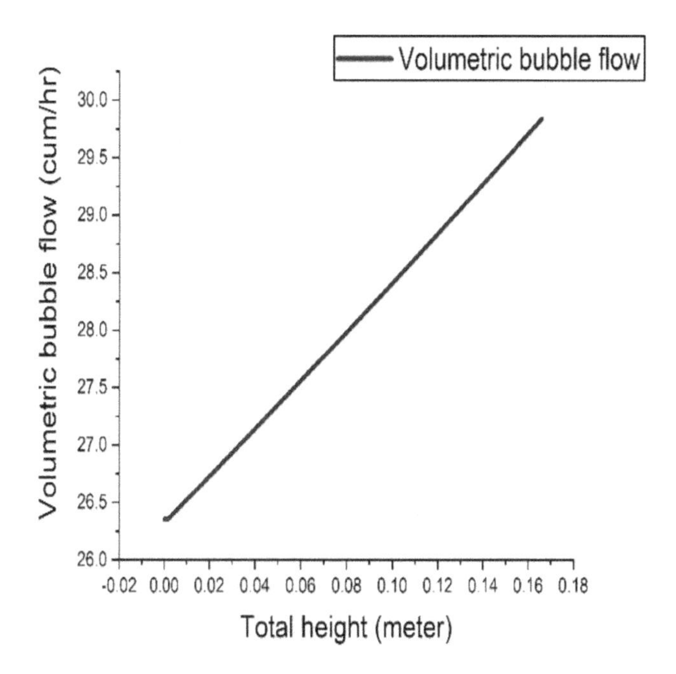

FIGURE 1.8 Total height vs. volumetric bubble flow.

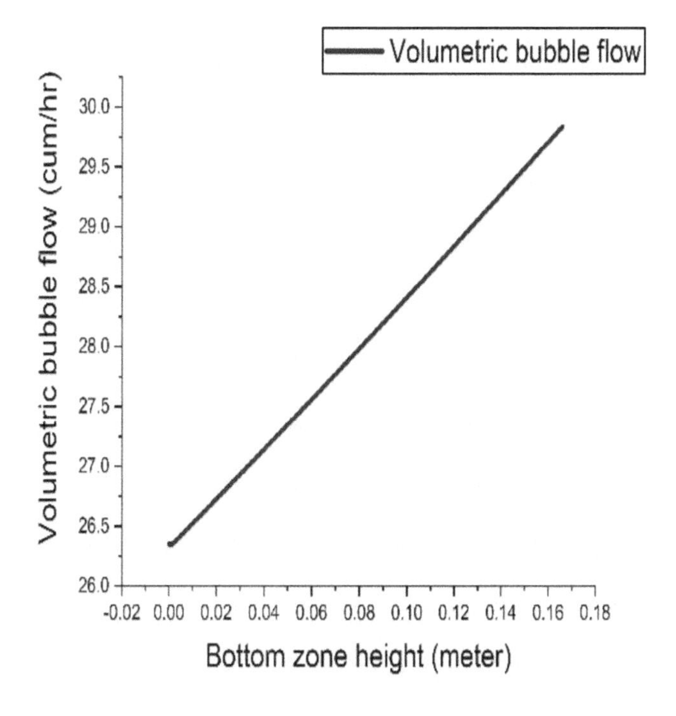

FIGURE 1.9 Bottom zone height vs. volumetric bubble flow.

1.4.3.6 Bubble Rise Velocity

The simulation analysis showed that the bubble rise velocity rises with total and bottom zone height. The bottom zone height rises when the incoming gas velocity approaches the minimum fluidization velocity and the voidage begins to rise, increasing the bubble rise velocity. The bubble rise velocity profile is shown in Figure 1.10 and Figure 1.11

1.4.3.7 Bubble Volume Fraction

Bubble volume fraction is the percentage of bubble volume in the fluidized bed's unit volume (Scholarship, Bai, and Zhu 2018). The bubble volume percentage decreases with overall height as well as bottom zone height above the gas distributor, as indicated by Aspen Plus. The profiles of bubble volume fraction may be found in Figures 1.12 and 1.13. Larger bubbles form at higher elevations, as seen in Figures 1.7 and 1.14, and the bubble rise velocity is faster.

1.5 PRESSURE DROP

Simulation has been used to examine pressure drop variation at different heights at varied inlet air velocities ranging from 0.0721 m/s to 0.55 m/s, as shown in Figure 1.15. The pressure drop diminishes with height and peaks at the bottom of the bed, where it is mostly determined by the incoming air flow rate. Also, in both Aspen and CFD analysis, the pressure drop at 0.55 m/s deviates the most. The computational analysis results are similar to those of the Aspen simulation. The distributor pressure drop,

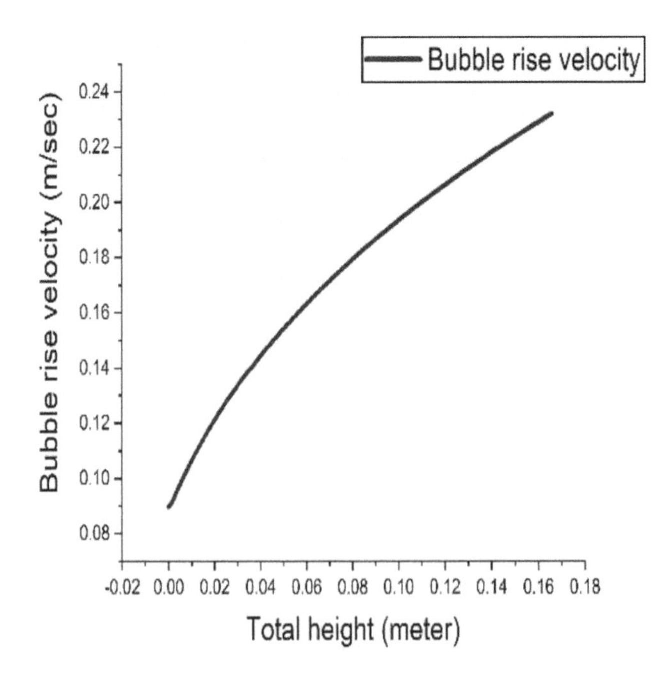

FIGURE 1.10 Total height vs. bubble rise velocity.

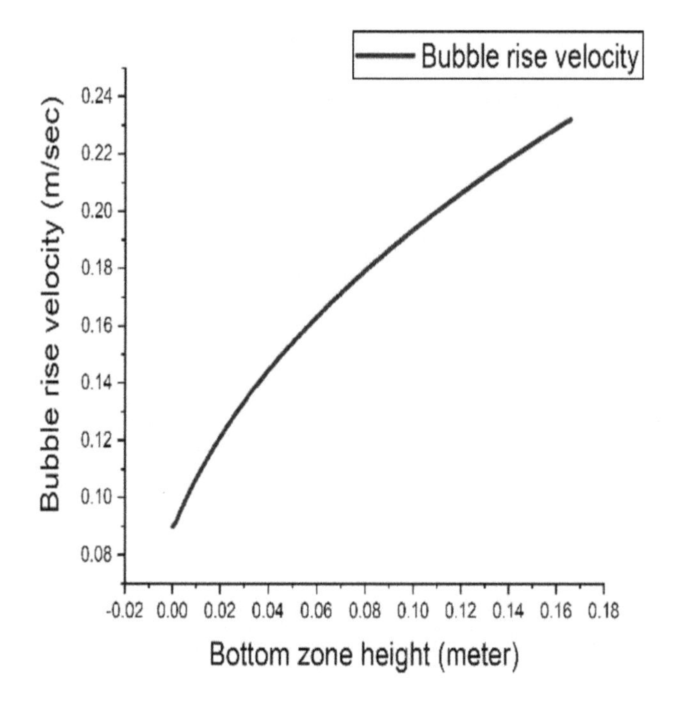

FIGURE 1.11 Bottom zone height vs. bubble rise velocity.

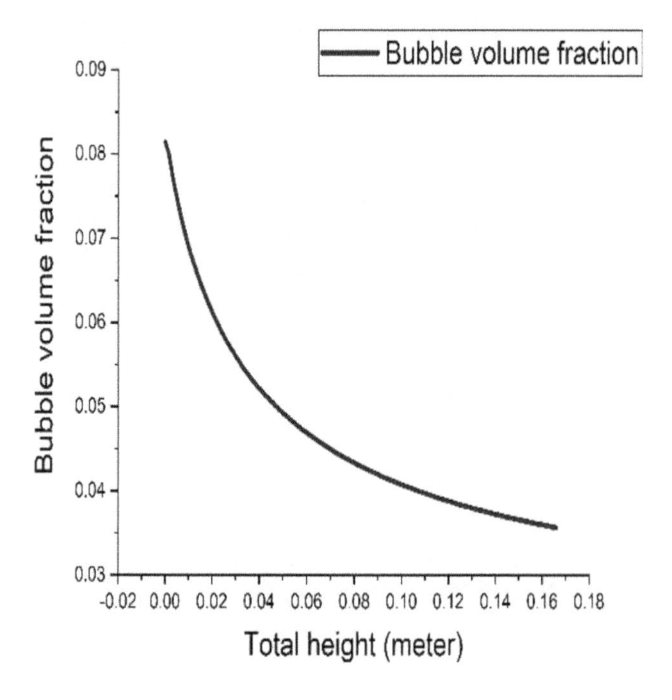

FIGURE 1.12 Total height vs. bubble volume fraction.

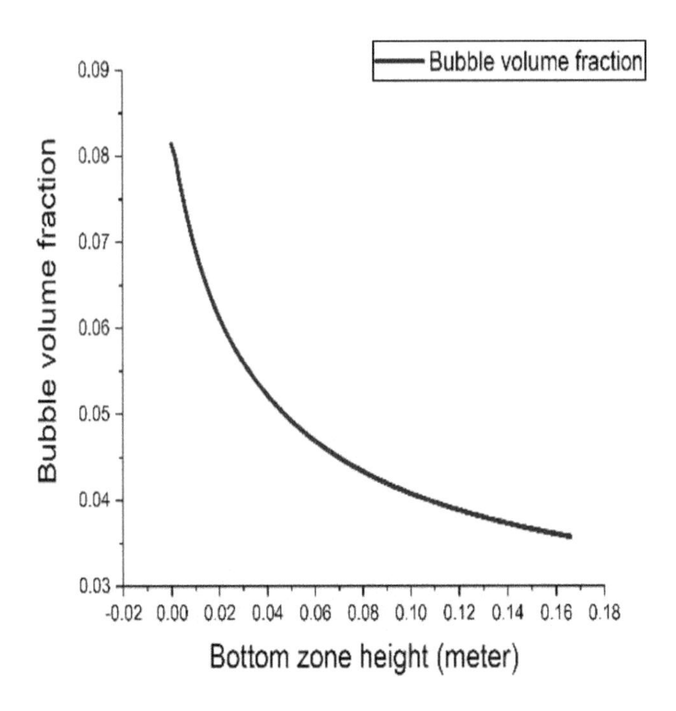

FIGURE 1.13 Bottom zone height vs. bubble volume fraction.

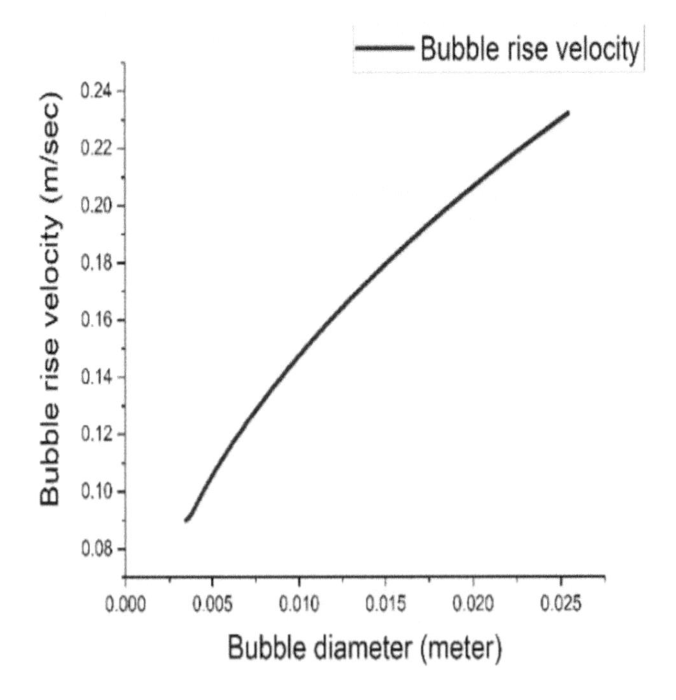

FIGURE 1.14 Bubble diameter vs. bubble rise velocity.

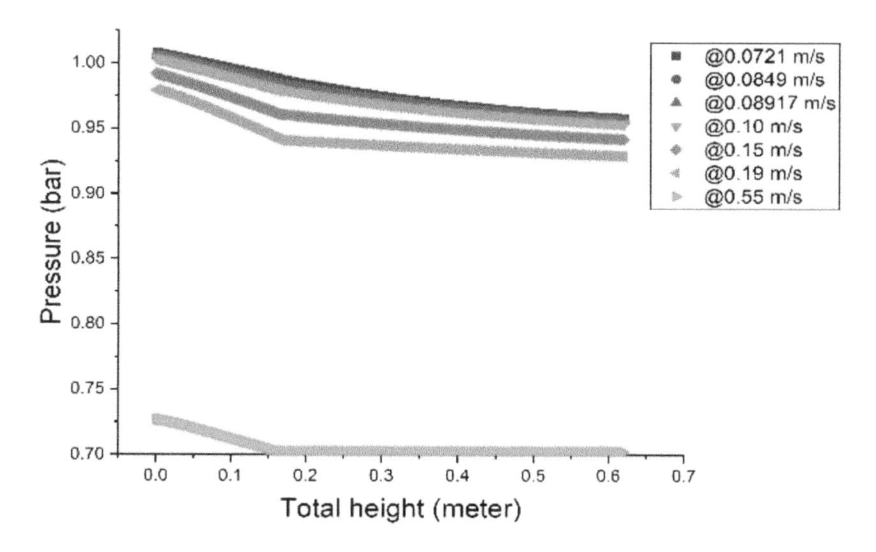

FIGURE 1.15 Variation of pressure drop across the column.

bottom zone pressure drop, freeboard pressure drop, fluidized bed pressure drop, and overall pressure drop were all observed to be 0.049 bar, 0.0199 bar, and 0.0299 bar, respectively, in the simulation.

1.6 DISCUSSION

The previous work predicted various heights and minimum fluidization velocity with the contour of biomass volume fraction. It was difficult to anticipate accurate velocities and bottom zone heights, whereas Aspen Plus analyzed to provide comprehensive data. It has been found that the TDH and freeboard height are 0.3230 m and 0.453 m, respectively, for an inlet velocity of 0.0721 m/s. It has also calculated various pressure drops, such as freeboard pressure drop, fluidized bed pressure drop, etc. The surface area of fluidization has been calculated as 31.938 sq^2. The comparative understanding of simulation results and results from the previous study is shown in Table 1.4.

1.7 CONCLUSION

The parameters from the prior experimental and CFD studies were used to develop a typical fluidized bed using Aspen Plus. The findings were more exact and comprehensive. The current research revealed a wide range of fluidized bed profiles with total height and bottom zone height, allowing for a better knowledge of hydrodynamic characteristics. It has been found that the minimum fluidization velocity is 0.612 m/s. Fluidization at a flow rate of 34 lpm has better efficiency with the lowest elutriation. The simulation evaluated the bottom zone height, which corresponds to the bed expansion in the computational analysis. The study also analyzed the

TABLE 1.4
Comparison between CFD and Aspen Results in Terms of MFV and OCBO

Parameters	Experimental Setup and Ansys Fluent CFD Package	Aspen Plus Simulation
Minimum fluidization velocity (MFV)	Predicted between 0.15m/s and 0.19 m/s	Exact value found out to be 0.612 m/s
Optimal condition for bed operation (OCBO)	0.19 m/s to 0.55 m/s	0.0721 m/s to 0.08917 m/s

TABLE 1.5
Comparison between CFD and Aspen Results in Terms of Inlet Air Velocity, Flow Rate, and Bed Parameters

		Results	
Inlet Air Velocity (m/s)	Flow Rate (lpm)	Experimental Setup and Ansys Fluent CFD Package	Aspen Plus Simulation
		Extended bed height (m/s)	Bottom zone height (m/s)
0.05	23.55	0.1519	–
0.0721	34	–	0.166010827
0.0849	40	–	0.192744424
0.08917	42	–	0.201508072
0.10	47.10	0.1860	0.223836095
0.15	70.650	0.2170	0.338665376
0.19	84.49	0.2387	0.453103602
0.55	259.05	0.4557	0.62

optimum condition for the fluidization, which is at an inlet air velocity of 0.0721 m/s to 0.08917 m/s, which is a good comparison to the prior study. The simulation also investigated the behavior of various fluidized bed profiles with total and bottom zone height. Such comprehensive data will aid in the future scaling-up of fluidized beds.

REFERENCES

Baron, T., Ecok Natima, C. L. Briens, P. Galtier, and M. A. Bergougnou. 1990. "Effect of Bed Height on Particle Entrainment Beds from Gas-Fluidized a New Model Which Accounts for 'Gulf Streaming' Effects Can Predict the Flux of Ejected Particles." *Powder Technology* 63.

Debnath, Shubhankar, Harjeet Nath, and Vishal Chauhan. 2019. "CFD Modeling of a Typical Fluidized Bed Column." *Materials Today: Proceedings* 46: 6178–6184. Elsevier Ltd. https://doi.org/10.1016/j.matpr.2020.04.079.

Olsson, S. E., J. Wiman, and A. E. Almstedt. 1995. "Hydrodynamics of a Pressurized Fluidized Bed with Horizontal Tubes: Influence of Pressure, Fluidization Velocity and Tube-Bank Geometry." *Chemical Engineering Science* 50.

Philippsen, Caterina Gonçalves, Antônio Cezar Faria Vilela, and Leandro Dalla Zen. 2015. "Fluidized Bed Modeling Applied to the Analysis of Processes: Review and State of the Art." *Journal of Materials Research and Technology*. Elsevier Editora Ltda. https://doi.org/10.1016/j.jmrt.2014.10.018.

Rudra Paul, Tamal, Harjeet Nath, Vishal Chauhan, and Abanti Sahoo. 2019. "Gasification Studies of High Ash Indian Coals Using Aspen Plus Simulation." *Materials Today: Proceedings* 46: 6149–6155. Elsevier Ltd. https://doi.org/10.1016/j.matpr.2020.04.033.

Scholarship, Western, Tianzi Bai, and Supervisor Zhu. 2018. "Bubble Dynamics and Dense Phase Composition in 2-D Binary Gas-Solid Fluidized Bed." https://ir.lib.uwo.ca/etd://ir.lib.uwo.ca/etd/5514.

Yogendrasasidhar, D., G. Srinivas, and Y. Pydi Setty. 2018. "Effect of Distributor on Performance of a Continuous Fluidized Bed Dryer." *Heat and Mass Transfer/Waerme-Und Stoffuebertragung* 54 (3): 641–649. https://doi.org/10.1007/s00231-017-2169-2.

2 Surfactant and Potable Water Recovery from Gray Water Using Integrated UF and RO Membranes – A Typical Method for Waste Recycling, Reusing, and Recovery of Value-Added Products

Harjeet Nath and Sukanta Reang

2.1 INTRODUCTION

Access to safe and clean drinking water is a fundamental human right, essential for sustaining life, promoting health, and ensuring overall well-being. However, the global demand for drinking water has been steadily increasing, driven by factors such as population growth, urbanization, industrialization, and climate change (Das et al. 2019). WHO reports that by 2025, half of the population will not be able to have a source of water (Sundaramahalingam and Chandrasekaran 2019). It has to be used optimally and should not be wasted. Most of the water available in our planet is salty (seawater) or trapped in ice caps and glaciers, and only 1% of the world's drinkable water is available for human consumption (Ahmad and EL-Dessouky 2008). Industrial growth places additional strain on water resources, leading to increased water stress in many regions. Industries require large amounts of water for manufacturing processes, cooling, and sanitation. This demand for water often competes with the needs of communities, agriculture, and ecosystems (Lenzen et al. 2013). The entire population of the world depends on 1% of water available for agriculture, industries, household, and other use. Therefore, it is required to conserve this for a sustainable future. As per estimates, a regular household uses approximately

DOI: 10.1201/9781003517108-2

80–100 gal of water per day, and majority of it is used for washing clothes, bathing, flushing toilet, etc. Urban wastewater, which includes majority of surfactants mixed with dirt and the like, essentially obtained from bathrooms, hand basins, washing machine, dishwashers, kitchen sinks, etc. is called gray water (Li, Wichmann, and Otterpohl 2009). Gray waste water obtained from such sources is generally high in total dissolved solids, total suspended solids, turbidity, as well as oxygen demand (Reang and Nath 2021). It is important that wastewater be treated in order to save water as a future source and protect the environment from pollution (Saha, Nath, and Das 2023; Sophia, Harjeet, and Praneeth 2016; Das et al. 2022). While gray water is relatively less contaminated than blackwater (which contains fecal matter), it still contains some organic matter, chemicals, and potential pathogens. Releasing untreated gray water into the environment can lead to several environmental issues: (a) *Nutrient overload.* Gray water contains nutrients like phosphorus and nitrogen, which can promote excessive growth of algae and aquatic plants when released into water bodies. This overgrowth, known as eutrophication, can lead to oxygen depletion in the water, harming fish and other aquatic organisms. (b) *Contamination of water sources.* If untreated gray water enters groundwater or surface water bodies, it can contaminate these sources, making them unsafe for drinking, irrigation, or recreational use. (c) *Soil degradation.* Gray water that is high in certain chemicals or detergents can harm soil quality and plant life when used for irrigation without treatment. Over time, the accumulation of chemicals can disrupt the soil's natural balance and affect plant growth. (d) *Spread of pathogens.* While gray water does not typically contain high levels of pathogens, there is still a risk of microbial contamination. Direct exposure to untreated gray water or consumption of contaminated water can lead to the spread of diseases and pose a health risk to humans and animals (Li, Wichmann, and Otterpohl 2009; Gross et al. 2005). Gray water can be treated for removing physical impurity using coarse sand and soil filtration, membrane filtration, etc. (Li, Wichmann, and Otterpohl 2009; Das et al. 2019; Sophia, Harjeet, and Praneeth 2016; Das et al. 2019; Reang and Nath 2021). Nowadays, reverse osmosis (RO)–based treatment has become the leading method for purifying and cleaning wastewater so that it can be reused (Gao et al. 2016). Reuse of wastewater for recovering water in addition to nutrients is becoming an important research area. The RO system uses a high-pressure pump that forces water across the semi-permeable membrane to separate ions, unwanted molecules, and larger particles from initial water. It has a membrane pore size of >2 nm or less. UF is the pre-treatment before RO to remove bacteria, viruses, parasites, suspended solid, and solute of high molecular weight. The pore size of the UF membrane is 2–100 nm; according to this pore size, contaminants are removed. RO membrane can reduce BOD and COD (Said, Rozaimakh Sheikh Abdullah, and Wahab Mohammad 2016). Gray water, which is of less TDS, can be directly cleaned by RO membrane. Combination of a UF membrane with an RO membrane has been recently tested to have reduced BOD by approximately 98% (Li, Wichmann, and Otterpohl 2009). The main challenge is to regenerate the membrane from fouling, which is the main reason for decreasing the permeate flow rate, which necessitates high-pressure requirement, thereby leading to high energy consumption and reduced membrane life (Wu et al. 2022). So in order to regenerate the membrane lifespan, backwash and backwash–backflush techniques may be adopted (Reang and Nath 2021; Venkatesh and Senthilmurugan 2017). Therefore, the following study will focus primarily on studying the process of recovering the valuable surfactant solution from surfactant and dirt containing gray water by adapting suitable membrane operations along with proper backwash and/or backwash–backflush methods for improving membrane lifespan. The process will therefore lead to the recovery of valuable surfactant/detergent which can be reused again and simultaneously recover the potable water, which can again be reused, thereby making the process sustainable and environment-friendly.

2.2 MATERIALS AND METHODS

2.2.1 Materials

Washing machine discharge obtained in the form of gray water is used as the primary feed to be treated. For preparing the detergent solution, 40 ml of liquid detergent is mixed with 40 L of fresh tap water (Reang and Nath 2021) and fed to a batch of soiled clothes inside a typical washing machine. The total dissolved solid (TDS) and turbidity are measured using Eutech CON 700 conductivity meter and Thermo Fisher turbidity meters, respectively. The experiment consists of utilizing a hollow fiber UF membrane (HFUF) (Wellon brand) and a spiral wound RO membrane (SWRO) (DOW Filmtech) in order to recover the surfactant and water out of the dirty water. Pressure dampers are used for stable pressure gauge measurements coming out of the diaphragm water booster pumps. The flow rate is measured using flow meters/rotameters. The flow rates and pressures are manipulated using control valves. Figure 2.1 shows the complete flow diagram of the recovery process.

2.2.2 Experimental Method For HFUF and SWRO Membrane

Gray water in the form of washing machine discharge is collected and stored in a feed tank (feed solution) which is initially pumped by a diaphragm booster pump at a flow rate of 40 lph and 5 psig pressure to a hollow fiber UF membrane (HFUF membrane) similar as elaborated by *S.* Reang et al. (Reang and Nath 2021). The SWRO module is operated at a pressure of 60 psig. The entire process is described in Figure 2.1 (Reang and Nath 2021). After a continuous operation which lasted for 60 min, a total of 23 L of UF product water is obtained and stored in UF product water tank, and a total of 7.2 L of potable water is obtained as permeate from SWRO membrane module. Approximately 15.8 L of concentrated surfactant solution were recovered in the UF product water tank. This recovered solution is essentially surfactant solution, which can be reused for cleaning clothes later. Membrane fouling effects are observed hereafter, which further required appropriate cleaning methodologies.

2.3 RESULT AND DISCUSSION

2.3.1 HFUF Membrane Performance

In 40 L of feed gray water, a total of 23 L of HFUF product water is obtained upon maintaining a continuous pressure of 5 psig. The process continued for a duration of approximately 60 min. It can be observed from Figure 2.2 that the initial condition of gray water was found to be around 114 NTU, and after passing through the HFUF

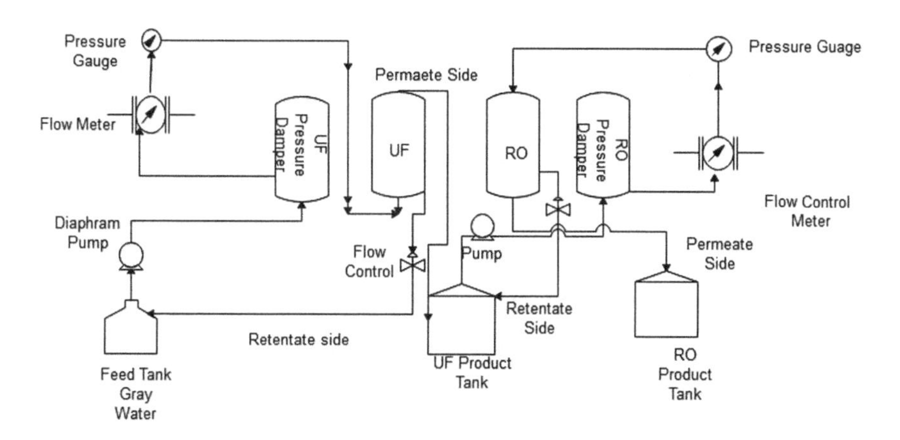

FIGURE 2.1 View of the experimental setup.

Source: Figure taken with permission from Reang and Nath (2021).

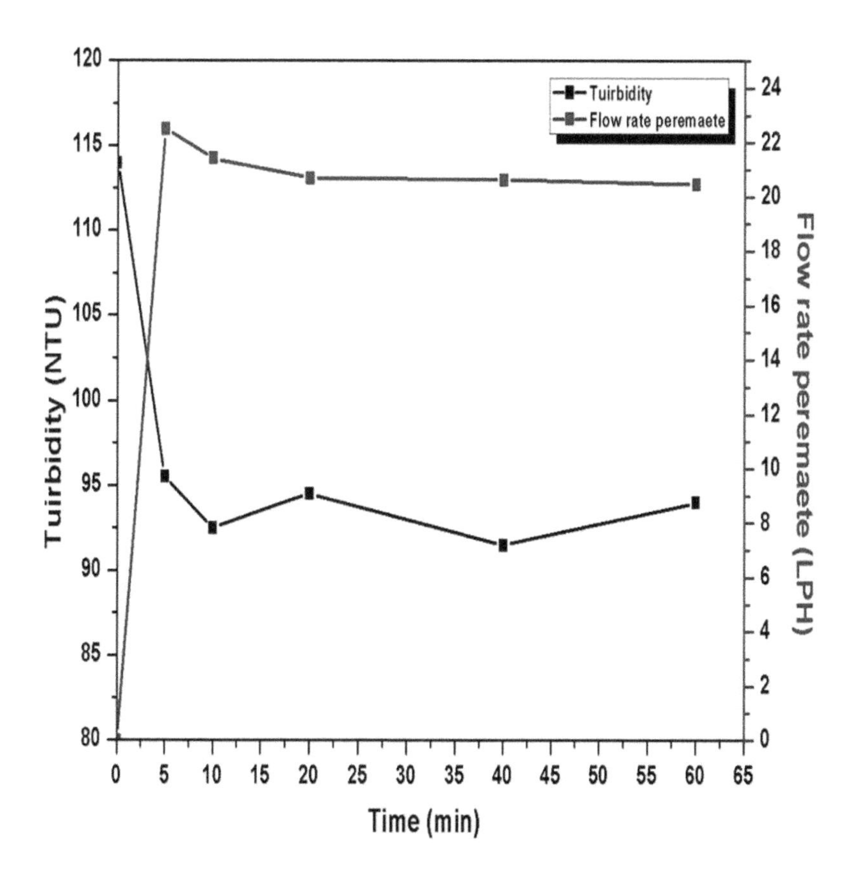

FIGURE 2.2 HFUF performance with respect to time vs. turbidity and flow rate of permeate.

setup, its turbidity decreased to 94 NTU. The BOD of gray water showed a slight decrease in its amount from 32.4 mg/L to 29.55 mg/L. From Figure 2.2, it can be observed that flow rate of HFUF permeate decreased with time mainly due to fouling (Reang and Nath 2021). The fouling phenomenon was later reduced by adapting backwash methodology (Venkatesh and Senthilmurugan 2017). The feed and permeate TDS remained almost constant at around 132 ppm throughout the process; however, the turbidity of the feed water tank increased to 164 NTU.

2.3.2 SWRO Membrane Performance

The UF product tank containing 23 L of product solution is taken as a feed for RO membrane, delivering a total of 7.2 L of pure potable water as RO membrane permeate and 15.8 L of concentrated surfactant solution as reject, which is sent back to the UF product tank. From Figure 2.3, it can be seen that the feed solution TDS decreased from 132 ppm to 4.6 ppm. Table 2.1 contains the detailed value of each tested parameter, where the turbidity factor was mostly tackled by both the HFUF and SWRO membranes, whereas the TDS could only be tackled via the SWRO membrane owing to its pore size (Yeom et al. 1998; Qin, Li, and Lee n.d.; Huang and McCutcheon 2015). Occurrence of membrane fouling was later analyzed and tackled using appropriate membrane backwash or backwash–backflush methods.

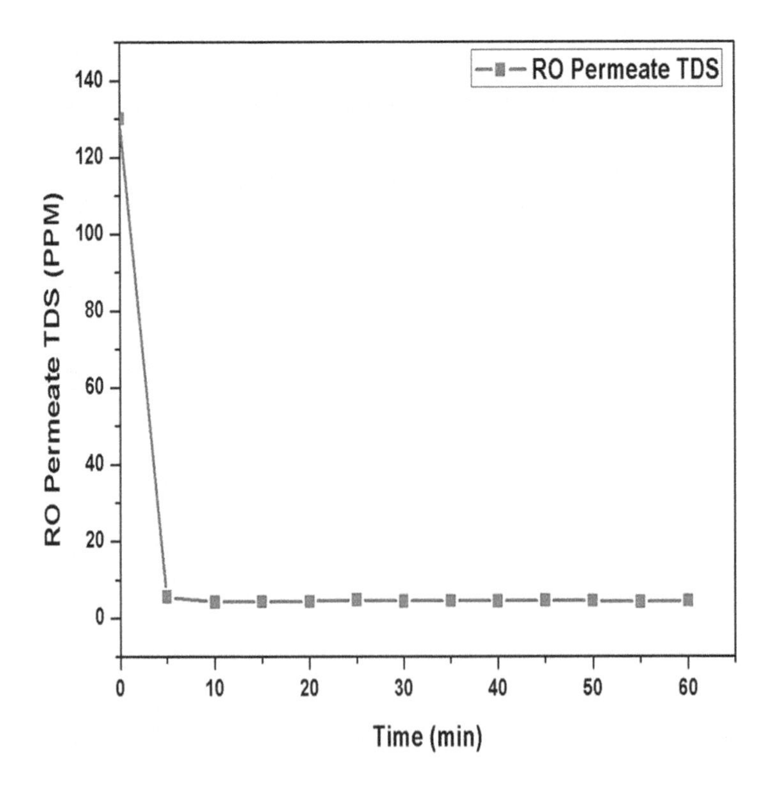

FIGURE 2.3 RO performance with respect to time vs. TDS of permeate.

TABLE 2.1
Comparison of Water Parameters for Initial and Final Product Tank and RO Permeate Water

Parameters	UF Product Tank (Initial)	UF Product Tank (Final)	RO Permeate Water
TDS	130 ppm	169 ppm	4.6 ppm
Turbidity	94 NTU	96 NTU	0.48 NTU
BOD	29.55 mg/l	30.1 mg/l	10.05 mg/l

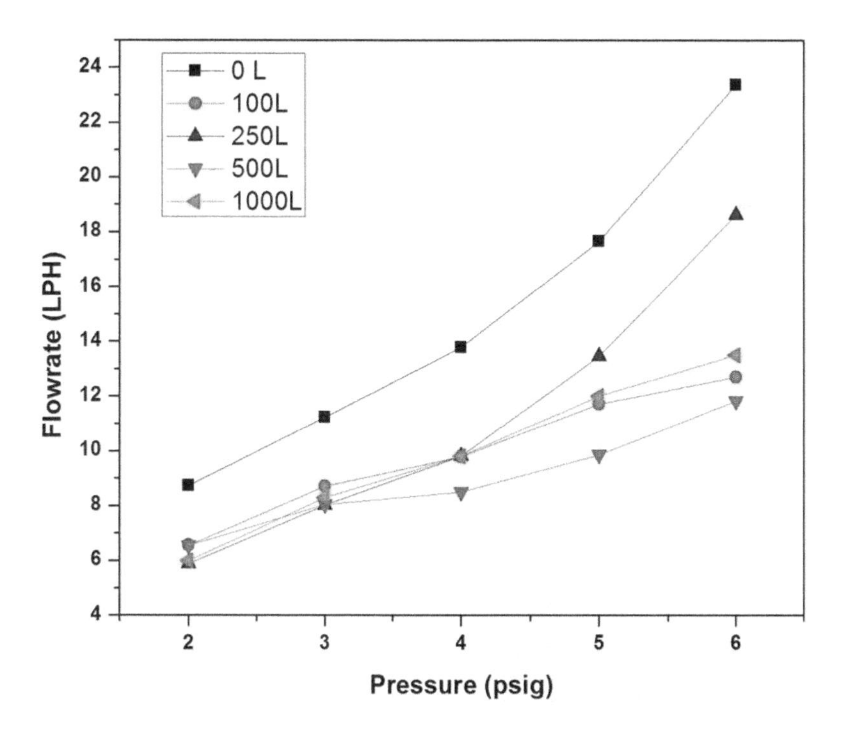

FIGURE 2.4 Performance of HFUF with respect to pressure.

2.3.3 HFUF Membrane Life Cycle Test

The fouling characteristics of HFUF membrane can be evaluated by checking the HFUF permeate flow rate at different pressures. The process adapted by S. Reang et al. (Reang and Nath 2021) was followed for checking the membrane life cycle test. For checking the occurrence of any fouling activity, the membrane is operated for batches of 100 L, 250 L, 500 L, and 1,000 L of gray water containing surfactant and dirt. The flow rates of the water at various pressures were evaluated after each batch, which is represented by Figure 2.4. Figure 2.4 thus proved the hypothesis of fouling due to the decrement of the permeate flow rate at varying pressures for the same membrane operated for various batches of gray water.

2.3.4 Backwash of HFUF

HFUF membrane was backwashed using pure water (0.35 NTU and 10 ppm) and at 1 psig pressure, as shown in Figure 2.5. From the turbidity readings, as elaborated in Figure 2.6, it is evident that the initial turbidity of the backwashed water was very high due to the presence of the fouling materials which were initially trapped on the membrane surface and later released during backwashing. The membrane required a total of around 4.33 L of pure water and 60 min for cleaning the membrane. The recovered HFUF membrane was then tested for its performance by passing pure water (0.35 NTU and 10 ppm) through it and comparing with the data obtained while the HFUF membrane was new (compared with 0 L flow from Figure 2.4) and represented by Figure 2.7. From Figure 2.7 it can be found that the recovered HFUF membrane is performing almost as was when it was new.

2.3.5 SWRO Membrane Life Cycle Test

The fouling characteristics of SWRO membrane can be evaluated by analyzing the SWRO permeate flow rate at varying pressures. In order to check membrane fouling behavior after the treatment of 100 L, 250 L, 500 L, and 1,000 L of HFUF-treated surfactant solution, the procedure adapted by S. Reang et al. (Reang and Nath 2021) was followed and represented in Figure 2.8. The figure clearly showed the decrement in membrane performance with more and more feed treatment, which is actually due to fouling.

2.3.6 Backwash of SWRO

Figure 2.9 shows the SWRO backwash methodology (Reang and Nath 2021). Figure 2.10 shows the presence of fouling materials in the SWRO membrane due to

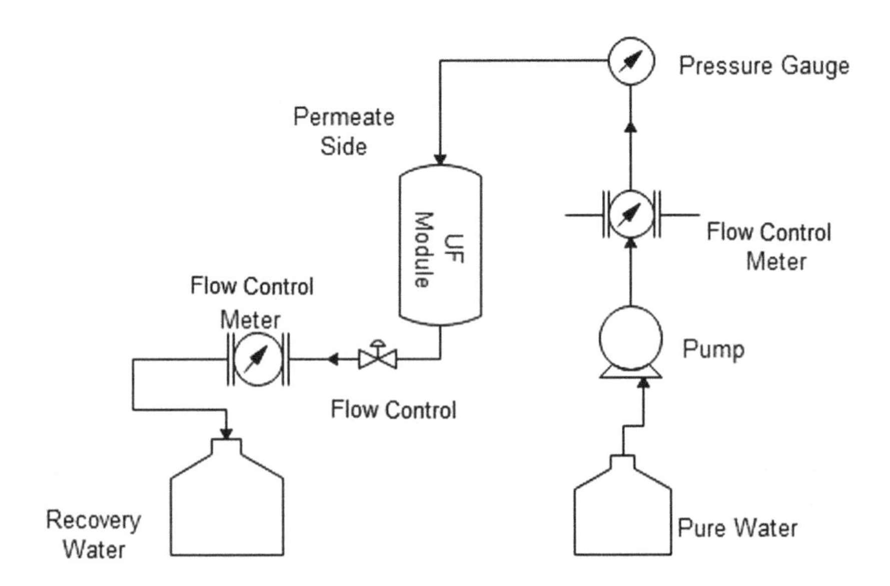

FIGURE 2.5 Flowchart of HFUF backwash method.

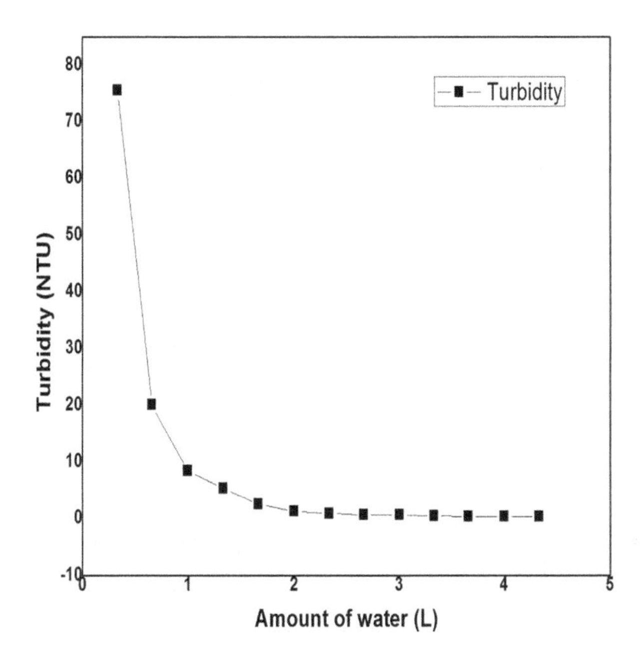

FIGURE 2.6 Amount of water with respect to turbidity used for UF membrane in backwashing operation.

Source: Figure taken with permission from Reang and Nath (2021).

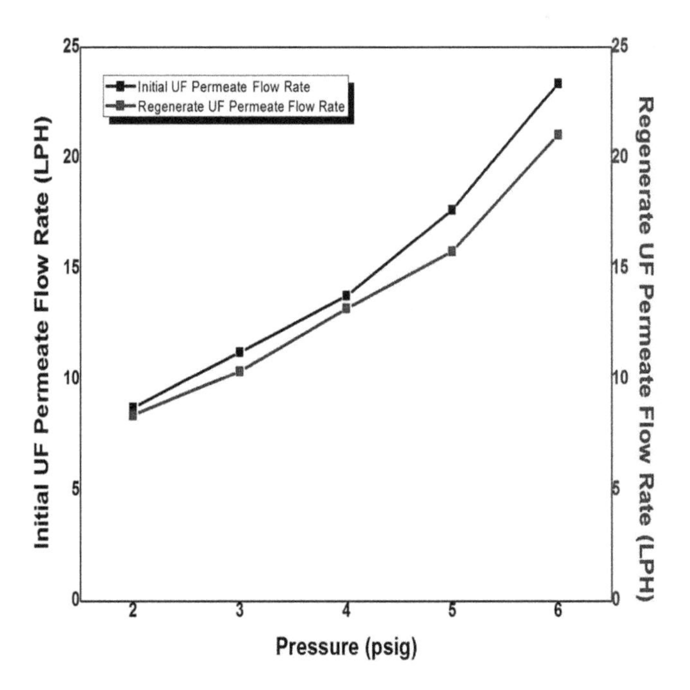

FIGURE 2.7 Performance of the regenerated HFUF membrane (comparison with 0 L).

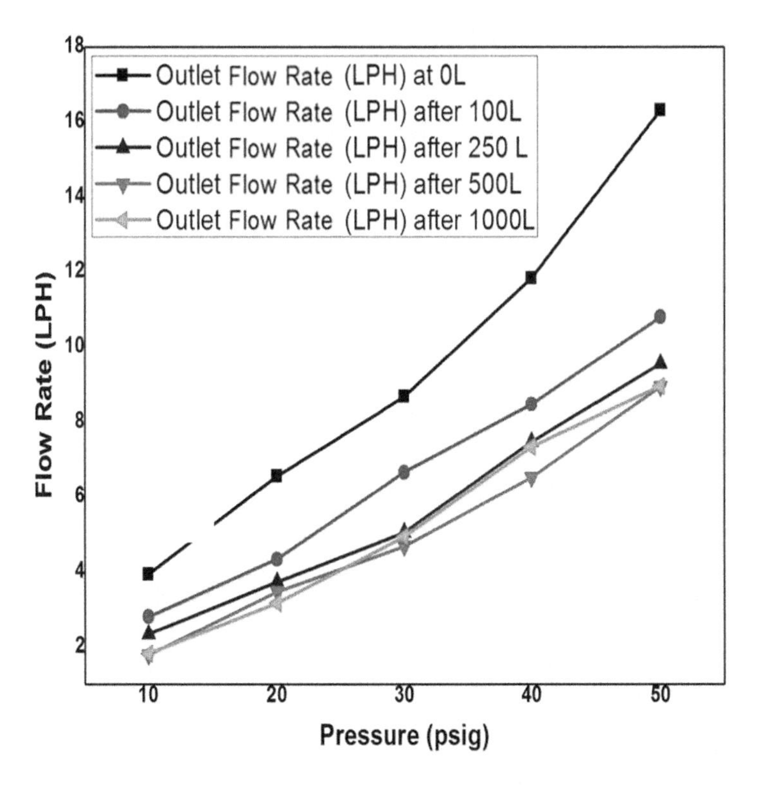

FIGURE 2.8 Life cycle of SWRO pressure vs. flow rate at different amount of pressure.

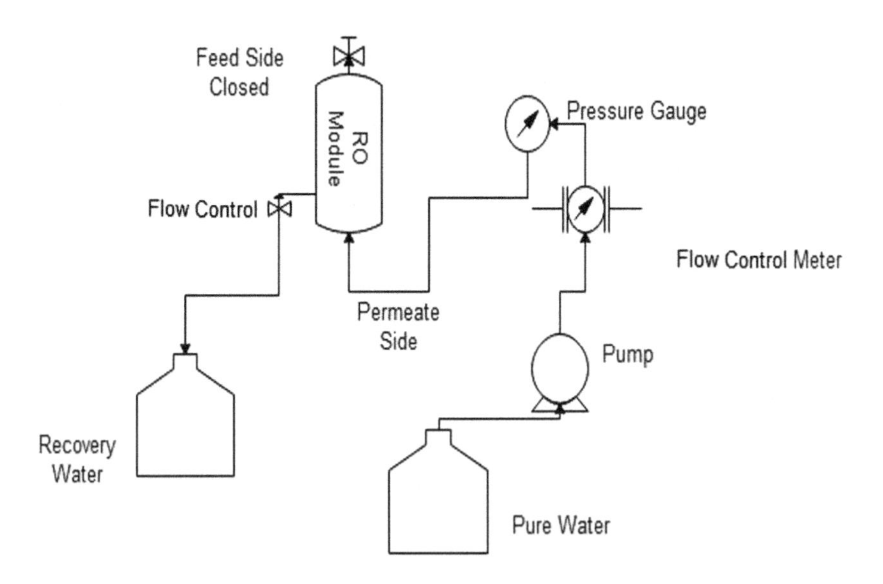

FIGURE 2.9 Flowchart of RO backwash method.

Source: Figure taken with permission from Reang and Nath (2021).

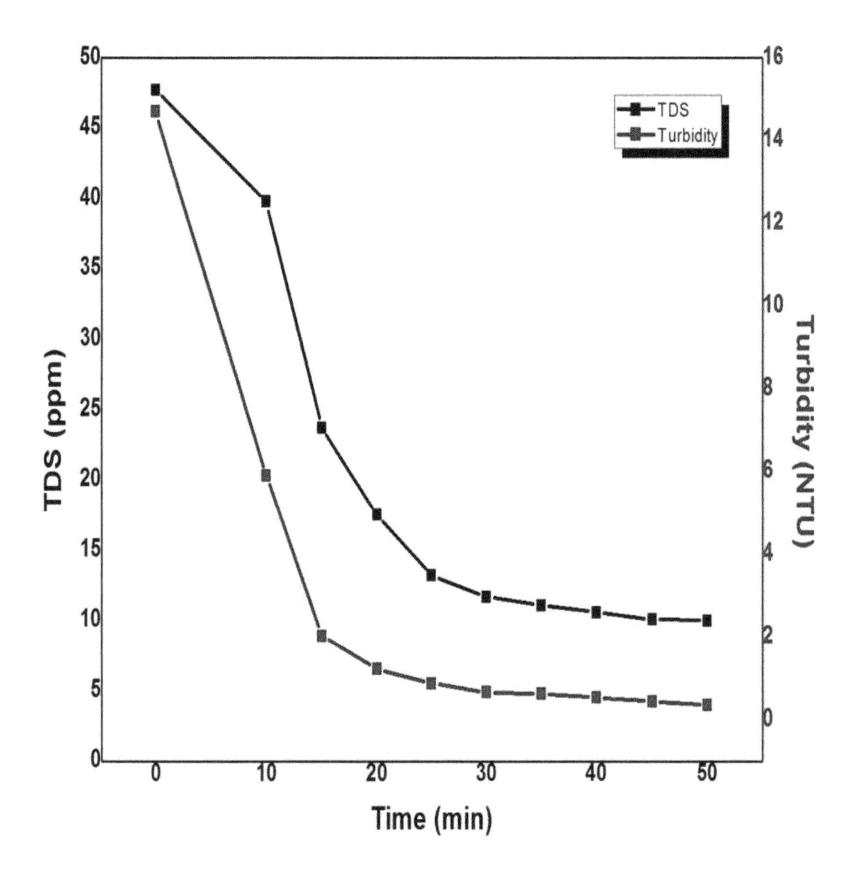

FIGURE 2.10 TDS and turbidity with respect to amount of time used for RO membrane in backwashing operation.

the presence of high TDS and turbidity which were released with the backwashed water. The membrane needs a total of around 3.33 L of pure water and 60 min in order to clean the membrane, and the performance of the recovered SWRO membrane was evaluated again by passing pure water (0.35 NTU and 10 ppm) through it and comparing with the data obtained while the RO membrane was new (compared with 0 L flow from Figure 2.8), represented in Figure 2.11. Figure 2.11 confirms the SWRO regeneration process.

2.3.7 Backwash–Backflush of SWRO

The SWRO process led to the production of 7.2 L of potable water and 15.8 L of concentrated surfactant solution. SWRO backwash–backflush method as adapted by Sukanta et al. (Reang and Nath 2021) was followed, as shown in Figure 2.12. Here the backwash pressure is kept at 70 psig, and the recovered product is pumped to the feed side of RO module at a pressure of 40 psig. Approximately 2.64 L of pure water were required for backwash–backflush. In Figure 2.13 it is shown that TDS and

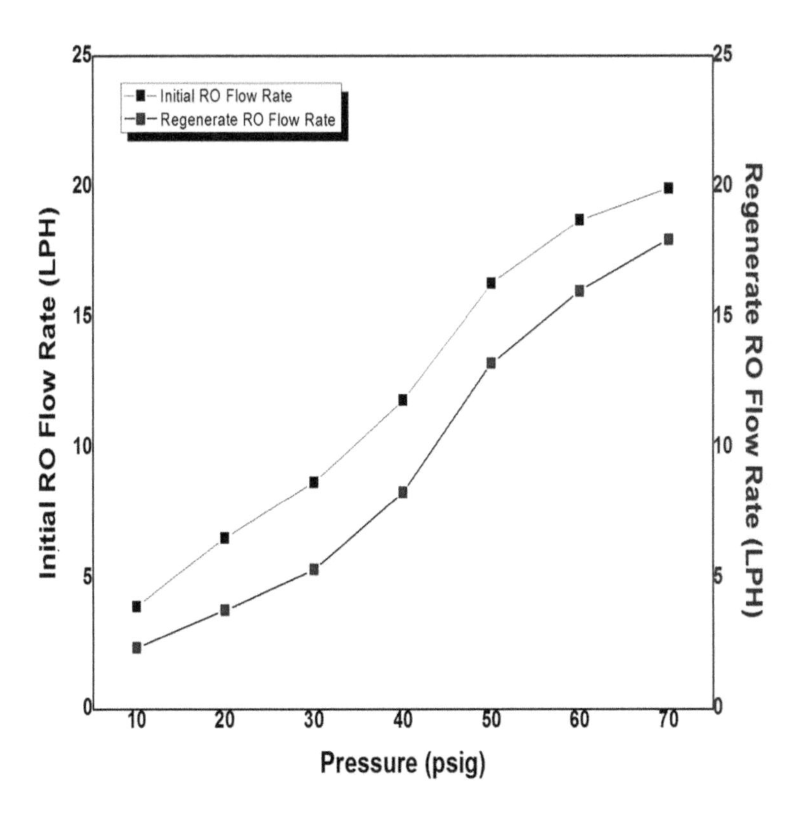

FIGURE 2.11 Performance of the regenerated SWRO membrane (comparison with 0 L).

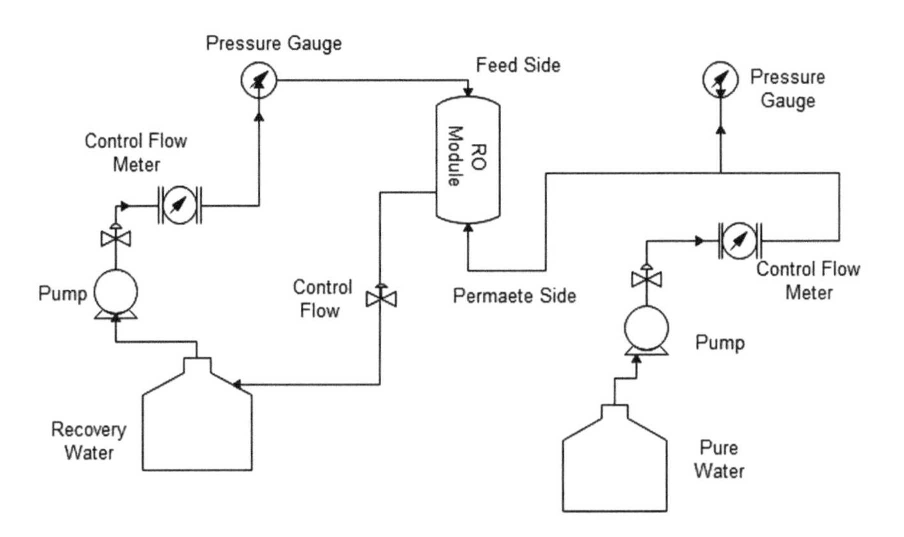

FIGURE 2.12 Flowchart of backwash–backflush of SWRO.

Source: Figure taken with permission from Reang and Nath (2021).

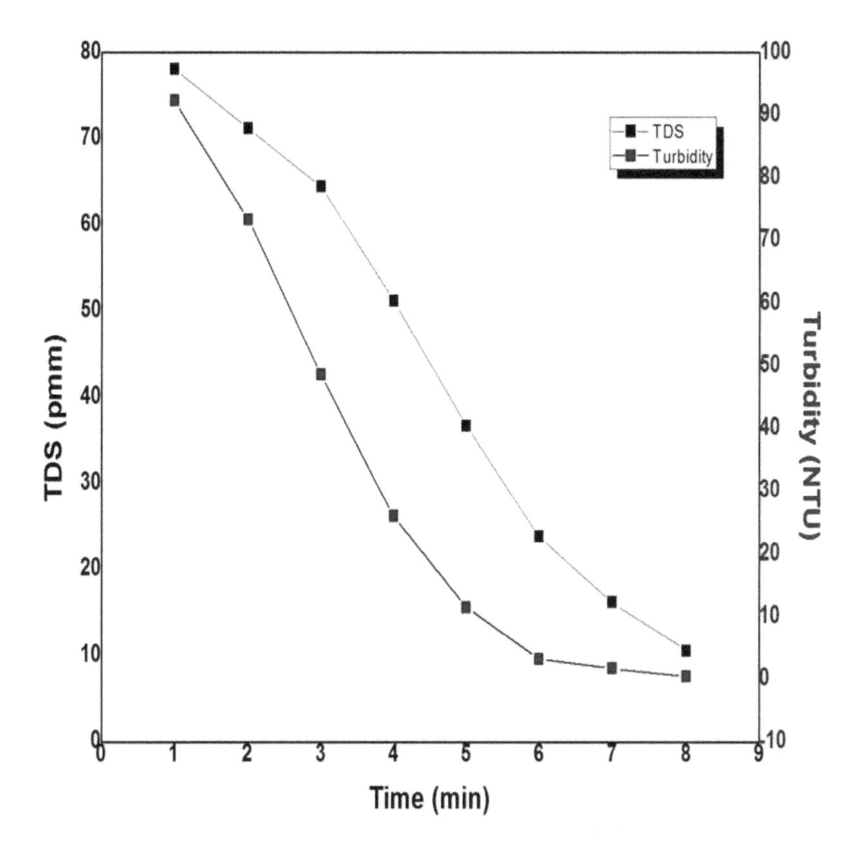

FIGURE 2.13 Time with respect to TDS and turbidity used for RO membrane in backwash–backflush operation.

turbidity decrease after certain amount of water is used, which is less as compared to backwash. Only 8 min are needed for the completion of backwash–backflush, as shown in Figure 2.14. TDS and turbidity decrease with increases in time, which is less as compared to backwash, as represented in Figure 2.15.

2.4 CONCLUSION

The study provided a sustainable, environment-friendly method to treat gray water coming out as washing machine discharge. The process leads to separating the dirt using the HFUF process. The surfactant was separated from water using SWRO process. Both surfactant and water quality were estimated to be of very potable nature and can be easily reused. In the experiment, HFUF backwash and SWRO backwash–backflush methods were found to have increased the overall efficiency of the process. From the membrane life cycle test, the decline of the flow rate can be seen in all the membranes, clearly indicating that fouling is present, but by backwash and backwash–backflush methods, especially for the SW membranes, the membrane can

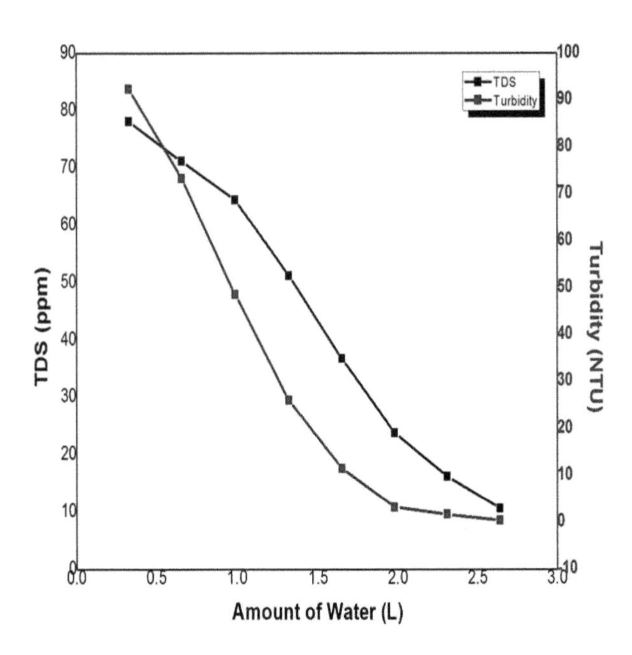

FIGURE 2.14 TDS and turbidity with respect to amount of water used for RO membrane in backwash–backflush operation.

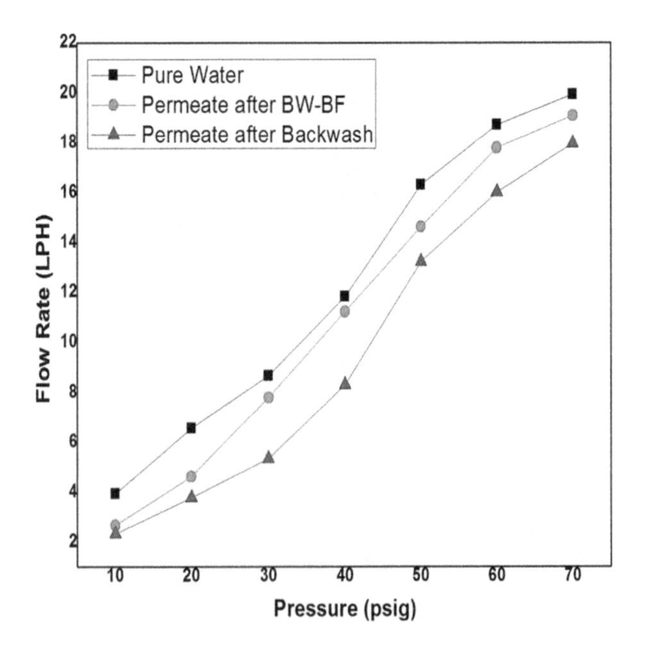

FIGURE 2.15 Comparison of regenerated RO membrane after backwash and backwash–backflush.

be regenerated easily. It is found that the backwash–backflush method is more efficient than the backwash method for spiral wound membranes. This combination can be adapted for the sustainable and environmentally friendly recovery of surfactants and potable water.

REFERENCES

Ahmad, Jamil, and Hisham EL-Dessouky. 2008. "Design of a Modified Low Cost Treatment System for the Recycling and Reuse of Laundry Waste Water." *Resources, Conservation and Recycling* 52 (7): 973–978. https://doi.org/10.1016/j.resconrec.2008.03.001.

Das, Joydeep, Chandrani Debnath, Harjeet Nath, and Rishabh Saxena. 2019. "Antibacterial Effect of Activated Carbons Prepared from Some Biomasses Available in North East India." *Energy Sources, Part A: Recovery, Utilization, and Environmental Effects* 0 (0): 1–11. https://doi.org/10.1080/15567036.2019.1656305.

Das, Joydeep, Rahul Saha, Harjeet Nath, Abhijit Mondal, and Soma Nag. 2022. "An Eco - Friendly Removal of Cd (II) Utilizing Banana Pseudo - Fibre and Moringa Bark as Indigenous Green Adsorbent and Modelling of Adsorption by Artificial Neural Network." *Environmental Science and Pollution Research.* https://doi.org/10.1007/s11356-022-21702-z.

Gao, Larry X., Anditya Rahardianto, Han Gu, Panagiotis D. Christofides, and Yoram Cohen. 2016. "Novel Design and Operational Control of Integrated Ultrafiltration – Reverse Osmosis System with RO Concentrate Backwash." *Desalination* 382: 43–52. https://doi.org/10.1016/j.desal.2015.12.022.

Gross, A., N. Azulai, G. Oron, Z. Ronen, M. Arnold, and A. Nejidat. 2005. "Environmental Impact and Health Risks Associated with Greywater Irrigation: A Case Study." *Water Science and Technology: A Journal of the International Association on Water Pollution Research* 52 (8): 161–169.

Huang, Liwei, and Jeffrey R. McCutcheon. 2015. "Impact of Support Layer Pore Size on Performance of Thin Film Composite Membranes for Forward Osmosis." *Journal of Membrane Science* 483 (June): 25–33. https://doi.org/10.1016/j.memsci.2015.01.025.

Lenzen, Manfred, Daniel Moran, Anik Bhaduri, Keiichiro Kanemoto, Maksud Bekchanov, Arne Geschke, and Barney Foran. 2013. "International Trade of Scarce Water." *Ecological Economics* 94: 78–85. https://doi.org/10.1016/j.ecolecon.2013.06.018.

Li, Fangyue, Knut Wichmann, and Ralf Otterpohl. 2009. "Review of the Technological Approaches for Grey Water Treatment and Reuses." *Science of the Total Environment* 407 (11): 3439–3449. https://doi.org/10.1016/j.scitotenv.2009.02.004.

Reang, Sukanta, and Harjeet Nath. 2021. "Grey Water Treatment with Spiral Wound UF and RO Membranes." *Materials Today: Proceedings* 46 (January): 6253–6259. https://doi.org/10.1016/j.matpr.2020.04.781.

Saha, Aritrika, Harjeet Nath, and Rahul Das. 2023. *Sustainable Chemical, Mineral and Material Processing. Lecture Notes in Mechanical Engineering*. Edited by Eswaraiah Chinthapudi, Suddhasatwa Basu, and Bhaskar Narayan Thorat. Singapore: Springer Nature. https://doi.org/10.1007/978-981-19-7264-5.

Said, Muhammad, Siti Rozaimakh Sheikh Abdullah, and Abdul Wahab Mohammad. 2016. "Palm Oil Mill Effluent Treatment Through Combined Process Adsorption and Membrane Filtration." *Sriwijaya Journal of Environment* 1 (2). https://doi.org/10.22135/sje.2016.1.2.36-41.

Sophia, A. Carmalin, Nath Harjeet, and Praneeth, N. V. S. 2016. "Synthesis of Nano-Porous Carbon from Cellulosic Waste and Its Application in Water Disinfection Synthesis of Nano-Porous Carbon from Cellulosic Waste and Its Application in Water Disinfection." *Current Science* 111 (October): 1377–1382. https://doi.org/10.18520/cs/v111/i8/1377-1382.

Sundaramahalingam, Balaji, and Karthikeyan Chandrasekaran. 2019. "Design and Development of Nanoceramic Filter as Point of Use Water Filter." *Advances in Natural Sciences: Nanoscience and Nanotechnology* 10 (4). https://doi.org/10.1088/2043-6254/ab527b.

Venkatesh, T., and S. Senthilmurugan. 2017. "Grey Water Treatment and Simultaneous Surfactant Recovery Using UF and RO Process." *Separation Science and Technology (Philadelphia)*. https://doi.org/10.1080/01496395.2016.1273244.

Wu, Jishan, Bongyeon Jung, Arezou Anvari, Sung Ju Im, Mackenzie Anderson, Xiaoyu Zheng, David Jassby, et al. 2022. "Reverse Osmosis Membrane Compaction and Embossing at Ultra-High Pressure Operation." *Desalination* 537. https://doi.org/10.1016/j.desal.2022.115875.

Yeom, C. K., C. U. Kim, B. S. Kim, K. J. Kim, and J. M. Lee. 1998. "Recovery of Anionic Surfactant by RO Process. Part I. Preparation of Polyelectrolyte-Complex Anionic Membrane." *Journal of Membrane Science* 143 (1–2): 207–218. https://doi.org/10.1016/S0376-7388(98)00026-X.

3 Withania Somnifera
A Rasayana Herb for Sustainable Management of Human Health and Overall Well-Being

Jackson Sugunakara Chary and Anuradha Sharma

3.1 INTRODUCTION

The ancient medicinal system of India, Ayurveda, emphasizes on a holistic treatment approach which defines health as the state of the soul, body, and mind together and suggests that overall well-being of all the three components can be achieved only after being in harmony with nature (Shi et al., 2021). In addition to India, other Asian countries, including China, Thailand, Japan, and South Korea, also practice herbal medicine as a traditional medicinal system. These days, it has also gotten famous in Western countries as well in terms of aromatherapy or essential oil therapy as an alternative treatment strategy. Along with the health benefits suggested by Ayurveda and current research, herbal medicine is also associated with some side effects, like drug toxicity, which are comparatively less than synthetic medicines. Herbal medicine is an affordable medicinal system, especially to the poor population of India and other developing countries, which meets the target of sustainability – medicine for every individual. There are several herbs listed in Ayurveda for treatment of various ailments, and *Withania somnifera* is one of the famous Rasayana herbs.

W. somnifera, also known as winter cherry or Indian ginseng, is known for its myriad health benefits for ages. This Rasayana herb has been used as tonic for children and as a rejuvenator for middle- and older-aged individuals. It acts as a potent adaptogen by enhancing the body's capacity to endure stress. It is an excellent memory enhancer and also suggested to improve health and functioning of central as well as peripheral nervous system. This herb also promotes a balanced sexual and reproductive system by improving the reproductive system. By enhancing cell-mediated immunity, *W. somnifera* boosts the body's natural ability to fight against diseases. Additionally, it possesses wonderful antioxidant properties that scavenge free radicals, ameliorate oxidative stress, and delay aging. Owing to these valuable properties, *W. somnifera* is used to cure a variety of ailments, including cancer, neurological and cardiovascular disorders, hepatotoxicity, arthritis, hyperlipidemia, impotence, dementia, anxiety, and others (Bharti et al., 2016).

DOI: 10.1201/9781003517108-3

Various studies based on phytochemical characterization enlisted various phytochemicals like alkaloids, steroidal lactones, flavonoids, saponins, terpenoids, and tannins as phytoconstituents of *W. somnifera*. Alkaloids and steroidal lactones are majorly considered responsible for the pharmacological properties of this herb. Withanolides and withaferins have shown the most promising bioactivities, specifically anti-cancer properties. Phytoconstituents of *W. somnifera* are listed as in Table 3.1.

TABLE 3.1
Various Phytochemicals of *W. somnifera* and Their Beneficial Activities

S. No.	Type/Class	Phytochemicals	Plant Parts	Pharmacological Activity
1.	Alkaloids	Choline, cuscohygrine, dl-isopelletierine, somniferine, nicotine, somniferinine, withanine, withananine, pseudowithanine, tropine, pseudotropine, and 3-tigloyloxytropane	Roots	Pervasive bradycardia, hypotension, and respiratory stimulant effect
2.	Glycosides	Sitoindosides (VII, VIII)	Roots	Anti-stress and adaptogenic activity
3.	Glycowithanolide	Sitoindosides VII–X and withaferin-A	Roots	Attenuate cerebral functional deficits, including amnesia in geriatric patients, neuroprotective activity against Alzheimer's disease
–		Glycowithanolides and sitoindosides IX and X	Whole plant	Deficits significant stress-reduction efforts
4.	Steroidal lactones	Withaferin A	Roots and leaves	Antiproliferative, anti-tumor, anti-inflammation, anti-angiogenesis, cell cycle arrest, NF-kB inhibiting activity, chemotherapeutic, immunomodulatory, nephroprotective
		Withanolides (ashwagandhanolide, withanoside I, withanoside, L-asparaginase, withanoside IV, and withanoside VI, withanolide A, withanolide D)	Fruits, root, leaves, whole plants	Anti-inflammatory, neuroprotective, memory-enhancing, radio-sensitization, chemoprevention
5.	Less-common alkaloids	Isopelletierine, anahygrine, anaferine, and cuseohygrine	Root and stem	Ant-cancer, effective for cramps in diarrhea, antimicrobial
6.	Phenols	Gallic, syringic, vanillic, benzoic, p-coumaric, and syringic acids	Root and stem	Antioxidative agents
7.	Flavonoids	Catechin, kaempferol, and naringenin	Root and stem	Inhibit tumor growth promoters

3.2 *W. SOMNIFERA* IN MANAGEMENT OF DISEASE AND OVERALL HEALTH

3.2.1 Communicable Diseases and *W. somnifera*

Diseases caused by infectious pathogens are known as infectious diseases or communicable diseases. Different microbes, including bacteria, viruses, fungi, and parasites, are responsible for the onset of communicable diseases, and these diseases are typically treated with specific antibiotic, antiviral, antifungal, or antiparasitic medications, depending on the pathogenic agent. The onset and spread of communicable diseases affect the economic growth by exponentially raising the healthcare expenditures and reducing productivity worldwide. We have recently witnessed the COVID-19 pandemic, and every one of us can judge the need to tackle communicable diseases for the sustainable and healthy development of a nation.

Various studies have suggested that extracts from ashwagandha roots, leaves, and berries exhibit antibacterial activity against different gram-positive and gram-negative bacteria. Methanolic extract of *W. somnifera* exhibited antibacterial effects against methicillin-resistant *Staphylococcus aureus* (MRSA) and *Enterococcus* sp. obtained from soft tissue infection (Bisht and Rawat, 2014). Butanol subfraction of methanolic extract exhibited the maximum efficacy against experimental salmonellosis, as suggested by the largest zone of inhibition. Additionally, this extract exhibited safety to other living cells in comparison to standard antibiotic chloramphenicol, which induces lysis of erythrocytes (Kishanrao and Mandge, 2020). Khanchandani et al. (2019) reviewed an extensive list of bacteria against which different extracts of *W. somnifera* showed potential antibacterial activity.

During the COVID-19 era, this plant was most explored for its antiviral properties and was suggested as a potential candidate for antiviral drug development and therapeutic management of viral diseases. Withaferin A and withanone, steroidal lactones from ashwagandha, suppress the inflammatory cytokines production, strengthen the immune system and protects multiple organs from stress and toxicity, which made these compounds the potential drug targets for antiviral treatment therapy (Singh et al., 2022). A computational study by Cai et al. (2015) also suggested the potential antiviral effect of withaferin A against influenza viruses by targeting neuraminidase. Withaferin A containing herbal drugs also show higher binding affinity to herpes simplex virus in computation studies, suggesting it as a candidate for anti–herpes simplex drug development as well.

W. somnifera and its different extracts also exhibited antifungal effect against different species, including *Candida albicans*, *Trichophyton rubrum*, *Aspergillus flavus*, *Fusarium oxysporium*, *Fusarium moniliformis*, and *Aspergillus niger*. Hexane, ethyl acetate, and aqueous extracts of *W. somnifera*, when tested on the mentioned fungal species except *C. albicans* and *Trichophyton rubrum* using disk diffusion assay, showed differential antifungal activity to different fungus. A cream formulation containing nanoparticles synthesized using *W. somnifera* extracts inhibited the growth of *C. albicans*, suggesting it as an eco-friendly alternative of antibiotic creams (Marslin et al., 2015).

Parasitic diseases such as malaria, leishmaniasis, schistosomiasis, pediculosis, chagas disease, etc., which, although need a medium like vectors or water source, if spread can contribute to great economic and productivity loss too. A study by Dikasso et al. (2006) reported the anti-plasmodial effect of *W. somnifera* as suggested by dose-dependent drop of parasitemia of malarial parasite in Swiss albino mice. The effect of *W. somnifera* against visceral leishmaniasis, a parasitic infection which targets macrophages and induces immunosuppression, was also studied, where treatment with *W. somnifera* significantly reduced parasitic (*L. donowani*) infection and successfully normalized the immune response in infected BALB/c mice (Kaur et al., 2014).

3.2.2 Non-Communicable Diseases and *W. Somnifera*

Non-communicable diseases are responsible for approximately 74% of global deaths. Cardiovascular diseases, cancers, lung disease, diabetes, strokes, neurodegenerative diseases, and mental ailments are the top listed NCDs as per the World Health Organization. Control of NCDs onset and reduction of NCDs-associated socioeconomic burden are crucial.

The cardioprotective activity of Ashwagandha has been suggested by some reports, including the studies on myocardial infarction, cardiorespiratory endurance, atherogenesis, etc. A clinical study on healthy adults showed the improvement of various parameters related to cardiovascular endurance, including power, velocity, and VO_2 max, in healthy individuals who consumed *W. somnifera* root extract (Sandhu et al., 2010). Another randomized clinical trial in healthy athletes also supported the findings of a previous study and showed significant improvement in different parameters of cardiorespiratory endurance (Tiwari et al., 2021). Withanone and withaferin A were shown to support the repair of muscles and muscular activity (Wang et al., 2021). Additionally, a preclinical study suggested a significant cardioprotective effect of *W. somnifera* in experimental model of myocardial infarction, the most lethal manifestation of heart diseases. Consumption of *W. somnifera* powder by Wistar albino rats significantly improved the hemodynamic parameters and raised the levels of antioxidants, thus resulting in protection from myocardial infarction induced by isoprenaline (Mohanty et al., 2004).

Diabetes increases the risk of getting cardiovascular diseases by twofold and also predisposes to other non-communicable diseases. Some polyherbal formulations containing *W. somnifera* as a component, such as Trasina or Dianix, possess anti-diabetic properties. Consumption of Ashwagandha powder is shown to stabilize blood sugar index and insulin sensitivity in both preclinical and clinical studies. Consumption of *W. somnifera* extracts is demonstrated to enhance glucose uptake by skeletal myocytes or adipocytes and to reduce the blood sugar levels in urine as well as blood (Anwer et al., 2008; Paul et al., 2021). Additionally, treatment with *W. somnifera* extracts protected prepubertal rats from diabetes-associated testicular impairments, which were induced by injecting streptozotocin to the animals (Kyathanahalli et al., 2014).

Beneficial effects of *W. somnifera* in various neurological and neuropsychiatric disorders have also been suggested by a plethora of studies. *W. somnifera* is famous for its stress-relieving property due to its adaptogenic potential. Preclinical and clinical studies witness the anti-stress, anti-anxiety, anti-depression, and sleep promoting effects of this herb. Consumption of *W. somnifera* commercial capsules suppressed the symptoms of depression in schizophrenic individuals; however, the studies were associated with significant limitations. Alcoholic/hydroalcoholic extracts of *W. somnifera*, which were rich in withanolides, withanones, and withaferins, exhibited improvement in sleep quantity and quality. Further, use of *W. somnifera* along with anti-depressant drugs showed a synergistic effect and potentiated the overall anti-depressive outcome (Speers et al., 2021). Stress- and depression-related memory decline and impaired cognitive ability were also found to be improved in clinical subjects upon consumption of *W. somnifera* for 30 days' period (Remenapp et al., 2022).

This herb is also indicated to have potential therapeutic potential in neurodegenerative diseases, as evidenced by studies suggesting regeneration of axons, synaptic plasticity improvement, and clearance of Aβ plaques in brain of mice upon treatment with *W. somnifera* extracts in different preclinical models of Alzheimer's disease. Different phytoconstituents of Ashwagandha are also indicated to exert neuroprotection individually, such as withanoside IV administration for 21 days promoting axonal regeneration, resulting to motor function recovery post–spinal cord injury (Kuboyama et al., 2014). Further, improvement of locomotor activity, cholinergic function, amelioration of oxidative stress, and mitochondrial dysfunction has also been demonstrated by treatment with different extracts or phytoconstituents of *W. somnifera* in Parkinson's and Huntington's disease models as well (Dar, 2020).

As discussed, cancer is also among the top debilitating non-communicable diseases, which requires prompt development of alternative treatment therapies. Phytoconstituents of *W. somnifera* are among the potential targets for plant-based chemotherapeutic agents. Withaferin A and different extracts induced apoptosis in breast cancer cells (MCF-7) as well as altered the gene expression in triple-negative breast cancer cells. Some studies suggest increased ROS production leading to apoptosis upon treatment with this herb. Anti-cancer properties of *W. somnifera* have also been indicated against colorectal cancers, brain cancers, lung cancers, ovarian cancers, gastric carcinoma, renal carcinoma, and others. Use of *W. somnifera* along with traditional chemotherapy to mitigate the side effects and improve quality of life was also suggested by some preclinical studies; however, detailed exploration of the concept as dietary supplement still needs to be explored (Dutta et al., 2019).

3.2.3 Other Health-Affecting Factors and *W. somnifera*

W. somnifera and its different formulations are further reported to have protective effects against traumatic brain injury, renal disorders, and wound healing properties against chronic injuries. A case report supports the wound healing claim of this herb by successfully managing the ulcerative wound in toe upon application of Ashwagandha-containing herbal paste, which was otherwise suggested to be amputated by the doctors (Sheoran et al., 2020). Withaferin A was found to exert a healing effect against bone injury and promoting the mechanical strength of bone (Khedgikar et al., 2013). Some studies also highlight the potential of this herb in managing substance abuse and addiction. An *in vitro* study by Bansal and Banerjee (2016) studied the effect of *W. somnifera* and *A. punjabianum* against anxiety associated with alcohol withdrawal. Swiss albino mice consuming alcohol for 21 days (lab model for chronic alcohol consumption) fed with either alone or a combination churna of *W. somnifera* and *A. punjabianum* exhibited significantly reduced alcohol dependency and associated anxiety. Adverse symptoms associated with chronic alcohol withdrawal, like seizures, depression, impaired cognition, and motor coordination, were significantly attenuated by *W. somnifera*, suggesting this herb as one of the potential targets for anti-additive therapy (Ruby et al., 2012).

It is well-known for its aphrodisiac property, highlighting its utility in the management of sexual and reproductive health. A recent randomized clinical trial of eight weeks where adults with lower sexual desire were given 300 mg root extract of Ashwagandha daily and reported, in the evaluation after eight weeks, a significant increase in DISF (Derogatis Interview for Sexual Functioning – Male) scores and testosterone levels, suggesting the improvement of sexual well-being (Chauhan et al., 2022). Further, *W. somnifera* is suggested to enhance reproductive health by increasing semen quality, reducing oxidative stress, and enhancing reproductive enzymes activity by different preclinical and clinical studies (Nasimi Doost Azgomi et al., 2018).

Obesity epidemic, which is one of the prime risk factors for various co-morbid conditions (hypertension, cardiovascular diseases, diabetes, dementia, neurodegenerative diseases), affects human health at a global level. *W. somnifera* extracts and phytoconstituents also have anti-adipogenic activity, suggesting it as a potential intervention for obesity treatment (Lee et al., 2021). Kaur and Kaur (2017) reported that Ashwagandha extract inhibited obesity-related neurological ailment, including anxiety and neuroinflammation, in Wistar albino rats.

Overall, this valuable herb "Indian ginseng" is suggested to have the potential to treat communicable and non-communicable diseases as well as to improve overall health and well-being. However, to devise the effective formulations and to test their efficacy requires detailed studies for extensive implementation to bedside.

3.3 DIFFERENT PHARMACOLOGICAL ACTIVITIES OF *W. SOMNIFERA*

The therapeutic potential of *W. somnifera* against various ailments and in promoting overall well-being is attributed to the various pharmacological properties of this herb, such as adaptogenic, anti-ulcer, antioxidative, anti-inflammatory, neuroprotective, chemopreventive, and immunomodulatory properties (Figure 3.1), which are further suggested to be exerted by the different phytoconstituents of this plant. Various pharmacological activities of this herb are summarized in the following table (Table 3.2).

3.4 *W. SOMNIFERA* IN SUSTAINABLE DEVELOPMENT

Goal 3 of Sustainable Development Goals aims to improve overall health and well-being by targeting to reduce neonatal death, reduce mortality and morbidity during pregnancy, treat and prevent different communicable and non-communicable diseases, reduce global mortality due to injuries and accidents, reduce substance

FIGURE 3.1 Various pharmacological activities of *W. somnifera*.

abuse, enhance access to reproductive/sexual healthcare, enhance overall affordability of health service and training of healthcare professionals for disease-associated risk management (Mamidi & Thakar, 2011). As discussed in detail in the current chapter, *W. somnifera* possesses different pharmacological properties, rendering it beneficial for the treatment of various communicable and non-communicable disease and injuries, improvement of sexual and reproductive health, reduction of addiction, and promotion of overall health. Additionally, herbal medicine is the part of ancient traditional medicine which is cost-effective and accessible to majority of the population, including tribal people in developing/economically weaker populations. So incorporating this herb in the healthcare system would significantly contribute to the holistic health approach, which focuses on preventive natural medicine. It will aid in the reduction of dependency on palliative care and fuel the sustainable practices of healthcare (Figure 3.2).

Additionally, Sustainable Development Goal 15 also covers medicinal plants, suggesting their cultivation, preservation, and enhancement of biodiversity. *W. somnifera*, along with other traditional herbs, could be planned as part of sustainable development, which will help protect the heritage of a nation, maintain biodiversity,

TABLE 3.2
Pharmacological Activities of *W. somnifera*

Property	Plant Part Used and Reference	Description	Results of the Study
Adaptogenic property	Root powder (Priyanka et al., 2020)	Kathiawari horses were given Ashwagandha root powder and exposed to different types of stressors, like exercise, noise, and separation.	Indicators of stress, that is, increased cortisol and epinephrine and decreased serotonin, were reversed significantly by consumption of *W. somnifera* root powder as compared to control horses, which were only given the normal diet.
	Root extract (Bhattacharya and Muruganandam, 2003)	Wistar albino rats were given Ashwagandha root extract 1 h prior to foot shock–induced stress for 21 days.	Stress due to foot shock induced different impairments in immune system, blood glucose levels, stress regulation, impaired sexual performance, and depression; however, administration of *W. somnifera* root extract before giving foot shock significantly suppressed these impairments and showed anti-stress or adaptogenic activity.
	Standardized ashwagandha extract (Shoden) Lopresti et al., 2019	60 healthy adults were randomly given 240 mg of extract or placebo, and stress-related parameters, including hormones and Hamilton anxiety scale, were evaluated.	Significant decrease in HAS score and morning cortisol in the individuals who consumed ashwagandha extract. No adverse effects were reported.
	Water extract (Kaur et al., 2001)	Anti-stress effect of aqueous suspension, and a phytoconstituent X was examined in CHR (cold, hypoxia, and restraint) model of stress in rodents.	*W. somnifera* and phytoconstituent X consumption significantly ameliorated the adverse effects exerted by low temperatures, hypoxic conditions, as suggested rapid recovery of normal temperature after cold shock as compared to control animals.

Antioxidant property	Root powder	Kathiawari horses were given Ashwagandha root powder and exposed to different type of stressors, like exercise, noise, and separation.	Post-stress evaluation of antioxidant profile suggested improved glutathione (GSH) and superoxide dismutase (SOD) activity and significantly reduced TBARS (thiobarbituric acid reactive substances) levels in animals consuming *W. somnifera* powder, indicating the antioxidative potential of *W. somnifera*.
	Root and leaf powder (Singh et al., 2010)	Root and leaf extracts were administered to diabetic-prone rats, whereas glibenclamide was provided as a control.	Significant reduction in the TBARS levels and hyperglycemia. Increased expression of glycogen, vitamins C and E, SOD, CAT, GPx, GST, and GSH.
	Glycowithanolides (Bhattacharya et al., 2001).	In a chronic foot shock stress paradigm, rats were given glycowithanolides once daily for 21 days, and effect on foot shock–induced oxidative stress was evaluated.	Increased SOD activity and lipid peroxidation were reversed in animals treated with glycowithanolides, along with enhanced activity of CAT and GPx as compared to shock control animals.
	Ethanolic root extract (Suganaya et al., 2020)	Wistar albino rats were given ethanolic root extract of *W. somnifera* for 30 days and checked oxidative status post–sleep deprivation for one week.	Significant increase in antioxidant enzymes, including GSTx, SOD, CAT, was observed in animals treated with *W. somnifera* extract, along with sleep deprivation, as compared to sleep-deprived control animals. Additionally, the consumption of extract significantly reversed the sleep deprivation–induced lipid peroxidation and free radical production.
	Ethanolic root extract (Birla et al., 2019)	Swiss albino mice exposed to bisphenol A (BPA) were given Ashwagandha extract, and different parameters related to oxidative stress as well as cognition were studies.	*W. somnifera* extract treatment ameliorated the BPA-induced oxidative stress, as suggested by significant increase in CAT, SOD, and significant reduction in lipid peroxidation, as compared to only BPA-treated animals.

(Continued)

TABLE 3.2 *(Continued)*
Pharmacological Activities of *W. somnifera*

Property	Plant Part Used and Reference	Description	Results of the Study
Immunomodulatory potential	Immunomodulatory potential	Kathiawari horses were given Ashwagandha root powder and exposed to different type of stressors, like exercise, noise, and separation.	Significant reduction in interleukin-16 (IL-16) post-stress in the animals which consumed diet containing *W. somnifera* root powder as compared to control animals.
	Withanolide glycoside–enriched extract of roots and leaves (Tharakan et al., 2021)	Immunomodulatory effect was accessed in a blinded clinical trial where healthy adults were given the mentioned extract or placebo and were assessed for different immunological parameters.	Consumption of *W. somnifera* extract significantly increased the levels of different immune components, like Igs, immune cells (including T, B, and NK), and level of cytokines as compared to placebo group.
	Leaf extract (Khan et al., 2009)	Role of *W. somnifera* in immune system activation was studied using splenocytes as in vitro model system and using BALBc mice as in vivo system.	*W. somnifera* extracts treatment enhances the proliferation and activation of T cells, upregulates the secretion of integrins and different co-stimulatory molecules required for T cell function.
	Aqueous root extract (Khan et al., 2019)	Animals with collagen-induced arthritis were given root extract of *W. somnifera*, and effects on different inflammatory-, oxidative-, and immune system–related parameters were studied.	This extract regulated the inflammatory component of immune system and significantly downregulated pro-inflammatory cytokines, which, in turn, provides relief from inflammation in arthritis.
Anti-inflammatory property	Hydroalcoholic extract of the plant (Chandra et al., 2012)	The protein denaturation bioassay was chosen in the current investigation to evaluate the hydroalcoholic extract of ashwagandha's anti-inflammatory effects in vitro. Different concentrations of ashwagandha extract were mixed with egg albumin, and parameters related to inflammation were studied.	The hydroalcoholic extract of ashwagandha (HAWS), when incubated with albumin, inhibited protein denaturation significantly and was more effective as compared to standard drug diclofenac sodium.
	Hydroalcoholic root extract (Joshi et al., 2020)	Anti-inflammatory potential of *W. somnifera* and its polyherbal formulation was studied against lipopolysaccharide-induced inflammation in RAW264.7 macrophages.	*W. somnifera* extracts inhibited LPS-induced TNF alpha cytokine production.

	W. somnifera root powder (Gupta and Singh, 2014)	In this study, *W. somnifera* root powder (at dose levels of 600 and 800 mg kg1) is administered to normal control, arthritic control, and arthritic rat models.	*W. somnifera* root powder (600 mg kg1) administered to the arthritic rats lessened the severity of the condition by successfully suppressing the symptoms of arthritis and enhancing the functional recovery of motor coordination and radiological score.
Antimicrobial activity	Methanolic leaf extract (Bisht and Rawat, 2014)	The study conducted to examine the antimicrobial potential of leaf extract of *W. somnifera* against gram-positive cocci ($n = 20$) from pus samples of patients suffering with infection of soft tissues.	Methanolic extract of *W. somnifera* showed potent antibacterial activity against gram-positive clinical isolates that have been reported to be more resistant against routinely used first line of antibiotics, such as ampicillin, co-trimoxazole, and erythromycin, at the concentration of 2 mg/ml.
	Leaf and root extract (Arora et al., 2004)	*W. somnifera* leaves and roots were extracted using methanol, hexane, and diethyl ether and tested for antibacterial activity against *S. typhimurium* and *E. coli* along with standard antibiotics rifampicin and isoniazid.	Methanol and hexane extracts of *W. somnifera* (leaf as well as root) show significant antibacterial activity, as suggested by agar plate disc-diffusion assay.
	Roots and leaves extract (Owais et al., 2005	The antibacterial activity of the extracts (from different parts of the plant) against salmonellosis was determined using agar well diffusion method.	It found that both aqueous as well as methanolic extracts of leaves and roots were successful in killing the *S. typhimurium* bacteria in a dose-dependent manner. Butanol subfraction of methanol extract was the most effective in inhibiting salmonellosis in BALB/c mice.
Chemoprotective	Roots, stems, and leaves (Dutta et al., 2019)	The anti-cancer efficacy of using *W. somnifera* in MCF-7 breast cancer cell lines was studied.	Reactive oxygen species (ROS) produced by *W. somnifera* extracts have been proven to apoptosis and kill cancer cells, but not normal cell growth. It was found that the phytochemical downregulated the estrogen receptor-α (ER-α) protein in MCF-7 cells.
	Root, stem, and leaves (Dutta et al., 2019)	The anti-cancer efficacy using *W. somnifera* against HCT-116, SW-480 colorectal cell lines was studied.	The extracts have reduced transcriptional activity of STAT3 (signal transducer and activator of transcription 3 on colorectal cancer cell lines' migration and proliferation).
	(Mehta et al., 2021)	The cancer chemoprevention efficacy using *W. somnifera* in U-937 leukemia cancer cell lines was studied.	L-Asparaginase purified from ethanolic *W. somnifera* fruit extract showed antiproliferative activity in acute lymphoblastic leukemia cells isolated from patient blood.

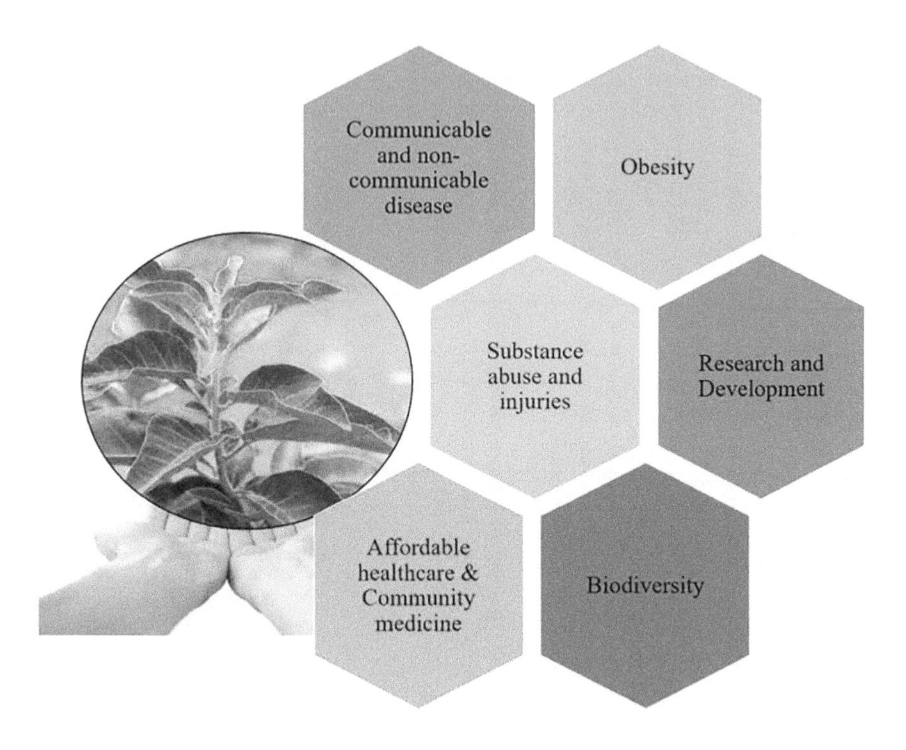

FIGURE 3.2 Ashwagandha, health management, and sustainable development.

and ensure the long-term availability of this medicinal plant. Further, cultivation of this herb and related trade could prove as an employment opportunity, specifically for rural communities. The development of value-added products, such as powders, extracts, ointments, supplements, or medicinal formulation, from *W. somnifera* would provide a source of improving livelihood as well as serve to meet the community healthcare demand.

A significant amount of research is being conducted globally on *W. somnifera* related to different aspects, including propagation, cultivation and genetic improvement, different processing approaches or quality control, identification of therapeutic phytochemical candidates, and development of therapeutic formulations, followed by looking for treatment regimens. This research could provide help in developing evidence-based complementary medicine accessible to majority of the population, along with the development of high-quality products, while ensuring sustainability.

3.5 SUMMARY

W. somnifera possesses various medicinal properties which could be beneficial in the treatment of various infections, heart problems, and neurological disorders and injuries and the improvement of sexual and reproductive health as well as overall well-being. It is an indispensable component of traditional community medicine of India as well as some other Asian countries. Integration of *W. somnifera* into sustainable development projects can promote a harmonious coexistence of environmental preservation, cultural preservation, and human well-being. *W. somnifera* can support a more sustainable and inclusive future through valuing traditional knowledge, fostering economic opportunities, and implementing eco-friendly practices.

REFERENCES

Anwer, T., Sharma, M., Pillai, K. K., & Iqbal, M. (2008). Effect of Withania somnifera on insulin sensitivity in non-insulin-dependent diabetes mellitus rats. *Basic & Clinical Pharmacology & Toxicology*, *102*(6), 498–503.

Arora, S., Dhillon, S., Rani, G., & Nagpal, A. (2004). The in vitro antibacterial/synergistic activities of Withania somnifera extracts. *Fitoterapia*, *75*(3), 385–388.

Bansal, P., & Banerjee, S. (2016). Effect of Withinia somnifera and shilajit on alcohol addiction in mice. *Pharmacognosy Magazine*, *12*(Suppl 2), S121.

Bharti, V. K., Malik, J. K., & Gupta, R. C. (2016). Ashwagandha: multiple health benefits. In *Nutraceuticals* (pp. 717–733). Academic Press.

Bhattacharya, A., Ghosal, S., & Bhattacharya, S. K. (2001). Anti-oxidant effect of Withania somnifera glycowithanolides in chronic footshock stress-induced perturbations of oxidative free radical scavenging enzymes and lipid peroxidation in rat frontal cortex and striatum. *Journal of Ethnopharmacology*, *74*(1), 1–6.

Bhattacharya, S. K., & Muruganandam, A. V. (2003). Adaptogenic activity of Withania somnifera: an experimental study using a rat model of chronic stress. *Pharmacology Biochemistry and Behavior*, *75*(3), 547–555.

Birla, H., Keswani, C., Rai, S. N., Singh, S. S., Zahra, W., Dilnashin, H., . . . Singh, S. P. (2019). Neuroprotective effects of Withania somnifera in BPA induced-cognitive dysfunction and oxidative stress in mice. *Behavioral and Brain Functions*, *15*(1), 1–9.

Bisht, P., & Rawat, V. (2014). Antibacterial activity of Withania somnifera against grampositive isolates from pus samples. *Ayu*, *35*(3), 330.

Cai, Z., Zhang, G., Tang, B., Liu, Y., Fu, X., & Zhang, X. (2015). Promising anti-influenza properties of active constituent of Withania somnifera ayurvedic herb in targeting neuraminidase of H1N1 influenza: computational study. *Cell Biochemistry and Biophysics*, *72*, 727–739.

Chandra, S., Chatterjee, P., Dey, P., & Bhattacharya, S. (2012). Evaluation of anti-inflammatory effect of Ashwagandha: a preliminary study in vitro. *Pharmacognosy Journal*, *4*(29), 47–49.

Chauhan, S., Srivastava, M. K., & Pathak, A. K. (2022). Effect of standardized root extract of Ashwagandha (Withania somnifera) on well-being and sexual performance in adult males: a randomized controlled trial. *Health Science Reports*, *5*(4), e741.

Dar, N. J. (2020). Neurodegenerative diseases and Withania somnifera (L.): an update. *Journal of Ethnopharmacology*, *256*, 112769.

Dikasso, D., Makonnen, E., Debella, A., Abebe, D., Urga, K., Makonnen, W., . . . Guta, M. (2006). Anti-malarial activity of Withania somnifera L. Dunal extracts in mice. *Ethiopian Medical Journal*, *44*(3), 279–285.

Dutta, R., Khalil, R., Green, R., Mohapatra, S. S., & Mohapatra, S. (2019). Withania somnifera (Ashwagandha) and withaferin A: potential in integrative oncology. *International Journal of Molecular Sciences*, *20*(21), 5310.

Gupta, A., & Singh, S. (2014). Evaluation of anti-inflammatory effect of Withania somnifera root on collagen-induced arthritis in rats. *Pharmaceutical Biology*, *52*(3), 308–320.

Joshi, P., Yadaw, G. S., Joshi, S., Semwal, R. B., & Semwal, D. K. (2020). Antioxidant and anti-inflammatory activities of selected medicinal herbs and their polyherbal formulation. *South African Journal of Botany*, *130*, 440–447.

Kaur, P., Mathur, S., Sharma, M., Tiwari, M., Srivastava, K. K., & Chandra, R. (2001). A biologically active constituent of Withania somnifera (Ashwagandha) with antistress activity. *Indian Journal of Clinical Biochemistry*, 16, 195–198.

Kaur, S., Chauhan, K., & Sachdeva, H. (2014). Protection against experimental visceral leishmaniasis by immunostimulation with herbal drugs derived from Withania somnifera and Asparagus racemosus. *Journal of Medical Microbiology*, *63*(10), 1328–1338.

Kaur, T., & Kaur, G. (2017). Withania somnifera as a potential candidate to ameliorate high fat diet-induced anxiety and neuroinflammation. *Journal of Neuroinflammation*, *14*(1), 1–18.

Khan, M. A., Ahmed, R. S., Chandra, N., Arora, V. K., & Ali, A. (2019). In vivo, extract from Withania somnifera root ameliorates arthritis via regulation of key immune mediators of inflammation in experimental model of arthritis. *Anti-Inflammatory & Anti-Allergy Agents in Medicinal Chemistry (Formerly Current Medicinal Chemistry-Anti-Inflammatory and Anti-Allergy Agents)*, *18*(1), 55–70.

Khan, S., Malik, F., Suri, K. A., & Singh, J. (2009). Molecular insight into the immune up-regulatory properties of the leaf extract of Ashwagandha and identification of Th1 immunostimulatory chemical entity. *Vaccine*, *27*(43), 6080–6087.

Khanchandani, N., Shah, P., Kalwani, T., Ardeshna, A., & Dharajiya, D. (2019). Antibacterial and antifungal activity of Ashwagandha (Withania somnifera l.): a review. *Journal of Drug Delivery and Therapeutics*, *9*(5-s), 154–161.

Khedgikar, V., Kushwaha, P., Gautam, J., Verma, A., Changkija, B., Kumar, A., . . . Trivedi, R. (2013). Withaferin A: a proteasomal inhibitor promotes healing after injury and exerts anabolic effect on osteoporotic bone. *Cell Death & Disease*, *4*(8), e778.

Kishanrao, K. B., & Mandge, S. V. (2020). Studies on antimicrobial & antioxidant properties of Ashwagandha around Nanded districts of Maharashtra. *European Journal of Molecular & Clinical Medicine*, *7*(11). ISSN 2515-8260.

Kuboyama, T., Tohda, C., & Komatsu, K. (2014). Effects of Ashwagandha (roots of Withania somnifera) on neurodegenerative diseases. *Biological and Pharmaceutical Bulletin*, *37*(6), 892–897.

Kyathanahalli, C. N., Manjunath, M. J., & Muralidhara. (2014). Oral supplementation of standardized extract of Withania somnifera protects against diabetes-induced testicular oxidative impairments in prepubertal rats. *Protoplasma*, *251*, 1021–1029.

Lee, B. S., Yoo, M. J., Kang, H., Lee, S. R., Kim, S., Yu, J. S., . . . Kim, K. H. (2021). Withasomniferol D, a new anti-adipogenic Withanolide from the roots of Ashwagandha (Withania somnifera). *Pharmaceuticals*, *14*(10), 1017.

Lopresti, A. L., Smith, S. J., Malvi, H., & Kodgule, R. (2019). An investigation into the stress-relieving and pharmacological actions of an Ashwagandha (Withania somnifera) extract: a randomized, double-blind, placebo-controlled study. *Medicine*, *98*(37).

Mamidi, P., & Thakar, A. B. (2011). Efficacy of Ashwagandha (Withania somnifera Dunal. Linn.) in the management of psychogenic erectile dysfunction. *AYU (An International Quarterly Journal of Research in Ayurveda)*, *32*(3), 322.

Marslin, G., Selvakesavan, R. K., Franklin, G., Sarmento, B., & Dias, A. C. (2015). Antimicrobial activity of cream incorporated with silver nanoparticles biosynthesized from Withania somnifera. *International Journal of Nanomedicine*, *10*, 5955.

Mehta, V., Chander, H., & Munshi, A. (2021). Mechanisms of anti-tumor activity of Withania somnifera (Ashwagandha). *Nutrition and Cancer*, *73*(6), 914–926.

Mohanty, I., Arya, D. S., Dinda, A., Talwar, K. K., Joshi, S., & Gupta, S. K. (2004). Mechanisms of cardioprotective effect of Withania somnifera in experimentally induced myocardial infarction. *Basic & Clinical Pharmacology & Toxicology*, *94*(4), 184–190.

Nasimi Doost Azgomi, R., Zomorrodi, A., Nazemyieh, H., Fazljou, S. M. B., Sadeghi Bazargani, H., Nejatbakhsh, F., . . . Ahmadi AsrBadr, Y. (2018). Effects of Withania somnifera on reproductive system: a systematic review of the available evidence. *BioMed Research International*, *2018*, 4076430.

Owais, M., Sharad, K. S., Shehbaz, A., & Saleemuddin, M. (2005). Antibacterial efficacy of Withania somnifera (Ashwagandha) an indigenous medicinal plant against experimental murine salmonellosis. *Phytomedicine*, *12*(3), 229–235.

Paul, S., Chakraborty, S., Anand, U., Dey, S., Nandy, S., Ghorai, M., . . . Dey, A. (2021). Withania somnifera (L.) Dunal (Ashwagandha): a comprehensive review on ethnopharmacology, pharmacotherapeutics, biomedicinal and toxicological aspects. *Biomedicine & Pharmacotherapy*, *143*, 112175.

Priyanka, G., Anil Kumar, B., Lakshman, M., Manvitha, V., & Kala Kumar, B. (2020). Adaptogenic and immunomodulatory activity of Ashwagandha root extract: an experimental study in an equine model. *Frontiers in Veterinary Science*, *7*, 541112.

Remenapp, A., Coyle, K., Orange, T., Lynch, T., Hooper, D., Hooper, S., . . . Hausenblas, H. A. (2022). Efficacy of Withania somnifera supplementation on adult's cognition and mood. *Journal of Ayurveda and Integrative Medicine*, *13*(2), 100510.

Ruby, B., Benson, M. K., Kumar, E. P., Sudha, S., & Wilking, J. E. (2012). Evaluation of Ashwagandha in alcohol withdrawal syndrome. *Asian Pacific Journal of Tropical Disease*, *2*, S856–S860.

Sandhu, J. S., Shah, B., Shenoy, S., Chauhan, S., Lavekar, G. S., & Padhi, M. M. (2010). Effects of Withania somnifera (Ashwagandha) and Terminalia arjuna (Arjuna) on physical performance and cardiorespiratory endurance in healthy young adults. *International Journal of Ayurveda Research*, *1*(3), 144.

Sheoran, S., Khanam, B., Mahanta, V., & Gupta, S. K. (2020). Efficacy of Ashwagandha [Withania somnifera (Linn.) Dunal] leaf paste in the management of chronic non-healing wound: a case report. *Journal of Ayurveda Case Reports*, *3*(3), 95.

Shi, Y., Zhang, C., & Li, X. (2021). Traditional medicine in India. *Journal of Traditional Chinese Medical Sciences*, *8*, S51–S55.

Singh, M., Jayant, K., Singh, D., Bhutani, S., Poddar, N. K., Chaudhary, A. A., . . . Khan, S. (2022). Withania somnifera (L.) Dunal (Ashwagandha) for the possible therapeutics and clinical management of SARS-CoV-2 infection: plant-based drug discovery and targeted therapy. *Frontiers in Cellular and Infection Microbiology*, 1098.

Singh, S. P., Tanwer, B. S., & Khan, M. (2010). Antifungal potential of Ashwagandha against some pathogenic fungi. *International Journal of Biopharmaceutics*, *1*(2), 72–74.

Speers, A. B., Cabey, K. A., Soumyanath, A., & Wright, K. M. (2021). Effects of Withania somnifera (Ashwagandha) on stress and the stress-related neuropsychiatric disorders anxiety, depression, and insomnia. *Current Neuropharmacology*, *19*(9), 1468.

Suganya, K., Kayalvizhi, E., Yuvaraj, R., Chandrasekar, M., Kavitha, U., & Suresh, K. K. (2020). Effect of Withania somnifera on the antioxidant and neurotransmitter status in sleep deprivation induced Wistar rats. *Bioinformation*, *16*(8), 631.

Tharakan, A., Shukla, H., Benny, I. R., Tharakan, M., George, L., & Koshy, S. (2021). Immunomodulatory effect of Withania somnifera (Ashwagandha) extract – a randomized, double-blind, placebo controlled trial with an open label extension on healthy participants. *Journal of Clinical Medicine*, *10*(16), 3644.

Tiwari, S., Gupta, S. K., & Pathak, A. K. (2021). A double-blind, randomized, placebo-controlled trial on the effect of Ashwagandha (Withania somnifera Dunal.) root extract in improving cardiorespiratory endurance and recovery in healthy athletic adults. *Journal of Ethnopharmacology*, *272*, 113929.

Wang, J., Zhang, H., Kaul, A., Li, K., Priyandoko, D., Kaul, S. C., & Wadhwa, R. (2021). Effect of Ashwagandha withanolides on muscle cell differentiation. *Biomolecules*, *11*(10), 1454.

4 Microbial Nexus for Converting Dairy Wastewater into Value-Added Products

Shaon Ray Chaudhuri, Mandakini Gogoi, Ajoy Modak, and Sujan Das

Milk processing plants generate a copious amount of nutrient-rich wastewater which needs energy, labor, and land-intense treatment before getting discharged. Elaborate treatment with seven- to nine-unit operations generates a large volume of treated water with an enormous quantity of sludge that, in turn, needs energy-intense treatment. The existing practices are beyond the reach of small- and middle-scale enterprises while making the effluent treatment operation a drain on its revenue for large installations. An eco-friendly and economical alternative to this energy-intense operation would be the conversion of the wastewater into value-added products that can earn revenue for the dairy industry using selectively enriched microbes from nature. The first product could be an ammonia-rich liquid biofertilizer that could replace fresh water and chemical fertilizer use during irrigation (agriculture consumes 89% of fresh water daily). This biofertilizer could enhance the yield of economic crops per unit of land compared to chemical fertilizer–based cultivation without wasting fresh water for agriculture. This conversion takes place in a single unit, saving investment on land (80%) and energy (89%) by the industry while generating revenue from the product's sale. The pilot-scale (11 m^3/day) biofilm-based wastewater treatment system could be installed quickly and continue functioning for more than four years under ambient conditions without additional microbial charging, with little maintenance. The biofertilizer production took 16 h of hydraulic retention time compared to 105 h for conventional dairy effluent treatment. This system suits rural dairies with extensive lands near the effluent treatment plant. In urban dairies, the nutrient-rich wastewater could be converted into biofuel/bioalcohol feedstock within 68 h, along with the generation of nutrient-free treated water suitable for non-potable application. The development of such value-added products necessitates a proper understanding of the pollutant nature, the possible product that could be generated from them, and the selection of the right microbial candidate from an environmental origin for the formation of tailor-made consortium for wastewater-specific bioconversion/biotransformation, leading to an eco-friendly and economical solution to industrial pollutants, promising a green future for the generations to come.

 DOI: 10.1201/9781003517108-4

4.1 THE WASTEWATER SCENARIO

Fresh waste scarcity and copious quantities of wastewater generation have become a global phenomenon. This situation becomes even more severe in densely populated countries. The existing reserves of fresh water are dwindling and need to be prevented from getting contaminated by seepage of wastewater. The average daily requirement of fresh water globally is about 10^6 m^3, of which a major share is on account of virtual water (water consumption on account of the products we use, hence indirect consumption). Food production claims about 3.5 m^3 of water per day per person (FAO UN, 2017), indicating the quantum of water required for the growing population's food production and processing. Agriculture uses between 30% (high-income countries) and 90% (low- to middle-income countries) of fresh water drawn every day (Saha et al., 2018; FAO UN, 2021) from various water sources. The problem of water scarcity is expected to worsen with the passing of time (CPCB, 2021). The increasing population calls for the intensification of ongoing industrial activities. That, in turn, would generate large volumes of wastewater. The relative quantum of wastewater generation and the existing treatment capacity reveal a huge gap that must be addressed by adopting an innovative approach that would lead to adequate, rapid, and affordable wastewater treatment.

The volume of sewage production is directly related to the population size. As per the recent report of the Central Pollution Control Board (Environmental Protection Agency equivalent in India) (WWTT UN, 2003), there is a gap of 78.7% between the amount of wastewater generated and the existing facilities. Considering that sewage treatment plant (STP) construction is planned/initiated for additional 1,742.6 (MLD) million liters per day capacity, this unmet need will reduce to 72.7%. A major cause of this gap is the large capital investment in the installation of STP, technical expertise, as well as investment on its operation and maintenance. The land requirement for one MLD STP varies between 0.2 and 1 ha, explaining further the huge capital investment in setting up of the STP/effluent treatment plant (ETP). Unless treated adequately, wastewater seepage would pollute existing freshwater sources, creating further environmental issues. The treatment involves steps like screening, grit removal, primary sedimentation, aeration tank, secondary sedimentation, sludge treatment, tertiary treatment, disinfection, and discharge/reuse of treated water. The secondary treatment methods would depend upon the nature of the pollutant, the volume of the wastewater, the availability of land, and the climate of the region having the STP installation. The secondary treatments could be carried out using the activated sludge-based method, trickling bed filters, rotating biological contractors, membrane bioreactors, moving bed biofilm reactors, waste stabilization ponds, aeration lagoons, up-flow anaerobic sludge blanket (UASB) process, photobioreactors, biofilm reactor, and so on (Gogoi, Biswas, et al., 2021). However, the need of the hour is developing a process requiring less space, time, labor, and energy for wastewater treatment to make it affordable for a more extensive user base.

The tailor-made bacterial formulation developed using selected bacteria from an environmental origin in a definite proportion leads to the setting up of a stable, economical, biofilm-based, single-unit operation with faster performance than the conventional systems for different types of wastewater (Saha et al., 2018;

Gogoi et al., 2021b; Biswas et al., 2021, 2022; Ray Chaudhuri et al., 2017). This approach reduces the land requirement, energy requirement, routine maintenance, and labor involvement in running the effluent treatment plant. In certain cases, the reusable pollutants are recovered for recycling, while the treated water was made suitable for non-potable applications, like agriculture and aquaculture (depending on the influent wastewater composition and the extent of treatment). Adopting this approach (tested at a pilot scale) makes effluent treatment economical and eco-friendly, hence adaptable by small- and middle-scale installations.

Water and food are two of the major interconnected necessities of human civilization. While food requirement increases due to the ever-increasing population, its production accounts for the consumption of 89% of the daily fresh water withdrawn in agriculture-based economies, resulting in fast depletion of the available freshwater reserves, pushing the population towards water scarcity. Agricultural sustenance also uses chemical fertilizer, a major (70%) portion of which leaches into agricultural runoff, polluting the available freshwater reserves. Some of the principal pollutants in agricultural runoff, food industries wastewater, and sewage are common (nitrates and phosphates), which are essential for plant growth. The production of nitrogen and phosphorous for agricultural use involves immense energy expenditure (for the production of urea) and the use of non-renewable resources (rock phosphate). An eco-friendly bio-remedial approach could be used to recover these essential nutrients for reuse while decreasing the pollution (Saha et al., 2018) load in the treated water. In some instances, the pollutant might be converted to a form preferred by the plants over the form present in the influent (Gogoi et al., 2021b). Some bacterial consortia ensure that the plant growth nutrients applied to the soil are restricted in the soil root zone (Press Information Bureau, 2022), preventing their leaching and promoting plant growth through easy uptake of the nutrients.

4.2 DAIRY WASTEWATER AND THEIR TREATMENT

In Indian agricultural history, the decade of the 1970s was magnificent, with the onset of white revolution under the leadership of Dr. Verghese Kurian. That beginning resulted in today's India being the highest milk producer and one of the top exporter countries (Press Information Bureau, 2022), with 887 active dairy processing plants having an annual production of 2.1×10^5 kg of processed milk (Press Information Bureau, 2022; Dairy Directory, 2023). This enormous amount of milk processing generates a large amount of dairy wastewater (0.2–10 L per liter of milk processes) (Mansoorian et al., 2016; Akansha et al., 2020). This wastewater contains different types of nutrients which act as pollutants for the surface water. The dairy industry includes different upstream operations, like cheese and butter manufacturing, condensed milk production, and ice cream production, other than pasteurized milk packaging (Britz et al., 2006). In India, whole milk contains 80% casein, 3.2% protein, >3.9% fat, 5.1% lactose, 9.5% total solid, and 8.7% organic solids, while maintaining a pH of 7.2. The composition of skim milk, buttermilk, and whey includes 80% casein, along with 3.3%, 3.4%, and 0.9% protein; 0.1%, 0.4%, and 0.3% fat; 5.3%, 4.3%, and 4.9% lactose; 9.4%, 6.9%, and 12.9% total solids; 8.7%, 6.3%, and 12.2% organic solids, while maintaining a pH of 7.2, 3.5, 10.3, respectively (Mohanrao & Subramanyam, 1972).

There is a high amount of pollutant discharged from the different segments/units of dairy processing plant, including dissolved organic substances, like lactose, different mineral salts, and protein suspensions in colloidal form (Charalambous et al., 2020). In addition, some non-edible inorganic substances such as phosphates are used as deflocculants and emulsifiers in cleaning compounds. Chlorine is used as disinfectant, while detergents and nitrogen are used as wetting agents. These agents increase the incidences of eutrophication in water bodies receiving the pollutants and increase human health risk (Ahmad et al., 2019). Therefore, the treatment of dairy wastewater is important to get a clean environment. The conventional approach involves high operational cost through physicochemical followed by biological methods (Charalambous et al., 2020; Gogoi et al., 2021b). The government imposes strict regulations for discharging the effluent to protect the environment. This becomes a burden for the industries to adopt, as conventional treatment facilities need quite an ample space and energy, with no monetary benefit in the process. Therefore, industry owners often release effluent directly or after giving some primary treatment due to a lack of funds and a dearth of awareness regarding environmental issues (Dairy Directory, 2023) (Figure 4.1).

The biodegradable component in dairy wastewater gets consumed by microbes, depleting the dissolved oxygen with the generation of BOD and COD. The nitrogen in protein is converted into inorganic form (ammonia, ammonium, nitrite, and nitrate). All these forms of nitrogen are toxic at elevated levels to the biota. Nitrate above 10 mg/L in drinking water causes methemoglobinemia through conversion to nitrite and conversion of hemoglobin to methemoglobin in the blood that fails to carry oxygen. The nitrite above 0.02 mg/L causes brown blood disease in fish, in which the gills of fish become pale tan or brown-colored. Ammonia above 0.5 mg/L causes tissue toxicity in fish through inhibition of excretion, causing lethargy and, subsequently, death. The prescribed limit for inorganic nitrogen and phosphorus is 30–100 mg/m^3

FIGURE 4.1 Partially treated or untreated dairy wastewater being discarded into the environment due to expensive, elaborate conventional effluent treatment methods.

and 15–30 mg/m^3 in effluent water. Higher concentrations cause eutrophication in water bodies, leading to further deterioration of the environment (Shete & Shinkar, 2013). The permitted limit of discharge of the different components in the dairy effluent as per Central Pollution Control Board, India (The Environment (Protection) Rules, 1986) are as follows: total solids <2,200 mg/L, total dissolved solids <2,100 mg/L, suspended solids <100 mg/L, total chloride <600 mg/L, sulfates <1,000 mg/L, phosphate <5 mg/L, oil and grease <10 mg/L, chemical oxygen demand <360 mg/L, biological oxygen demand <30 mg/L, nitrate <10 mg/L, pH between 5.4 and 9.1 (Dongre et al., 2020). The conventional process of dairy effluent treatment involves physicochemical methods, namely, coagulation/flocculation, adsorption, electrocoagulation, electro flocculation, to name a few (Yonar et al., 2018). These steps involve large land and energy requirement. Biological treatment, on the other hand, includes aerobic and anaerobic treatment. Aerobic treatment for high organic wastewater removal involves aerated lagoons, oxidation ponds, activated sludge processes, and sequencing batch reactors, generating enormous quantities of sludge that need elaborate treatment. The other systems based on attached bacterial growth are tricking bed filters and rotating biological contractors. Energy consumption is the major bottleneck in these cases. Anaerobic filters, upflow anaerobic sludge blanket (UASB) reactor, continuous stirred tank reactor (CSTR), anaerobic contact processes (ACP), anaerobic fluidized bed reactors (AFBR), and anaerobic packed-bed digester (APBR) are used for anaerobic treatment (Joshibaa et al., 2018). The advantages of using an immobilized system compared to its suspended form are as follows: less cumbersome to remove biomass from treated effluent, less to no sludge (Gogoi et al., 2021b) generation, better performance than suspended cells as cell density is higher upon immobilization (Gogoi et al., 2022), and the cells are resistant to external perturbations, with a reduced discharge of microbes into the environment (Mohebrad et al., 2022). In most of these cases, large space and energy requirement along with maintenance cost is involved.

About 1 L of milk processing generates between 0.2 and 10 L of wastewater (Biswas et al., 2019) based on the upstream operation and the technology used for processing of milk. The global milk production in 2018 was 76.48×10^{10} kg, in 2019 was 77.29×10^{10} kg, while in 2020 was 78.12×10^{10} kg. It is expected (OCED-FAO, 2020) to increase to 90.45×10^{10} kg by 2029 (Feed Additive, 2021). From the preceding figures, the minimum and maximum amount of milk processing wastewater generation could be predicted to be 76.48×10^7 to 76.48×10^8 m^3 for 2018, 77.29×10^7 to 77.29×10^8 m^3 for 2019, 78.12×10^7 to 78.12×10^8 m^3 for 2020, while 90.45×10^7 to 78.12×10^8 m^3 for 2029. Processing of dairy wastewater takes 105 to 120 h for a given volume in a 7- to 8-step operation (Biswas et al., 2019). Dairy wastewater treatment is known to consume energy to the tune of 2.29 kW per m^3 (Żyłka et al., 2021). However, the energy required for the actual operation of the conventional dairy wastewater treatment process at OMFED Bhubaneswar (India) was calculated to be 0.1 kW per m^3 of wastewater. Based on the lowest energy consumption data, treatment of 600 m^3 of dairy wastewater as per conventional procedure would require 60 kW of energy, resulting in 475.92 t per year of CO_2 equivalent gas emission (Gogoi et al., 2021b). Adapting the conventional treatment technology involves using large space and sludge generation, which needs to be treated separately, with associated huge energy expenditure which further translates to large amounts of CO_2 gas emission, hence a concern for the climate impact.

Different groups are exploring the option of recovering essential materials from dairy wastewater to recycle and reuse them while reducing pollution. New products are generated in the course of action (Conidi et al., 2023; Gomes da Cruz et al., 2023). Dairy wastewater was converted to ammonia-rich liquid biofertilizer, which could enhance the yield of economic crops while replacing the use of fresh water and chemical fertilizer for agriculture. This technology was suitable for rural dairies with nearby land (Gogoi et al., 2021b) that could use up the produced biofertilizer. The most significant part of the system was a drastic reduction in space requirement (less than 20% of the conventional space), and the entire wastewater could be converted into liquid biofertilizer within 16 h of incubation. The system could be retrofitted with existing dairy effluent treatment plants, enhancing their wastewater processing capacity. For urban dairies, which do not have space in the surrounding, a combination of bacterial and microalgal treatment in two units could convert wastewater into treated water suitable for non-potable application within 68 h (105 to 120 h for a conventional system), while generating algal biomass with enhanced carbohydrate and lipid contents (Biswas et al., 2022).

Dairy wastewater can produce protease and lipase within 4 h under calcium alginate–immobilized conditions using *Pseudomonas aeruginosa* in the microbial electrochemical system (Mohebrad et al., 2022). This approach successfully coupled biotransformation with environmental sustenance at the laboratory scale with higher and faster production of these essential enzymes, indicating better performance by biofilms than the suspended cells (Mohebrad et al., 2022). The presence of stable, complex fats/lipids creates a problem in the operation of conventional treatment systems by forming a continuous layer on the microbial biofilms, blocking the access of pollutants by the microbes, hence negatively impacting (by reducing mass transfer efficiency) the system performance (Ferreira et al., 2021; Kim & Karthikeyan, 2021; Rahul et al., 2022; Egerland et al., 2021). Through adequate anaerobic degradation, these complex lipids produce biomethane (Sasidharan et al., 2023) and biohydrogen (Arun & Sivashanmugam, 2018).

4.3 DAIRY WASTEWATER CONVERSION TO BIOFUEL/BIO-OIL SUBSTRATE THROUGH MICROBIAL INTERACTION

Researchers have been actively involved in making dairy effluent treatment operations environmentally safe and economical. In order to do so, the effluent nature and the location of the effluent treatment plant are taken into consideration. For urban dairies, the treated product should be suitable for immediate discharge or reusable with a small storage space requirement. Biswas (Biswas et al., 2022) developed a combination of dairy effluent treatment processes with the 2-unit operation. The first unit was a biofilm-based bacterial treatment using environmental isolates (mostly from dairy-activated sludge) from the genus *Aeromonas*, *Acinetobacter*, *Thauera*, *Bacillus*, and uncultured bacterium. Each of the isolates, after detailed characterization, was found to be non-pathogenic, with the ability to remove carbohydrate, protein, lipid, chloride, calcium carbonate, nitrite, nitrate, and phosphate from dairy wastewater. Aeromonas from various environmental origins are reported to grow as structured biofilm fermenting carbohydrates with the production of lactic acid and reduction of nitrate (Biswas, 2020; Banerjee et al., 2023). These facultative anaerobes also produce lipase, amylase, catalase, oxidase, and protease, explaining their role in dairy wastewater treatment. *Acinetobacter* sp. are aerobic, biofilm-forming bacteria present in all types of environments and have nitrate-assimilating ability under low carbon conditions, as they lack nitrate dissimilatory pathway. They can reduce phosphate and promote plant growth, thus justifying their involvement in dairy wastewater bioremediation. The genus *Thauera* is known for its presence in wastewater and sludge. These are biofilm-forming denitrifiers with growth in oxic and anoxic environments. These properties justify their being involved in dairy wastewater bioremediation. Genus *Bacillus* from varied environments is known for its active involvement in nitrate removal (reduction and accumulation), essential extracellular enzyme production, and phosphate remediation (Wróbel et al., 2023; Islam et al., 2022).

In the current bacterial treatment of dairy wastewater, the following reduction was observed after 20 h of incubation in the biofilm reactor: 93% nitrate, 65% phosphate, 36% protein, and 72% chemical oxygen demand, getting the effluent close to discharge level prescribed by Environmental Protection Agency norms. However, with it being rich in carbon–nitrogen ratio, there was the production of ammonia (38%) during bioremediation. The treated water was found to promote mung bean cultivation/yield during pot trial (Biswas et al., 2021; Biswas, 2020). This treated wastewater had some pollutants (nitrate) within the discharge level, while others (ammonia and phosphate) were above the discharge level and had to be treated further before discharge (Biswas et al., 2019, 2022). Algae bacterial consortium enriched using inoculum from wastewater-fed fish pond could grow as attached biomass which, under ambient conditions, could get the rest of the pollutants (ammonia to 2.5 mg/L from 19.6 mg/L; nitrate and phosphate to below detectable level; COD to 229 mg/L from 3,481 mg/L) down to discharge level within 48 h while maintaining neutral pH. The total time of treatment becomes 68 h, along with the generation of 67% higher algae-bacterial biomass which is rich in lipid (42% enhancement) as well as carbohydrate (55% enhancement) content, within 48 h compared to algae-bacterial biomass grown on sewage water. The treated effluent was suitable for discharge or other applications. This was the highest reported from dairy wastewater (Biswas et al., 2022). Metagenomic analysis of the algae bacterial consortium was conducted to understand the underlying reason behind this biofuel/biodiesel substrate generation. Metagenomics revealed the presence of *Desulfovibrio vulgaris*, followed by *Azoarcus* sp., *Chloroflexus aurantiacus*, *Azospirillum bracsilense*, *Curvibacter*, *Pseudomonas*, and Cyanobacteria (Biswas et al., 2021; Banerjee et al., 2023). Through this approach, the three main problems associated with algal cultivation for biofuel production were addressed. The nutrient requirement for algal cultivation was met by utilizing the nutrients present in the dairy wastewater. The huge water requirement for algal cultivation could be met through the large volume of wastewater generated from the milk processing plant (about 10 L per liter of milk processed) without consuming fresh water for algal cultivation. The third bottleneck for algal cultivation is the energy required for harvesting and dewatering the algal biomass. In this case, algae are present as an attached growth, hence obviating the energy involved in harvesting, which accounts for about 30–40% of the total biodiesel production cost (Biswas, 2020). Through this, within 48 h, with mild aeration and light, the dairy wastewater could be treated with lipid- and carbohydrate-rich biomass enrichment that could be used for various by-product development, like animal and aquaculture feed and fertilizer, other than feedstock for bioalcohol and biodiesel. Metagenomics revealed 93% of the community to be of bacteria, and 7% was eukarya. Among the eukarya, 58% were algae. The bacterial population was dominated by Proteobacteria (71%) and Cyanobacteria (10%). Proteobacteria are known to be predominant in wastewater treatment plants (Banerjee et al., 2023). A comparison of the existing genus and their corresponding functions reported in the literature revealed the possible functions of the different members in the community (Banerjee et al., 2023). In order to understand the reason behind lipid and carbohydrate enrichment within the biomass, a detailed analysis of existing literature and a comparison with the community present was conducted. Higher algal biomass production with enhanced carbohydrate and lipid content is reported due to growth in nutrient-enriched medium, growth in defined medium with metal stress and cultivation in nutrient-rich wastewater, starvation (nitrate and phosphate) of the culture after growing in enriched condition, co-cultivation with selected varieties of bacteria, impact of secreted phytohormones, effect of light intensity (Biswas et al., 2022; Biswas, 2020; Banerjee et al., 2023), and CO_2 level (Banerjee et al., 2023). Increase in CO_2 concentration enhances lipid and carbohydrate concentration in the biomass in some cases, while in some cases, lower CO_2 concentration stimulates carbohydrate and lipid (Banerjee et al., 2023) accumulations. Hence, the innovation proposed by Biswas (Biswas et al., 2022) and Ray Chaudhuri (Ray Chaudhuri, 2023) saves time and space for the urban effluent treatment plant setup while producing value-added products.

4.4 DAIRY WASTEWATER CONVERSION TO BIOFERTILIZER AND ASSOCIATE TESTING AT FIELD SCALE

Dairy wastewater is rich in nitrogen (nitrate, nitrite, ammonia, protein, nitric acid used for cleaning), phosphorous (phosphate from detergent), carbon (from milk protein casein), lipids (milk fat), and calcium. These are removed through a combination of physicochemical followed by biological (aerobic and anaerobic) treatment in 8 to 9 steps. The existing dairy wastewater treatment process is elaborate, laborious, energy-intense, and space-consuming, which compels the small dairy processing plants to discard their effluent untreated, polluting the environment. The volume of wastewater generated from dairy installation is large, and the space requirement for effluent treatment plant installation is high. The pollutants in dairy wastewater are also essential for plant growth in specific forms. Gogoi (Gogoi, Bhattacharya, et al., 2021) developed a microbial biotransformation process that could convert the nitrogen present in different forms in dairy wastewater into ammonia, a form that plants prefer. In addition, the large volume of water could be completely used for agricultural purposes. Through this approach, fresh water could be prevented from being wasted for non-portable purposes. The final by-product is an ammonia-rich liquid fertilizer with polyphosphate and phosphatase. The total volume is converted into liquid biofertilizer within 4 h of incubation in the biofilm reactor. This biofertilizer has to be directly used for irrigation. This means that the process needs large cultivable land near the dairy effluent treatment plant. The situation in rural dairies matches the requirement, with large farmlands in its surrounding which produce fodder for dairy cattle grazing. Such a process is only suitable for rural dairies. In order to develop such a process, bacterial isolates were screened from the dairy-activated sludge (Biswas et al., 2019) and characterized at length. The isolates (67% *Aeromonas*, 16% *Bacillus*, and 17% *Acinetobacter*) were combined to develop a suitable consortium (Halder et al., 2020) that could reduce nitrate, nitrite, phosphate, calcium carbonate, chloride, protein, and carbohydrate from the dairy wastewater while producing ammonia under suspended and immobilized condition. After optimization of the consortium performance at the laboratory scale (Gogoi, Banerjee, et al., 2021), the process was scaled up to the level of a pilot plant at two dairy processing units in two states of India that are geographically distant (Gogoi, Biswas, et al., 2021). The reactor's design was made considering the minimum space requirement with a hydraulic retention time of 16 h (Figure 4.2a). The system upon installation continued to function for more than four years without any breakdown with little maintenance. With an initial fat removal system in place, the bioreactor functioned without sludge generation. The developed biofertilizer was tested on numerous varieties of 17 different types of economic crops, and the impact of the biofertilizer-based cultivation compared to the recommended dose of chemical fertilizer application is provided in Table 4.1.

In most of the cases, yield enhancement was at par with chemical fertilizer–based cultivation or improved over it. That indicates that the produced biofertilizer was efficient in replacing chemical fertilizer and fresh water use in agriculture. Literature shows that organic cultivation decreases yield (Mader et al., 2002), resulting in higher cost of organic food. This bottleneck was overcome through this study. Mader showed a 34–42% decrease in production of tuber crops upon organic cultivation. The work of Gogoi (Gogoi, 2021) reported a 30–37% decrease in yield of two varieties of tuber crops. Hence, the current biofertilizer could reduce the decrease in production due to organic cultivation to some extent. But the most significant part of the study was an increased or maintained protein content of all the tuber crops tested while a significant decrease in the carbohydrate content (14–55%) compared to chemical fertilizer–grown produce. So there was production of diet tubers without using genetically modified varieties. This finding correlates with the finding of Salunke (Salunke & Desai, 1988), which reported an increase in the protein–carbohydrate ratio of food upon nitrogen-rich fertilizer application. Through this approach, higher production per unit of land of healthy organic food could be obtained without the use of chemical fertilizer and fresh water, hence environmental protection. The environmental protection was because of the following:

1. Through prevention of leaching of unutilized chemical fertilizer.
2. Through minimization of discharge of untreated or improperly treated dairy wastewater into the environment that seeps into the freshwater reserves.

TABLE 4.1
Impact of Biofertilizer-Based Yield Enhancement Compared to Chemical Fertilizer–Based Cultivation or No Fertilizer Application (in case of Sorghum Sudan Grass)

S. No.	Crop	Fold Increase in Production
1	Mung bean (*Vigna radiata var.* MEHA)	1.56
2	Black gram (*Vigna mungo* var. Pant-U-31)	1.04
3	Maize (*Zea mays* var. Vijay)	1.19
4	Ramie (*Boehmeria nivea var.* R-1411 Hazarika)	1.39
5	Sorghum sudangrass (*Sorghum sudanense* var. (Piper) Stapf)	2.53
6	Scented rice (*Oryzae sativa* var. Kala Joha, var. Manikimadhuri)	1.14 (grain filling)
7	Aloe vera (*Aloe elongata* var. Ghikuari)	1.63
8	Lemon grass (*Cymbopogon citratus* var. Dhanitri and var. Krishna)	2.82
9	Potato (*Solenum tuberosum* var. MF1, MF2, TPS7, TPS13 and TPS67)	−1.44
10	Yam bean (*Pachyrhizus erosus* var. RM1)	−1.57
11	Colocasia (*Colocasia esculenta* var. Telia and var. Muktakeshi)	1.01
12	Sweet potato (*Ipomoea batatas* var. Krishna and Kanchangard)	1.44
13	Cassava (*Manihot esculenta* var. Srijaya, var. Srirekha, var. Sri Bishakha)	1.86
14	Yam (*Discorea alata* var. Srinidhi)	2.55
15	Elephant foot yam (*Amorphophallus paeoniifolius* var. Gajendra)	3.8
16	Field pea (*Pisum sativum var.* VL 142)	1.16
17	Sugarcane (*Saccharum officinarum)*	1.01

3. Preventing the misuse of fresh water for non-potable purposes like agriculture.
4. Decreasing the CO_2 equivalent gas emission due to lesser hydraulic retention time (4 h instead of 105 h) in the bioreactor for biofertilizer production and no sludge production.

As per the study of Gogoi, in a year, 475.92 t of CO_2 emission take place from one 600 m^3 per day dairy wastewater processing plant using a conventional system that uses 60 kW of energy. In the current microbial process, this emission could be reduced to 11.9 t of CO_2 equivalent gas annually, hence a major factor for environmental health. Another energy consumption point in a conventional plant is sludge management. This includes the dewatering (72–96 kW per day) and digestion (58–69 kW per day) of the generated sludge (Molinos et al., 2013), which generates about 2.55 to 3.26 t of CO_2 per day. The process of Gogoi generates no sludge, hence saves further release of CO_2 equivalent gas generation, if operated as per standard operating procedure, hence complete savings in CO_2 equivalent gas emission in the step of sludge treatment upon adoption of the current technology. This process is scalable as efficiency was maintained in 250 ml/day to 11 m^3/day processing capacity systems.

Hence, the significant impact of the developed process on the environment is clearly visible from the preceding findings.

Milk is a common component of the household, hence a component of every kitchen's wastewater. An attempt was made to design a domestic biofertilizer production reactor through 3D printing. The developed process for biofilm formation on the raschig rings was adopted to develop a stable biofilm on the reactor matrix (Figure 4.2b). Ammonia production was monitored at regular intervals. The comparative performance of the 0.32 L (large) and 0.12 L (small) reactor is shown in Figure 4.2c as fold increase in ammonia concentration compared to simulated wastewater. The system's performance in terms of ammonia production from three different volumes of packed-bed biofilm reactor varied between 734.60 mg/kg of matrix for 27 L, 933–992 mg/kg of matrix for 0.32 L, and 613–813 mg/kg of matrix for 0.12 L system. Hence, in each case, the consortium could perform the biotransformation of dairy wastewater. This finding also validates the performance of the designed reactor for the purpose of domestic biofertilizer production.

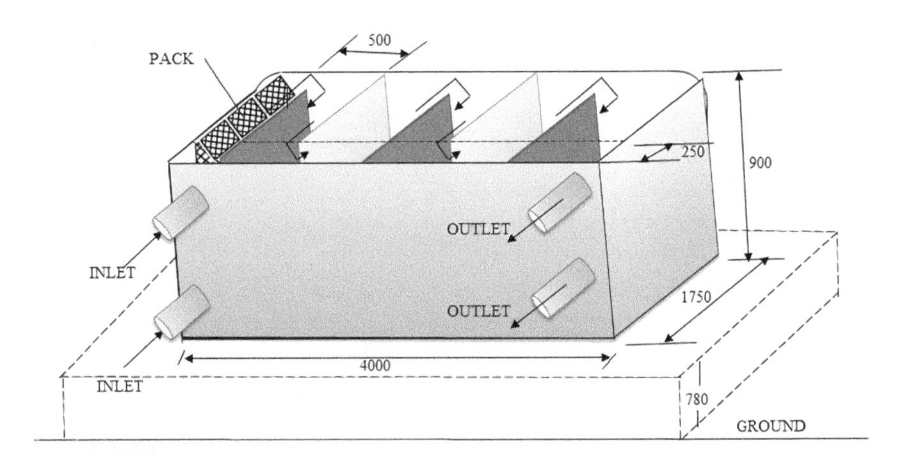

1. All dimensions are in mm.
2. Placement: The reactor is to be placed 780 mm above the ground.
3. Cover: GI sheet or FRP sheet (light cover) (4 M × 2 M).
4. Corners: 150-R rounding in four vertical corners; no rounding in bottom corners.
5. Packs: Pack dimension L × B × H = 800 × 500 × 750 mm. The pack comprises 25 number equispaced corrugated sheets having dimension 750 × 800 mm.
6. Inlet nozzles: 2 number 1 1/2″ with suitable flange.
7. Outlet nozzles: 2 number 1 1/2″ with suitable flange.

FIGURE 4.2 Bioreactor for conversion of dairy wastewater into liquid biofertilizer. From top to bottom: (a) Design of the pilot scale reactor with blocks of fiber-reinforced plastics as the immobilization matrix. The system has been performing bioconversion for more than four years during continuous and batch mode operation in India. (b) Schematic representation of the domestic rector with raschig rings and arrangement of granular activated charcoal for odor control. (c) Bar diagram representing fold increase in ammonia production in two domestic reactors with 18 raschig rings (small) and 66 raschig rings for immobilization of biofilm.

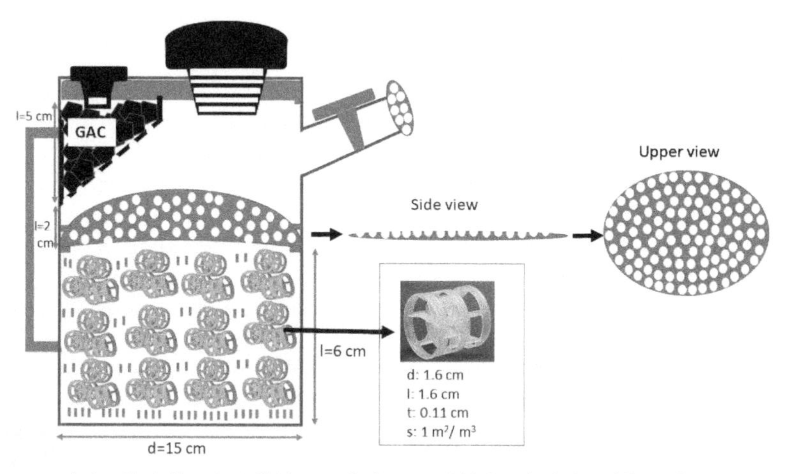

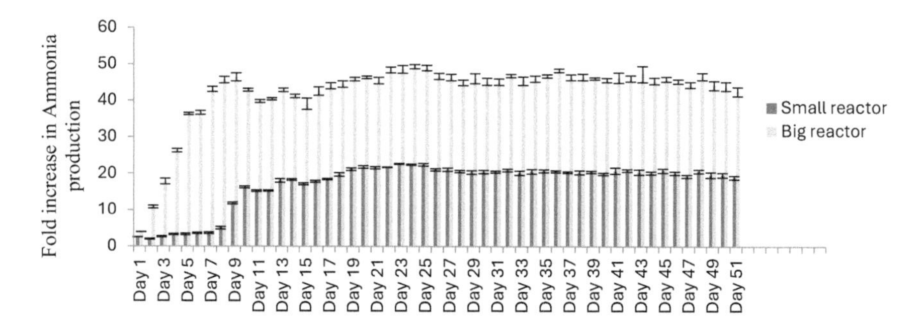

FIGURE 4.2 (Continued)

4.5 CONCLUSION

This chapter reported an approach of selecting the natural microbial population from an environmental origin for the development of a tailor-made consortium for dairy wastewater treatment with the generation of useful by-products like liquid biofertilizer, carbohydrate- and lipid-rich algal biomass, along with clean water suitable for non-potable application. This is an eco-friendly approach as it replaces the use of chemical fertilizer and fresh water with the large volume of nutrient-rich treated dairy wastewater, preventing wastage of valuable resources while treating the water adequately before discharge. Both algal cultivation and agriculture could be practiced safely without using fresh water and chemical fertilizer. It also makes the process of dairy wastewater treatment rapid, with no sludge generation and minimum CO_2 equivalent gas emission. This eco-safe process will reduce the discharge of untreated dairy wastewater into the environment while recovering all useful components in the wastewater for reuse.

ACKNOWLEDGMENTS

The authors acknowledge the funding support from the University Grant Commission (UGC), UGC Department of Atomic Energy (UGC DAE), Biotechnology Industry Research Assistance Council, Government of India. The authors acknowledge Tripura University for the laboratory infrastructure and the computational facility and the laboratory.

CRediT authorship contribution statement Shaon Ray Chaudhuri: conceptualization, manuscript preparation, data curation and interpretation, project administration, funding acquisition, supervision; Mandakini Gogoi: making the computer sketch of the domestic reactor based on the provided design, calculating the ammonia production per kilogram of matrix for the 28 L reactor; Ajoy Modak: data generation for the domestic reactor, formatting the references; Sujan Das: carrying out the literature survey.

REFERENCES

Ahmad, T., Aadil, R. M., Ahmed, H., Soares, B. C. V., Souza, S. L. Q., Pimentel, T. C., Scudino, H., Guimarães, J. T., Esmerino, E. A., Freitas, M. Q., Almada, R. B., Vendramel, S. M. R., Silva, M. C., Cruz, A. G., and U. Rahman. 2019. Treatment and utilization of dairy industrial waste: a review. *Trends Food Sci. Technol.* 88:361–372.

Akansha, J., Nidheesh, P. V., Ashitha, G., Anupama, K. V., and M. S. Kumar. 2020. Treatment of dairy industry wastewater by combined aerated electrocoagulation and phytoremediation process. *Chemosphere* 253:1–8.

Arun, C., and P. Sivashanmugam. 2018. Enhanced production of biohydrogen from dairy waste activated sludge pre-treated using multi hydrolytic garbage enzyme complex and ultrasound-optimization. *Energy Convers. Manag.* 164:277–287.

Banerjee, S., Pati, S., and S. Ray Chaudhuri. 2023. Algae bacterial mixed culture for waste to wealth conversation: a case study. In *Technological Advancement in Algal Biofuels Production*, ed. N. Srivastava and P. K. Mishra, 271–295. Springer Nature Singapore Pte Ltd.

Biswas, T. 2020. *Development of tailor-made consortia for efficient effluent treatment.* PhD Diss., Tripura University, India.

Biswas, T., Banerjee, S., Saha, A., Bhattacharya, A., Chanda, C., Gantayet, L. M., Bhadury, P., and S. Ray Chaudhuri. 2022. Bacterial consortium based petrochemical wastewater treatment: from strain isolation to industrial effluent treatment. *Environm. Adv.* 7:100132.

Biswas, T., Bhushan, S., Prajapati, S. K., and S. Ray Chaudhuri. 2021. An eco-friendly strategy for dairy wastewater remediation with high lipid microalgae-bacterial biomass production. *J. Environ. Manage.* 286:112196.

Biswas, T., Chatterjee, D., Barman, S., Chakraborty, A., Halder, N., Banerjee, S., and S. Ray Chaudhuri. 2019. Cultivable bacterial community analysis of dairy activated sludge for value addition to dairy wastewater. *Microbiol. Biotechnol. Lett.* 47(4):585–595.

Britz, T. J., Van, S. C., and Y. T. Hung. 2006. Treatment of dairy processing wastewaters. In *Handbook of Industrial and Hazardous Wastes Treatment*, ed. L. K. Wang, Y. T. Hung, H. H. Lo, and C. Yapijakis, 1–28. Taylor & Francis Group.

Central Pollution Control Board. 2021. *National Inventory of Sewage Treatment Plants.* https://cpcb.nic.in/openpdffile.php?id=UmVwb3J0RmlsZXMvMTIyOF8xNjE1MTk2MzIyX-21lZGlhcGhvdG85NTY0LnBkZg (accessed May 29, 2023).

Charalambous, P., Shin, J., Shin, S. G., and I. Vyrides. 2020. Anaerobic digestion of industrial dairy wastewater and cheese whey: performance of internal circulation bioreactor and laboratory batch test at pH 5–6. *Renew. Energy* 147:1–10.

Conidi, C., Basile, A., and A. Cassano. 2023. Food-processing wastewater treatment by membrane-based operations: recovery of biologically active compounds and water reuse. In *Advanced Technologies in Wastewater Treatment*, ed. A. Basile, A. Cassano, and C. Conidi, 101–125. Elsevier.

Dairy Directory. 2023. https://indiadairy.com/dairy-directory/dairy-plants-in-India/ (accessed March 11, 2023).

Dongre, A., Monika, S., and K. Sonu. 2020. Treatment of dairy wastewaters: evaluating microbial fuel cell tools and mechanisms. In *Environmental Issues and Sustainable Development*, ed. S. Sarvajayakesavalu and P. Charoensudjai. Intech Open. https://www.intechopen.com/chapters/73585

Egerland, B. B., Rosero, H. J., Rabelo, S., Gomes, T., Ribeiro, R., and G. Tommaso. 2021. Methane production from anaerobic digestion of dairy grease trap waste: effect of sugarcane bagasse addition. *Environ. Qual. Manag.* 31:73–83.

The Environment (Protection) Rules. 1986. (As amended to date). https://upload.indiacode.nic.in/showfile?actid=AC_MP_74_308_00003_00003_1543231806694&type=rule&filename=ep_rules_1986.pdf (accessed June 14, 2024).

FAO UN. 2017. *Water for Sustainable Food and Agriculture. A Report Produced for the G20 Presidency of Germany*. http://www.fao.org/3/i7959e/i7959e.pdf (accessed May 29, 2023).

Ferreira, T. F., Santos, P. A., Paula, A. V., de Castro, H. F., and G. S. S. Andrade. 2021. Biogas generation by hybrid treatment of dairy wastewater with lipolytic whole cell preparations and anaerobic sludge. *Biochem. Eng. J.* 169:107965.

Food and Agricultural Organization of the United Nations. 2021. *Land & Water*. http://www.fao.org/nr/water/docs/waterataglance.pdf (accessed May 28, 2023).

Global Dairy Industry and Trends. 2021. *International Magazine for Animal Feed and Additive Industries*. https://www.feedandadditive.com/global-dairy-industry-and-trends/ (accessed May 29, 2023).

Gogoi, M. 2021. *Development of microbial prototype for effluent treatment and reuse in dairy and aquaculture industry*. PhD Diss., Tripura University, India.

Gogoi, M., Banerjee, S., Pati, S., and S. Ray Chaudhuri. 2021. Microbial bioconversion of dairy wastewater in packed bed biofilm reactor into liquid biofertilizer. *GeoMicrobiology* 39(3–5):249–258.

Gogoi, M., Bhattacharya, P., Sen, S. K., Mukherjee, I., Bhushan, S., and S. Ray Chaudhuri. 2021. Aquaculture effluent treatment with ammonia remover *Bacillus albus* (ASSF01). *J. Environ. Chem. Eng.* 9(4):105697.

Gogoi, M., Biswas, T., Biswal, P., Saha, T., Modak, A., Gantayet, L. M., Nath, R., Mukherjee, I., Thakur, A. R., Sudarshan, M., and S. Ray Chaudhuri. 2021. A novel strategy for microbial conversion of dairy wastewater into biofertilizer. *J. Cleaner Prod.* 293:126051.

Gogoi, M., Mukherjee, I., and S. Ray Chaudhuri. 2022. Characterization of ammonia remover *Bacillus albus* (ASSF01) in terms of biofilm formation ability with application in aquaculture effluent treatment. *Environ. Sci. Pollut. Res.* 29(41):61838–61855. doi: 10.1007/s11356-021-16021-8.

Gomes da Cruz, A., Pimentel, T. C., Lippel Sant'Anna, G., and S. M. R. Vendramel. 2023. Advanced strategies for dairy wastewater treatment: a perspective. In *Advanced Technologies in Wastewater Treatment*, ed. A. Basile, A. Cassano, and C. Conidi, 275–310. Elsevier.

Halder, N., Gogoi, M., Sharmin, J., Gupta, M., Banerjee, S., Biswas, T., Agarwala, B. K., Gantayet, L. M., Sudarshan, M., Mukherjee, I., Roy, A., and S. Ray Chaudhuri. 2020. Microbial consortium-based conversion of dairy effluent into biofertilizer. *J. Hazardous Toxic Radioactive Waste* 24(1):04019039.

Islam, M. T., Rahman, M., and R. C. P. Pandey. 2022. *Bacilli in Agrobiotechnology: Plant Stress Tolerance, Bioremediation, and Bioprospecting*, ed. M. T. Islam, M. Rahman, and P. Pandey. Springer International Publishing.

Joshibaa, G. J., Kumara, P. S., Feminaa, C. C., Jayashreea, E., Racchanaa, R., and S. Sivanesan. 2018. Critical review on biological treatment strategies of dairy wastewater. *Desalin. Water Treat.* 160:94–109.

Kim, J. R., and K. G. Karthikeyan. 2021. Effects of severe pretreatment conditions and lignocellulose-derived furan byproducts on anaerobic digestion of dairy manure. *Bioresour. Technol.* 340:125632.

Mader, P., Fliessbach, A., Dubois, D., Gunst, L., Fried, P., and U. Niggli. 2002. Soil fertility and biodiversity in organic farming. *Science* 296:1694–1697.

Mansoorian, H. J., Mahvi, A. H., Jafari, A. J., and N. Khanja. 2016. Evaluation of dairy industry wastewater treatment and simultaneous bioelectricity generation in a catalyst-less and mediator-less membrane microbial fuel cell. *J. Saudi Chem. Soc.* 20(1):88–100.

Mohanrao, G. J., and P. V. R. Subramanyam. 1972. Sources, flows and characteristics of dairy waste water. *Indian J. Environ. Health* 14(3):207–217.

Mohebrad, B., Ghods, G., and A. Rezaee. 2022. Dairy wastewater treatment using immobilized bacteria on calcium alginate in a microbial electrochemical system. *J. Water Process Eng.* 46:102609.

Molinos, M., Hernández-Sancho, F., and R. Sala-Garrido. 2013. Cost modeling for sludge and waste management from wastewater treatment plants: an empirical approach for Spain. *Desalin. Water Treat.* 51:5414–5420.

OCED and Food and Agricultural Organization of the United Nations. 2020. *OCED-FAO Agricultural Outlook 2020–2029.* https://www.oecd-ilibrary.org/agriculture-and-food/oecd-fao-agricultural-outlook-2020-2029_1112c23b-en (accessed May 29, 2023).

Press Information Bureau. 2022. *Milk Production in India.* https://pib.gov.in/FeaturesDeatils.aspx?NoteId=151137&ModuleId%20=%202 (accessed March 11, 2023).

Rahul, K. B., Bhuvaneshwari, S., Majeed, F., Maneesha, M. M., Jose, E., and A. Mohan. 2022. Different treatment methodologies and reactors employed for dairy effluent treatment – a review. *J. Water Process Eng.* 46:102622.

Ray Chaudhuri, S. 2023. Are algal biofuels an answer to the petrochemical crisis? In *Environmental Sustainability of Biofuels*, ed. K. U. R. Hakeem, S. A. Bandh, F. A. Malla, and M. A. Mehmood, 299–312. Elsevier Inc. ISBN: 9780323911597.

Ray Chaudhuri, S., Mishra, M., De, S., Samal, B., Saha, A., Banerjee, S., Chakraborty, A., Chakraborty, A., Pardhiya, S., Gola, D., Chakraborty, J., Ghosh, S., Jangid, K., Mukherjee, I., Sudarshan, M., Nath, R., and A. R. Thakur. 2017. Microbe-based strategy for plant nutrient management. In *Waste Water Treatment and Resource Recovery*, ed. R. Farooq and Z. Ahmed, 3, 38–55. Intech.

Saha, A., Bhushan, S., Mukherjee, P., Chanda, C., Bhaumik, M., Ghosh, M., Sharmin, J., Datta, P., Banerjee, S., Barat, P., Thakur, A. R., Gantayet, L. M., Mukherjee, I., and S. Ray Chaudhuri. 2018. Simultaneous sequestration of nitrate and phosphate from wastewater using a tailor-made bacterial consortium in biofilm bioreactor. *J. Chem. Technol. Biotechnol.* 93:1279–1289.

Salunke, D. K., and B. B. Desai. 1988. Effects of agricultural practices, handling, processing, and storage on vegetables. In *Nutritional Evaluation of Food Processing*, ed. E. Karmas and R. S. Harris, 23–72. Springer.

Sasidharan, R., Kumar, A., Paramasivan, B., and A. Sahoo. 2023. Photocatalytic pretreatment of dairy wastewater and benefits of the photocatalyst as an enhancer of anaerobic digestion. *J. Water Process Eng.* 52:103511.

Shete, B. S., and N. P. Shinkar. 2013. Dairy industry wastewater sources, characteristics & its effects on environment. *Int. J. Current Eng. Technol.* 3(5):1161–1615.

United Nations. 2003. *Waste-Water Treatment Technologies: A General Review.* https://www.worldcat.org/title/waste-water-treatment-technologies-a-general-review/oclc/55489914 (accessed May 27, 2023).

Wróbel, M., Śliwakowski, W., Kowalczyk, P., Kramkowski, K., and J. Dobrzyński. 2023. Bioremediation of heavy metals by the genus Bacillus. *Int. J. Environ. Res. Public Health* 20(6):4964.

Yonar, T., Sivrioglu, O., and N. Ozengin. 2018. Physico-chemical treatment of dairy industry wastewaters: a review. In *Technological Approaches for Novel Applications in Dairy Processing*, ed. N. Koca, 9, 179–191. Intech Open.

Żyłka, R., Karolinczak, B., and W. Dąbrowski. 2021. Structure and indicators of electric energy consumption in dairy wastewater treatment plant. *Sci. Total Environ.* 782:146599.

5 Novel Microbial Techniques for Pollutant Environment

Their Principles, Advantages, Limitations, and Future Prospects

Vikas Kumar, Preeti Pallavi, and Sangeeta Raut

5.1 INTRODUCTION

Global industrialization has led to environmental, freshwater, and topsoil pollution. Due to human activities, such as mining and the eventual removal of toxic metal effluents from steel mills, battery manufacturers, and energy generation, water quality has become worse, raising serious environmental issues. The ecology is harmed by effluents like petroleum, PET, and trace metals. Heavy metals are contaminants that naturally occur in the crust of the Earth and are challenging to break down. They are extracted as minerals after existing as ores in rocks. Heavy metals may be released into the environment at high exposure levels. They remain hazardous for longer once they are in the environment. Synthetic biology involves the design and engineering of biological systems for specific purposes. In this context of bioremediation, synthetic biology can be used to develop new strategies for removing pollutants from the environment. Some strategies that are being explored in this field include the following.

Synthetic biology can be used to engineer microorganisms with specific capabilities for removing pollutants (1). For example, scientists can introduce genes into bacteria or fungi that allow them to produce enzymes that can break down specific pollutants. Engineered microorganisms can break down specific pollutants by producing enzymes or other metabolites that are capable of degrading the target contaminants. The process of engineering microorganisms for bioremediation involves the introduction of new genes into the microorganisms or the modification of existing genes to enhance the degradation process. The specific enzymes or metabolites produced by the engineered microorganisms depend on the type of pollutant being targeted (1). For example, some engineered microorganisms produce laccases, which are enzymes that can oxidize organic pollutants and break down toxic chemicals.

 DOI: 10.1201/9781003517108-5

Other engineered microorganisms produce proteases, which can break down proteins, and cellulases break down cellulose.

In the case of heavy metal pollution, engineered microorganisms can produce chelating agents, which can form stable complexes with the metal ions, effectively removing them from the environment. Some engineered microorganisms can also produce metalloproteases, which can break down metal–protein complexes, allowing the metal ions to be more easily removed. The process of pollutant removal by engineered microorganisms involves the introduction of microorganisms into the contaminated environment. The microorganisms then grow and produce the enzymes or metabolites that interact with the pollutants, breaking them down and removing them from the environment.

5.2 BIOSENSORS

Synthetic biology can be used to develop biosensors that can detect the presence of pollutants in the environment. These biosensors can then be used to trigger the degradation process by releasing specific enzymes or other metabolites that break down the pollutants (2). Nanoparticles can be used in biosensor applications and have appealing optical and electronic properties.

5.2.1 Development of Biosensors That Can Detect the Presence of Pollutants in the Environment

Biosensors are devices that can detect the presence of specific target molecules, such as pollutants, in the environment. In the context of bioremediation, biosensors can be used to monitor the presence of pollutants in real time and trigger the degradation process by releasing specific enzymes or other metabolites that break down the pollutants (3). The development of biosensors for pollutant detection involves the integration of biological recognition elements, such as antibodies or enzymes, with a transduction element that converts the biological response into an electrical or optical signal. Biosensors can be designed to detect a wide range of pollutants, including heavy metals, organic chemicals, and toxic gases (3). Developed biosensors for pollutant detection offer a number of advantages over traditional methods, including real-time monitoring, rapid response times, and the ability to detect low concentrations of pollutants.

5.2.2 Types of Biosensors for Pollutant Removal

Biosensors can be designed using enzymatic or aptamer-based approaches and can provide real-time monitoring and rapid response times for the detection of pollutants.

1. *Enzymatic biosensors.* These use enzymes that specifically bind to the target pollutant, resulting in a measurable change in the enzymatic activity. This type of biosensor can detect pollutants, such as heavy metals and organic chemicals.

2. *Aptamer-based biosensors.* These use aptamers, which are short RNA or DNA molecules that specifically bind to the target pollutant. The binding of the aptamer to the pollutant results in a conformational change that can be detected using optical or electrical methods.
3. *Antibody-based biosensors.* These use antibodies that specifically bind to the target pollutant. The binding of the antibody to the pollutant results in a measurable change in the electrical or optical properties of the biosensor.
4. *Bacterial biosensors.* These use bacteria that have been engineered to degrade specific pollutants. The bacteria can produce enzymes or other metabolites that break down the pollutants. The presence of the pollutants can be detected by monitoring the bacterial growth or the production of degradation products.
5. *Electrochemical biosensors.* These use electrodes that are modified with biological recognition elements, such as enzymes or antibodies, to detect the presence of pollutants. The electrical signals generated by the electrodes are proportional to the concentration of the pollutant in the environment.

5.2.3 Enzymatic Biosensors

Enzymatic biosensors are a type of biosensor that uses enzymes to detect the presence of specific target molecules, such as pollutants. The enzymatic biosensor works by using an enzyme that specifically binds to the target pollutant and generates a measurable signal, such as a change in pH or the production of a colored product. In enzymatic biosensors, the biological recognition element is the enzyme, and the transduction element is the chemical reaction catalyzed by the enzyme. The chemical reaction can result in a change in the pH, the production of a colored product, or the generation of an electrical signal (4). Enzymatic biosensors can be designed to detect a wide range of pollutants, including heavy metals, organic chemicals, and toxic gases. They offer a number of advantages over traditional methods, including real-time monitoring, rapid response times, and the ability to detect low concentrations of pollutants (5). One of the main advantages of enzymatic biosensors is their specificity. The enzyme used in the biosensor is chosen for its ability to specifically bind to the target pollutant, which minimizes the risk of false positive or negative signals.

5.2.4 Bacterial Biosensors

Bacterial biosensors use bacteria to detect the presence of specific target molecules, such as pollutants. The bacterial biosensor works by engineering bacteria to produce enzymes or other metabolites that degrade specific pollutants. The presence of the pollutant in the environment can be detected by monitoring the bacterial growth or the production of degradation products. In bacterial biosensors, the biological recognition element is the bacteria, and the transduction element is the metabolic response of the bacteria to the presence of the pollutant. The bacterial metabolic response can result in a change in the optical density, fluorescence, or conductivity of the bacterial culture. Bacterial biosensors can be designed to detect a wide range of pollutants, including heavy metals, organic chemicals, and toxic gases. They offer a number of advantages over traditional methods, including real-time monitoring, rapid response times, and the ability to detect low concentrations of pollutants (6). One of the main advantages of bacterial biosensors is their versatility. Bacteria can be engineered to degrade a wide range of pollutants, and the bacterial biosensor can be easily modified to target different pollutants by changing the bacteria or the metabolic pathway used to degrade the pollutant.

5.2.5 Aptamer-Based Biosensors

Aptamer-based biosensors use aptamers, which are short single-stranded RNA or DNA molecules, to detect the presence of specific target molecules (7), such as pollutants. The aptamer-based biosensor works by binding the aptamer to the target pollutant, which generates a measurable signal, such as a change in fluorescence or the production of a colored product. In aptamer-based biosensors, the biological recognition element is the aptamer or chemical antibodies (7), and the transduction element is the binding of the aptamer to the target pollutant. The binding of the aptamer can result in a change in the fluorescence, conductivity, or color of the aptamer. Aptamer-based biosensors can be designed to detect a wide range of pollutants, including heavy metals, organic chemicals, and toxic gases. They offer a number of advantages over traditional methods, including specificity, real-time monitoring, and the ability to detect low concentrations of pollutants. A number of biosensors, using aptamers as the bioreceptors, have been developed in recent years to detect environmental toxins. These biosensors include colorimetric, fluorescent, electrochemical, and surface-enhanced Raman spectroscopy (SERS) (8). Also, the creation of fresh nanomaterials demonstrated their enormous potential for the creation of groundbreaking aptasensors. The latter are maintained by aptamers due to their great biocompatibility. Aptamers can be selected for their ability to specifically bind to the target pollutant, which minimizes the risk of false positive or negative signals (9).

5.2.6 Antibody-Based Biosensors

Antibody-based biosensors use antibodies, which are proteins that recognize and bind to specific target molecules (10). The antibody-based biosensor works by binding the antibody to the target pollutant, which generates a measurable signal, such as a change in fluorescence or the production of a colored product. In antibody-based biosensors, the biological recognition element is the antibody, and the transduction element is the binding of the antibody to the target pollutant. The binding of the antibody can result in a change in the fluorescence, conductivity, or color of the antibody (11). *Immunosensors*, another name for antibody-based biosensors, are portable instruments that use a transducer to detect and measure the precise interaction between immunoglobulins and antigens (12). They offer a number of advantages over traditional methods, including specificity, real-time monitoring, and the ability to detect low concentrations of pollutants. Immunosensors possess the advantages of better selectivity and sensitivity than classical analytical methods (13).

Examples of antibody-based biosensors include electrochemical antibody-based biosensors, which use an electrochemical transduction mechanism to detect the binding of the antibody to the target pollutant; fluorescence antibody-based biosensors, which use a fluorescence transduction mechanism to detect the binding of the antibody to the target pollutant; and colorimetric antibody-based biosensors, which use a colorimetric transduction mechanism to detect the binding of the antibody to the target pollutant (14).

5.2.7 Enzymatic Pollutant Biomarkers

Enzymatic pollutant biomarkers are a type of biological marker that uses specific enzymes as indicators of exposure to environmental pollutants. The presence of specific enzymes can serve as a sensitive and specific marker of exposure to certain pollutants, such as heavy metals, pesticides, and toxic organic compounds. For example, the presence of increased levels of certain enzymes, such as acetylcholinesterase (AChE), in organisms can indicate exposure to organophosphate pesticides. Similarly, the presence of increased levels of metallothionein in organisms can indicate exposure to heavy metals, such as cadmium, lead, and zinc. Enzymatic pollutant biomarkers provide a rapid and cost-effective method for monitoring exposure to environmental pollutants and assessing the health of populations and ecosystems. The use of enzymatic biomarkers can also provide important information for risk assessment and management of pollutants, as well as guide the development of remediation and prevention strategies (15).

In conclusion, different types of biosensors can be used for pollutant removal, including enzymatic, aptamer-based, antibody-based, bacterial, and electrochemical biosensors. Each type of biosensor has specific advantages and disadvantages, and the choice of biosensor will depend on the type of pollutant being targeted and the specific requirements of the bioremediation process.

5.3 GENOME EDITING

Synthetic biology can be used to edit the genomes of microorganisms to enhance their ability to remove pollutants. This can involve the insertion or deletion of specific genes to modify the metabolic pathways of the microorganisms (16). The main gene editing tools are CRISPR-Cas, ZFN, and TALEN, which can possibly fulfil the preceding expectations (17).

There are several strategies to edit the genomes of microorganisms to enhance their ability to remove pollutants. These strategies involve the manipulation of specific genes and pathways within the microorganisms to increase their ability to degrade, detoxify, or transport pollutants. **Some of these strategies include:**

1. *Metabolic engineering.* This involves the modification of metabolic pathways in microorganisms to increase their ability to degrade pollutants. This can be accomplished through the overexpression of genes involved in degradation pathways, the introduction of new genes encoding for novel degradation pathways, or the deletion of genes involved in competing metabolic pathways.
2. *Synthetic biology.* This involves the design and construction of new genetic circuits within microorganisms to enhance their ability to degrade pollutants.

This can involve the insertion of new genes encoding for metabolic pathways, the manipulation of regulatory elements to control gene expression, or the construction of biosensors to detect and respond to pollutants.

3. *Phage-assisted gene transfer.* This involves the use of bacteriophages (viruses that infect bacteria) to introduce new genes into microorganisms. This allows for the efficient transfer of genes encoding for degradation pathways or other bioremediation functions, without the need for transformation or other genetic modification methods.
4. *RNA interference (RNAi).* This involves the use of small RNA molecules to silence specific genes within microorganisms. This can be used to reduce the expression of genes involved in competing metabolic pathways or to increase the expression of genes involved in degradation pathways.

These are just a few of the strategies used to edit the genomes of microorganisms to enhance their ability to remove pollutants. By manipulating specific genes and pathways within the microorganisms, it is possible to increase their efficiency and specificity in removing pollutants from the environment.

5.4 CHASSIS MICROORGANISMS

Synthetic biology can be used to create "chassis" microorganisms, which are microorganisms that have been modified to serve as a platform for the development of new bioremediation strategies. This can involve the engineering of specific metabolic pathways or the integration of multiple enzymes into a single microorganism. The concept of the chassis organism is based on the idea that biological systems can be engineered by combining functional genetic elements, such as promoters, genes, and regulatory circuits, to produce a desired biological function or product (18). Chassis microorganisms provide a foundation for this process by serving as a framework onto which genetic parts can be assembled for proper function. The chassis organism can be modified by adding, deleting, or substituting genetic parts to create novel biological functions, such as biosynthesis of biofuels, production of pharmaceuticals, bioremediation, and other applications. The use of chassis microorganisms has revolutionized the field of synthetic biology by providing a standardized platform for genetic engineering, allowing for rapid prototyping and testing of new biological systems. Chassis microorganisms are also critical for the development of safe and reliable engineered biological systems, as they can be thoroughly characterized and tested for biosafety and biocontainment.

In synthetic biology, a *chassis microorganism* refers to a genetically tractable host organism that is used as a platform for the construction of synthetic biological systems. The chassis organism serves as a framework onto which functional genetic elements such as genes, promoters, and regulatory circuits can be assembled to create a desired biological function or product (18). Chassis microorganisms are typically chosen for their well-characterized genetics, high transformation efficiency, and ability to grow and function under controlled laboratory conditions. Examples of commonly used chassis microorganisms include *Escherichia coli*, *Bacillus subtilis*, and *Saccharomyces cerevisiae*, among others. Chassis microorganisms can be modified by adding, deleting, or substituting genetic parts to create novel biological functions, such as the production of biofuels and pharmaceuticals or the biodegradation of environment. Chassis microorganisms can be used to facilitate the biodegradation of environmental pollutants through the construction of synthetic biological systems that degrade target compounds. This involves engineering the chassis organism to express genes that encode enzymes or pathways capable of breaking down the target pollutant, or to upregulate endogenous genes involved in pollutant degradation. For example, a synthetic biological system could be designed to use the chassis organism to produce enzymes that break down hydrocarbons in oil spills or to degrade toxic chemicals such as polychlorinated biphenyls (PCBs) or polycyclic aromatic hydrocarbons (PAHs).

One example of this is the use of *Escherichia coli* as a chassis organism for bioremediation of heavy metals. Researchers have engineered *E. coli* to express genes that encode for metal-binding proteins and transporters, which help the organism take up and detoxify heavy metals, such as mercury, cadmium, and lead, from contaminated soil or water (19). Another example is the use of *Pseudomonas putida* as a chassis organism for the biodegradation of herbicides, which involves engineering the bacterium to express genes encoding enzymes that break down the herbicide into non-toxic compounds (18, 20, 21). The use of chassis microorganisms for biodegradation of environmental pollutants holds great promise for the development of more efficient and sustainable methods for cleaning up polluted environments.

5.5 MICROBIAL FABRICATED NANOSYSTEMS

Microbial fabricated nanosystems refer to the production of nanoscale materials or structures through the use of microbial cells or their components. This process is also known as biomineralization or biofabrication. Microbes such as bacteria, fungi, and algae have the ability to create intricate structures and materials on a nanoscale level (22). These structures can have a wide range of properties and applications, including use in biomedicine, energy, electronics, and environmental remediation. For example, some bacteria are capable of producing magnetite nanoparticles that have potential applications in magnetic resonance imaging (MRI) and targeted drug delivery. Other microbes can produce metallic nanoparticles that have antimicrobial properties, making them useful in medical applications, such as wound healing. Microbial fabricated nanosystems have the potential to revolutionize various industries and contribute to the development of sustainable and environmentally friendly technologies.

Nanosystems have the potential to revolutionize environmental cleaning by providing efficient, cost-effective, and environmentally friendly solutions to a wide range of pollution problems. One example of nanosystems in environmental cleaning is the use of nanoparticles for water treatment. Nanoparticles can be used to remove pollutants such as heavy metals, organic compounds, and pathogens from contaminated water sources (23). They can also be used to filter out microplastics and other small particles that are difficult to remove using traditional methods. Another example is the use of nanoscale materials in air purification. Nanoparticles can be used to remove harmful pollutants from the air, including volatile organic compounds

(VOCs), particulate matter, and nitrogen oxides (NOx). These particles can be embedded in air filters or incorporated into building materials to improve indoor air quality. Nanosystems can be used to clean up contaminated soil and groundwater. Nanoparticles can be designed to bind with pollutants, making them easier to remove from the environment. They can also be used to break down complex organic compounds into simpler, less-harmful substances.

Nanosystems offer a promising solution for environmental cleaning and can contribute to the development of sustainable and eco-friendly technologies (24). However, it is important to consider the potential risks associated with the use of nanoparticles, including their potential impact on human health and the environment. Nanoparticles can be used to remove pollutants such as heavy metals from contaminated water sources. Heavy metals, such as lead, arsenic, mercury, and cadmium, are toxic pollutants that can have serious health impacts on humans and the environment. Nanoparticles can be designed to selectively bind with heavy metal ions, effectively removing them from water. These nanoparticles can be made from a variety of materials, including iron oxide, titanium oxide, and carbon. When the nanoparticles are added to contaminated water, they attract and bind with the heavy metal ions, forming aggregates that can be removed using conventional filtration methods (25). Nanoparticle-based water treatment methods have several advantages over traditional treatment methods. They are more effective at removing heavy metals, require less energy and chemicals, and produce less waste. Additionally, nanoparticle-based water treatment methods can be designed to be scalable and cost-effective, making them suitable for use in developing countries and remote areas. It is important to note that the use of nanoparticles for water treatment also raises concerns about their potential impact on human health and the environment. Research is ongoing to better understand the risks associated with the use of nanoparticles for water treatment and to develop safe and sustainable technologies.

Nanoparticles can be used to remove pathogens, such as bacteria and viruses, from contaminated water sources. Pathogens in water sources can cause serious illnesses, and traditional water treatment methods such as chlorination and filtration may not be effective at removing all types of pathogens and can be designed to attract and bind with pathogens, effectively removing them from water. For example, silver nanoparticles have been shown to be effective in removing bacteria from water sources. Other types of nanoparticles, such as titanium dioxide nanoparticles, can be used to inactivate viruses by breaking down their protein coats and disrupting their genetic material. However, it is important to note that the use of nanoparticles for pathogen removal in water sources also raises concerns about their potential impact on human health and the environment. Some nanoparticles may have toxic effects on humans or other organisms, and their long-term effects on the environment are not fully understood. Therefore, research is ongoing to better understand the risks associated with the use of nanoparticles for pathogen removal in water treatment and to develop safe and sustainable technologies.

Nanosystems have the potential to revolutionize environmental cleaning by providing more efficient and effective solutions to a wide range of pollution problems. *Nanosystems* refer to the use of nanotechnology, which involves the manipulation and control of matter at the nanoscale level. Nanosystems can be used for a variety of environmental cleaning applications, such as water treatment, air purification, and soil remediation. For example, nanoparticles can be used to remove pollutants from water sources, including heavy metals, organic compounds, and pathogens. In air purification, nanoparticles can be used to remove harmful pollutants, such as volatile organic compounds (VOCs) and particulate matter (26). In soil remediation, nanoparticles can be used to bind with and remove pollutants from contaminated soil. Nanosystems offer several advantages over traditional cleaning methods. They are more effective at removing pollutants, require less energy and resources, and produce less waste. Nanosystems can also be designed to be more selective, targeting specific pollutants while leaving other components of the environment intact. It is important to note that the use of nanosystems for environmental cleaning also raises concerns about their potential impact on human health and the environment. Therefore, it is important to carefully evaluate and mitigate any potential risks associated with the use of nanosystems in environmental cleaning. Overall, nanosystems offer a promising solution for environmental cleaning and can contribute to the development of sustainable and eco-friendly technologies.

5.6 NANOFIBROUS WEBS FILTERS FOR WASTEWATER TREATMENT

Filtration and pollutant trapping using membranes are two more technologies utilized for wastewater treatment. However, many studies have been done to develop reasonably priced, highly effective filters for treatment using nanotechnological techniques, such as the creation of electrospun nanofibers and nanowebs. The nanofibrous web filters have a large surface area and nanoscale porosity, making it suited for the filtration and trapping of micro-nanoscale contaminants (27). As a result, wastewater filtration and purification can be improved by combining nanofibrous matrix with microbial cells like bacteria or algae. Several applications have been investigated and shown to have significant environmental effects. *Clavibacter michiganensis*–, *Pseudomonas aeruginosa*–, and *Aeromonas eucrenophila*–immobilized bacterial strains were successfully used to extract 95% of the color molecules from wastewater using reusable electrospun CA-NFW.

In a study, nitrates have been eliminated from wastewater using a hybrid model of electrospun chitosan nanofiber mats (ECNMs)–immobilized algal cells. The ECNM–algal matrix exhibits up to 87% of nitrate removal from wastewater and is simpler to manage. It also requires little area for production. However, the ammonium-oxidizing bacteria *Acinetobacter calcoaceticus* STB1 cells were immobilized by electrospun cellulose acetate nanofibrous webs (CA-NFW) for ammonium removal from wastewater (98.5%). As a result, NFWs are a cost-effective reactor system for the remediation of heavy metals and hazardous contaminants since they have a smaller volume and a greater active surface area. Moreover, under environmental stress, the biofilm of biogenic nanocomposite exhibits excellent stability and notable activity (28).

5.7 MICROBIAL FUEL CELLS IN WASTEWATER TREATMENT

Microbial fuel cells (MFCs) are devices that use bacteria to convert organic matter into electricity. In wastewater treatment, MFCs can be used as an alternative to traditional treatment methods to generate electricity while removing contaminants from the water. The basic design of an MFC consists of an anode and a cathode separated by a membrane. The anode is typically made of a conductive material, such as carbon, and is colonized by bacteria that consume organic matter and release electrons. The electrons flow through an external circuit to the cathode, where they combine with oxygen to produce water. In wastewater treatment, MFCs can be used to remove organic matter and nutrients, such as nitrogen and phosphorus, from the water. The bacteria in the anode consume the organic matter, breaking it down into simpler compounds and releasing electrons in the process. This generates electricity that can be used to power the treatment process (20).

One advantage of MFCs over traditional wastewater treatment methods is that they can operate at low temperatures and do not require aeration, which can be energy-intensive. Additionally, MFCs can generate a significant amount of electricity, which can be used to offset the energy costs of the treatment process. Acetate, propionate, and butyrate are some examples of complicated compounds that the bacterial cell aids in decomposing into less-complex forms, such as H_2O and CO_2. So by utilizing the energy produced by MFCs, the demand for conventional technology for power production is decreased (20). Overall, MFCs have the potential to be a sustainable and efficient option for wastewater treatment, but further research is needed to optimize their design and performance.

Furthermore, the integration of wastewater treatment systems with advanced communication networks and IoTs (Internet of Things) can provide real-time monitoring and point-of-detection capabilities for improved efficiency and reliability. By using sensors and data analysis tools, operators can monitor the performance of the treatment process in real time, detecting potential issues before they become serious problems. For example, sensors can be used to monitor the levels of organic matter, nutrients, and pathogens in the wastewater, providing early warning of changes that could affect the efficiency of the treatment process. IoT-enabled wastewater treatment systems can be designed to automatically adjust treatment parameters based on real-time data, optimizing the process for maximum efficiency and reducing the risk of over-treatment or under-treatment (29). This can also help reduce energy and chemical consumption, which can lead to cost savings and environmental benefits. The connectivity of wastewater treatment systems with advanced communication networks and IoTs can facilitate remote monitoring and control, allowing operators to manage the treatment process from anywhere at any time. This can be particularly beneficial in situations where the treatment plant is located in a remote or inaccessible location, or where there are limited resources for on-site monitoring and maintenance.

Overall, the integration of wastewater treatment systems with advanced communication networks and IoTs can provide significant benefits for improved efficiency, reliability, and environmental performance. The combination of cutting-edge technologies, such as IoTs, artificial intelligence (AI), machine learning, cloud computing, and 5G communication, with bio-nanotechnology-based water purification strategies has the potential to transform the currently time- and resource-intensive approaches. AI can also be used to investigate different microbes and their bio-reductants in order to create different classes of NMs without the need for actual experimental research (30). Hence, the fusion of contemporary technologies with bio-nanotechnology and the Internet of Nanothings has the potential to transform current water treatment techniques in order to meet the demand for clean water and the aims of sustainable development (30). To identify more MTB possibilities and gain a deeper understanding of the creation of magnetosomes and their function in the removal of pollutants from water bodies, more study is required in this field. If these challenges are overcome, MTBs will undoubtedly drive future nanotechnology advancements.

5.8 FUNGI NANOMATERIALS

Fungi have been found to be a rich source of nanomaterials, due to their ability to produce a wide range of metabolites, including enzymes, pigments, and polysaccharides. Researchers have been able to extract these metabolites from fungi and manipulate them to create nanomaterials with unique properties. These include:

1. *Fungal pigments.* Melanin, for instance, a pigment produced by many fungi, can be used to create semiconducting nanoparticles for use in solar cells and other electronic devices.
2. *Fungal enzymes.* Fungal enzymes have been used to create biodegradable and biocompatible materials for use in medical and environmental applications. For example, chitinases and chitosanases, enzymes produced by certain fungi, can be used to create chitosan-based nanoparticles for drug delivery.
3. *Fungal polysaccharides.* Fungal polysaccharides can be used to create hydrogels and other materials with potential applications in tissue engineering and wound healing.

Fungal metabolites also have potential for use as antimicrobials, for instance, in food packaging to prevent contamination by pathogenic bacteria. Fungi are a promising source of nanomaterials with a wide range of potential applications in fields such as medicine, electronics, and environmental science.

5.9 FUNGAL NANOREMEDIATION

Fungal nanoremediation is the use of fungi and their metabolites to remove or degrade pollutants in the environment. This can include the use of fungal enzymes and pigments to break down toxic compounds, as well as the use of fungal biomass to adsorb pollutants from contaminated soils and waters. One example of fungal nanoremediation is the use of mycoremediation, which is the use of fungi to break down pollutants in soil and water. Fungi can produce enzymes that can degrade a wide range of pollutants, including polycyclic aromatic hydrocarbons (PAHs), polychlorinated biphenyls (PCBs), and pesticides. Another example is the use of fungal-based nanoparticles for the removal of heavy metals from contaminated water. Fungi can produce nanoparticles such as melanin or chitosan, which can adsorb heavy metals, such as lead, cadmium, and mercury, and remove them from water. Fungi can also be used to degrade or remove pollutants from air, as certain species can remove volatile organic compounds (VOCs) and other pollutants from the air. Overall, fungal nanoremediation is a promising approach for the cleanup of contaminated environments. The use of fungi and their metabolites can be effective, sustainable, and cost-efficient and can also help reduce the use of harsh chemicals in the remediation process.

Fungi-based nanoparticles have been researched for their potential use in cleaning the environment. Some methods include:

1. *Fungal biosorption.* Using fungal cells as natural adsorbents to remove heavy metals from contaminated soil or water.
2. *Fungal biodegradation.* Using fungi to break down and degrade toxic pollutants, such as oil spills or polycyclic aromatic hydrocarbons (PAHs).
3. *Fungal bioremediation.* Integrating fungal processes into bioremediation systems to improve their efficiency and effectiveness.
4. *Fungal phytoremediation.* Using fungi in conjunction with plants to remove pollutants from the environment, such as using mycorrhizal fungi to help plants absorb heavy metals.
5. *Fungal biotransformation.* Using fungi to convert pollutants into less-toxic forms.

These methods are still being researched and developed, and their effectiveness and practicality for large-scale environmental cleanup projects are still being evaluated. *Fungal biotransformation* refers to the use of fungi to convert toxic pollutants into less-harmful forms. This process occurs naturally, as fungi degrade organic matter, but can also be enhanced in controlled laboratory settings. The fungi secrete enzymes that break down complex toxic compounds into simpler, less-toxic molecules. Examples of fungal biotransformation include degradation of persistent organic pollutants (POPs), such as polychlorinated biphenyls (PCBs), and transformation of xenobiotic compounds, like aromatic hydrocarbons and pesticides. Fungal biotransformation is a promising area of research for environmental remediation, as it offers a natural, cost-effective alternative to traditional chemical treatments. However, more research is needed to optimize the process for large-scale application.

Nanomycoremediation is a term used to describe the use of fungal-based nanoparticles to degrade pollutants, such as textile dyes. The fungal nanoparticles are produced by growing fungi on a substrate and then isolating the fungal cells and breaking them down to produce nanoparticles. These fungal nanoparticles are able to degrade pollutants by releasing enzymes and other metabolic products that break down the pollutants. Research has shown that fungal nanoparticles can be effective in degrading textile dyes. For example, studies have used fungal nanoparticles produced from white-rot fungi, such as *Phanerochaete chrysosporium* and *Pleurotus ostreatus*, to degrade azo dyes. The fungal nanoparticles have been found to be able to degrade the dyes more efficiently than the bulk mycelia from which they were derived.

One of the advantages of using fungal nanoparticles for nanomycoremediation is that they can be easily transported and distributed in the environment, making them useful for on-site treatment of pollutants. Additionally, the fungal nanoparticles are able to persist in the environment for longer periods of time, which allows them to continue to degrade pollutants over time. It is worth noting that the field of nanomycoremediation is still a developing field and more research is needed to fully understand the potential of this approach. Factors such as the fungal species used, the method of production of fungal nanoparticles, and the optimal conditions for their use need to be further studied.

Fungi are known to produce a wide variety of metabolites, including enzymes, pigments, and antibiotics. Some of these metabolites can be used to create nanomaterials with unique properties. For example, certain fungal pigments can be used to create semiconducting nanoparticles for use in solar cells and other electronic devices. Fungal enzymes can also be used to create biodegradable and biocompatible materials for use in medical and environmental applications. Researchers are actively studying fungal metabolites for their potential use in creating new and innovative nanomaterials.

5.10 BIOSYNTHESIS OF NANOPARTICLES BY MICROORGANISMS

Inorganic nanoparticles like gold, silver, calcium, silicon, iron, gypsum, and lead are known to be produced by microorganisms like bacteria, cyanobacteria, actinomycetes, yeast, and fungus. They produce nanoparticles that are intra- and/or extracellular in nature due to their intrinsic potential (24). However, due to additional processing steps such as ultrasonication and therapy with appropriate detergents, it is difficult to extract the nanoparticles produced by intracellular biosynthesis. In order to prevent the extracellular production of nanoparticles, bacteria must be screened. Presently, metals, a few metal sulfides, and very few oxides are the only compounds that can be produced by microbes when it comes to variable compound nanomaterials. They are all limited to microorganisms with an earthy origin. To produce nanoparticles on a wide scale, it is required to standardize the culture conditions that control the biological synthesis of nanoparticles using microorganisms. It is acknowledged that many microbes may manufacture metallic nanoparticles with qualities similar to those of chemically synthesized nanomaterials, despite stringent control being used over the shape, size, and combination of the particles. The hydrolytic activity of the microorganisms is anticipated to result in the formation of additional metal oxides. Finally, microorganisms can produce nanosized materials under moderate pressures and temperatures. Furthermore, using microbial processes to produce nanomaterials is inexpensive, simple, efficient, energy-efficient, and environmentally friendly.

5.11 BIOSYNTHESIS OF NANOPARTICLES BY FUNGI

Eukaryotic fungi can be found in a wide range of typical settings and usually function as decomposer organisms. Of the estimated 1.5 million species of fungi on Earth, only 70,000 have been named. According to more recent data, approximately 5.1 million fungal species are discovered using high-throughput sequencing methods. These organisms have the capacity to digest extracellular food and to hydrolyze

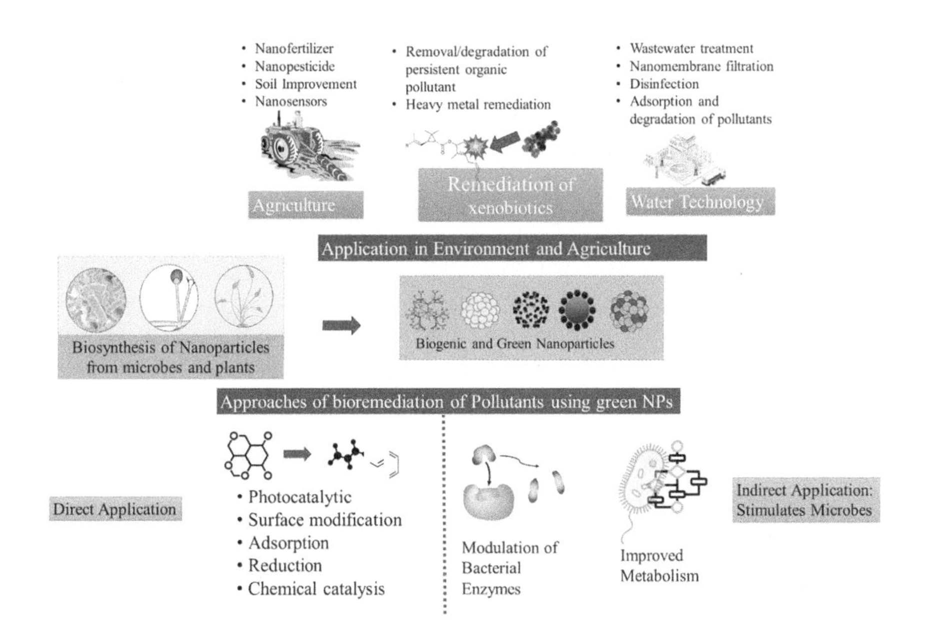

FIGURE 5.1 Applications of Nanoparticles in Environment and Agriculture.

complicated components into simpler molecules that can be ingested and used as an energy source. The ability of fungi to bioaccumulate metals has drawn increased interest in studies on the biological synthesis of metallic nanoparticles.

One clear benefit of employing fungi in the manufacture of nanoparticles is how easily they can be scaled up (e.g., utilizing a thin solid substrate fermentation technique). Because fungi are very effective secretors of extracellular enzymes, large-scale enzyme production is feasible (21). Another advantage of using a green approach mediated by fungal to synthesize metallic nanoparticles is the economic viability and ease of employing biomass. Furthermore, many species are easy to cultivate and maintain in a lab setting because they grow quickly (25). Most fungi have high wall-binding and intracellular metal uptake abilities (31). Fungi can produce metal nanoparticles/meso- and nanostructures by reducing enzyme intracellularly or extracellularly and using the biomimetic mineralization procedure.

5.12 REMOVAL OF HEAVY METALS BY FUNGI

Fungi have been researched for their ability to remove heavy metals from contaminated soil and water through a process known as biosorption. This process works by using the fungal cells as natural adsorbents, which can bind to heavy metals in the environment and effectively remove them. The mechanisms for heavy metal removal by fungi include:

Physical adsorption. The metal ions bind to the fungal cell surface through electrostatic interactions. The binding of metal ions to the fungal cell surface through electrostatic interactions is a mechanism of heavy metal removal by fungi through biosorption. This process occurs due to the difference in charge between the fungal cell surface and the metal ions. The fungal cell surface is negatively charged, while the metal ions are positively charged. This creates an attractive force between the two, allowing the metal ions to bind to the fungal cell surface through electrostatic interactions. Once the metal ions are bound to the fungal cell surface, they are effectively removed from the environment and can no longer cause harm. The fungal cells can then be harvested, and the metal ions can be recovered and reused or disposed of safely (32).

This process of heavy metal removal by fungi through electrostatic interactions is a natural and sustainable alternative to traditional chemical treatments. However, more research is needed to optimize the process for large-scale applications and to understand the mechanisms involved in greater detail.

Chemical adsorption. The mechanism by which metal ions form chemical bonds with functional groups on the fungal cell surface is another mechanism of heavy metal removal by fungi through biosorption. Functional groups, such as carboxyl and hydroxyl groups, are present on the fungal cell surface and can react with metal ions to form chemical bonds. These chemical bonds are much stronger than the electrostatic interactions that occur in physical adsorption, and as a result, the metal ions are more effectively removed from the environment. The specific type of chemical bond that forms between the metal ion and the fungal cell surface depends on the properties of the metal ion and the functional group (33). For example, carboxyl groups can form chelating bonds with metal ions, such as lead and cadmium, while hydroxyl groups can form coordination bonds with metal ions, such as copper and zinc.

This mechanism of heavy metal removal by fungi through chemical bonding is a natural and sustainable alternative to traditional chemical treatments. However, more research is needed to optimize the process for large-scale applications and to understand the mechanisms involved in greater detail.

Chelation. Fungal cells produce chelating agents, such as organic acids, that can bind to heavy metals and remove them from the environment. Fungal chelating agents are compounds produced by fungi that can bind to metal ions and effectively remove them from contaminated soil and water. The use of fungal chelating agents for pollutant removal is a type of bioremediation technology that takes advantage of the ability of fungi to produce these compounds and to remove heavy metals from the environment (34).

5.13 THE PROCESS OF POLLUTANT REMOVAL USING FUNGAL CHELATING AGENTS

Fungal cells are introduced into the contaminated environment.

The fungi secrete chelating agents, such as organic acids, into the environment.

The chelating agents bind to the metal ions in the contaminated soil or water, effectively removing them from the environment.

The metal ions are held in the chelating agents, which can be harvested and disposed of or recovered for reuse.

Fungal chelating agents have been shown to be effective for removing a range of heavy metals, including lead, cadmium, and mercury. Additionally, the use of fungal chelating agents for pollutant removal is a natural and sustainable alternative to traditional chemical treatments. However, more research is needed to optimize the process for large-scale applications and to understand the mechanisms involved in greater detail.

5.13.1 Mechanism of Fungal Chelation

The mechanism of fungal chelation for pollutant removal involves the binding of heavy metal ions by chelating agents produced by fungi. Chelating agents are compounds that can form complex structures with metal ions, effectively removing them from the environment (35). In fungal chelation, the fungi secrete chelating agents, such as organic acids, into the contaminated environment. The chelating agents then bind to the metal ions in the soil or water, forming complex structures that effectively remove the metal ions from the environment. The specific type of bond that forms between the metal ion and the chelating agent depends on the properties of the metal ion and the chelating agent (34). For example, carboxyl groups can form chelating bonds with metal ions, such as lead and cadmium, while hydroxyl groups can form coordination bonds with metal ions, such as copper and zinc.

Once the metal ions are bound to the chelating agents, they are effectively removed from the environment and can no longer cause harm. The chelating agents can then be harvested, and the metal ions can be recovered and reused or disposed of safely. This mechanism of heavy metal removal by fungi through chelation is a natural and sustainable alternative to traditional chemical treatments. However, more research is needed to optimize the process for large-scale applications and to understand the mechanisms involved in greater detail.

5.13.2 Chemical Interaction of Fungal Chelation

The chemical interaction between heavy metal ions and fungal chelating agents is the basis of fungal chelation for pollutant removal. The goal of fungal chelation is to form stable complexes between the metal ions and the chelating agents, effectively removing the metal ions from the environment. Chelating agents are compounds that can form multiple bonds with metal ions, effectively sequestering the metal ions and removing them from the environment. The specific type of bond that forms between the metal ion and the chelating agent depends on the properties of the metal ion and the chelating agent (36). For example, carboxyl groups can form chelating bonds with metal ions, such as lead and cadmium, while hydroxyl groups can form coordination bonds with metal ions, such as copper and zinc. The bond formation effectively removes the metal ions from the environment, reducing the risk of toxicity and other negative effects. The strength of the bond between the metal ion and the chelating agent determines the effectiveness of the fungal chelation process. Stronger bonds result in more effective removal of metal ions, while weaker bonds may result in the release of metal ions back into the environment. Overall, the chemical interaction between heavy metal ions and fungal chelating agents is a critical aspect of fungal chelation for pollutant removal. The goal is to form stable, strong bonds between the metal ions and the chelating agents, effectively removing the metal ions from the environment and reducing the risk of toxicity and other negative effects (36).

Fungal biosorption has been shown to be effective for removing a range of heavy metals, including cadmium, lead, mercury, and zinc. Additionally, fungal biosorption is a cost-effective and sustainable alternative to traditional chemical treatments for heavy metal removal. However, more research is needed to optimize the process for large-scale applications.

5.14 ENZYME'S ROLE IN POLLUTANT REMOVAL

Fungal enzymes play a crucial role in pollutant removal through bioremediation. *Bioremediation* is the use of microorganisms, such as fungi, to remove contaminants from the environment. The ability of fungi to produce a range of enzymes and other metabolites allows them to break down and remove a range of pollutants, including heavy metals, organic pollutants, and hazardous chemicals.

The specific enzymes produced by fungi and the mechanisms they use to remove pollutants depend on the type of pollutant and the species of fungus. For example, some fungi produce laccases, which are enzymes that can oxidize organic pollutants and break down toxic chemicals. Other fungi produce proteases, which can break down proteins, and cellulases, which can break down cellulose. In the case of heavy metal pollution, fungi can produce chelating agents, which can form stable complexes with the metal ions, effectively removing them from the environment. Some fungi can also produce metalloproteases, which can break down metal–protein complexes, allowing the metal ions to be more easily removed.

The process of pollutant removal by fungi involves the introduction of fungal spores or mycelium into the contaminated environment. The fungi then grow and produce enzymes and other metabolites, which interact with the pollutants to break them down and remove them from the environment. In total, fungal enzymes play a critical role in pollutant removal through bioremediation. By producing a range of enzymes and other metabolites, fungi can effectively break down and remove a range of pollutants, including heavy metals, organic pollutants, and hazardous chemicals.

5.14.1 Removal of Colorful Textile Dyes

The sophisticated oxidative wastewater pre-treatment techniques produce a number of hazardous by-products and costly mineralization techniques. Many of the colors released into the environment by the paper, paint, tanning, and cosmetics industries constitute toxic waste for aquatic and terrestrial life. Aromatic halogenated chemicals, which are non-biodegradable, poisonous, carcinogenic, and mutagenic, make up several significant groups of dyes (37). There have been several environmentally friendly procedures used in the past to quickly remove color from aqueous solutions without any hazardous consequences. One of several sustainable techniques is the use of biogenic PdNPs made by *Caldicellulosiruptor saccharolyticus*, a species of thermophilic, anaerobic cellulolytic bacterium, to remove dye (methyl orange [MO] and diatrizoate) from water bodies that show noticeably greater breakdown efficiency (36). As compared to PdNPs made chemically, the rate of color deterioration is twice as quick with biogenic PdNPs. In addition, the biogenic PdNPs readily absorb the dye breakdown product that is produced, reducing the harmful impact (25). *Caldicellulosiruptor saccharolyticus* produces and disperses PdNPs, which promote the breakdown of pollutants in water. Similarly, the ability of chemically produced and naturally occurring zinc oxide nanoparticles (ZnONPs) to remove dyes such as reactive black-5 and methylene blue from polluted water was tested. In comparison to chemically generated ZnONPs 85.4%, *Pseudochrobactrum* sp. C5-produced ZnONPs showed removal efficiencies up to 95% for a variety of dyes, including methylene blue, brilliant blue R, reactive black 5, reactive red 120, and brilliant yellow. Due to its small particle size and higher surface area, biogenic ZnONPs have a substantially higher catalytic effectiveness for the efficient removal and remediation of dye from aquatic systems. Silver nanoparticles (AgNPs) were created biologically in order to turn wastewater into a synthetic textile dye by adding functional groups to the NPs' caps and revealing the crystal structure.

5.14.2 Removal of Pharmaceutical and Healthcare Wastewater Pollutants

It was recently discovered that medical waste and pharmaceutical industrial effluents include various harmful xenobiotics and drug compounds that reach the aquatic system either directly or through contaminated sites. Several conventional methods are used to evaluate a small number of common contaminants, either qualitatively or quantitatively, but they are not appropriate for the evaluation of complex hazardous pollutants and their derivatives (18). The use of nanotechnology in the detection, analysis, and remediation of such contaminants has become increasingly practical. Biogenic manganese oxide NPs (MnONPs) produced utilizing *Pseudomonas putida* to extract steroids, estrone, and 7-ethinylestradiol from aqueous solution show the highest removal efficiency (37). Diclofenac, an anti-inflammatory halogenated medication discovered in biomedical waste, is degraded using ozonation procedures. This drug is found in polluted areas. Unfortunately, only a small portion of the drug's components are really removed, and a sizable amount of toxic and mutagenic by-products are also released.

Nanoparticles effectively adhere to the cell membrane of microbial contaminants by damaging the cellular DNA (38). They enhance cell permeability, which leads to the formation of free radicals and reactive oxygen species. Furthermore, these nano-filters are relatively inexpensive to produce and have a high efficacy in eliminating harmful microbial load in water. Nevertheless, the clearance rate of the microbial load was restricted to 4 min after the samples were treated. In another experiment, a cement–silver nanocomposite concrete pebble with 99% microbial load reduction was created. Green nanoparticles are derived from natural sources or synthesized using environmentally friendly methods and have been investigated as a potential solution for wastewater treatment. One approach is the use of green NPs as catalysts for advanced oxidation processes (AOPs) in wastewater treatment. AOPs are a group of treatment methods that use reactive species such as hydroxyl radicals to degrade organic pollutants in wastewater. Green NPs, such as iron oxide, silver, and copper NPs, have been shown to be effective catalysts for AOPs, due to their high surface area, reactivity, and stability.

Another approach is the removal of contaminants from wastewater using green NPs as adsorbents. As possible adsorbents for heavy metals, dyes, and organic contaminants in wastewater, green NPs made of chitosan, cellulose, and starch-based materials have been studied (31). These green NPs have several advantages over traditional adsorbents, including their biodegradability, low cost, and high adsorption capacity. Green NPs can be used in combination with other treatment methods, such as membrane filtration and bioreactors, to enhance their performance. For example, green NPs can be used to enhance the fouling resistance and antimicrobial properties of membrane filtration systems, or as a source of electron donors for microbial fuel cells. Hence, the methodologies outlined earlier demonstrated the significance of green NPs in wastewater treatment.

5.15 RECENT ADVANCEMENT AND CHALLENGES

5.15.1 Database Biology Approach

The goal of bioremediation is to interpret the underlying degradation mechanism carried out by a specific organism for a specific pollutant by utilizing information from different biological sources, including databases of chemical structure and composition, RNA/protein expression, organic chemicals, catalytic enzymes, microbial degradation pathways, and comparative genomics (15). A variety of bioinformatics tools is used to examine all these sources in order to research bioremediation and create more efficient environmental cleaning systems. A dearth of information on the factors influencing the growth and metabolism of microorganisms with potential for bioremediation has led to a dearth of bioremediation applications (39). These bioremediation-capable bacteria have been characterized, and their mineralization routes and processes have been mapped out using bioinformatics. Proteomic methodologies such as two-dimensional polyacrylamide gel electrophoresis, microarrays, and mass spectrometry are also important in the assessment of bioremediation methods and technologies. According to the researchers, it considerably enhances the structural characterization of microbial proteins with contaminant-degradable characteristics. The structural makeup of microbial proteins that may break down contaminants has significantly improved. This study ties together biology and computer science. For instance, data pertaining to the DNA, RNA, and proteins in the genome are stored, changed, and retrieved using computers.

5.15.2 Omics-Based Bioremediation Tools

Omics-based bioremediation tools involve the use of various high-throughput technologies to analyze the genetic, metabolic, and functional aspects of microbial communities that can degrade contaminants in the environment (40). These tools enable researchers to understand the mechanisms and pathways involved in bioremediation and identify key genes, enzymes, and metabolic pathways that can be manipulated to enhance the degradation of pollutants (41). Tools are genomics, metagenomics, metatranscriptomics, metaproteomics, metabolomics (42).

5.15.2.1 Omics-Based Approach: Several Applications in Environmental Pollution Techniques

1. *Identification of key genes and metabolic pathways.* Metatranscriptomics can help identify the genes and metabolic pathways that are actively involved in the degradation of pollutants. This information can be used to design targeted bioremediation strategies and optimize environmental conditions to enhance the activity of these pathways.
2. *Monitoring of bioremediation processes.* Metatranscriptomics can be used to monitor the progress of bioremediation processes by tracking changes in the expression of key genes and metabolic pathways (43). This information can be used to evaluate the effectiveness of bioremediation strategies and adjust them as necessary.
3. *Assessment of the health of microbial communities.* Metatranscriptomics can provide information about the health of microbial communities in contaminated environments. Changes in gene expression patterns can indicate stress on the microbial community, which can be used to identify potential problems and take corrective actions.
4. *Discovery of novel enzymes and metabolic pathways.* Metatranscriptomics can reveal new enzymes and metabolic pathways that are involved in the degradation of pollutants. These enzymes and pathways can be used to develop more effective bioremediation approaches.

By using these omics-based bioremediation tools, researchers can identify key microbial species and metabolic pathways involved in the degradation of contaminants. The use of genomes, transcriptomics, metabolomics, and proteomics in bioremediation research can be beneficial. This method facilitates in the assessment of the *in situ* bioremediation process by correlating DNA sequences with the amount of metabolites, proteins, and mRNA (44). This knowledge can then be used to develop strategies to enhance bioremediation processes, such as genetic engineering of microorganisms, optimization of environmental conditions, and selection of microbial consortia with enhanced degradation capabilities.

5.15.2.2 Genomics

The study of bioremediation bacteria is an emerging area in genomics. This tactic is predicated on microorganisms' capacity to fully comprehend their genetic data inside the cell. Microorganisms of all kinds are used in bioremediation (45). Genomic methods such as PCR, isotope distribution analysis, DNA hybridization, molecular connectivity, metabolic foot printing, and metabolic engineering are utilized to better understand the biodegradation process. PCR-based techniques for genotypic fingerprinting include amplified ribosomal DNA restriction analysis (ARDRA), amplified fragment length polymorphisms (AFLP), automated ribosomal intergenic spacer analysis (ARISA), terminal-restriction fragment length polymorphism (T-RFLP), single-strand conformation polymorphism (SSCP), randomly amplified polymorphic DNA (RAPD) analysis, and length heterogeneity of PCR amplicons (LH-PCR) (46). These techniques allow for the identification of specific microbial populations involved in biodegradation and the tracking of changes in microbial community structure over time.

Isotope distribution analysis, such as stable isotope probing (SIP) (31), is a technique that can be used to identify the microorganisms that are actively involved in the biodegradation of specific contaminants. This technique involves labelling the contaminant with a stable isotope (e.g., 13C), which is then taken up and incorporated into the DNA, RNA, or proteins of the microorganisms that are actively degrading the contaminant. The labelled DNA, RNA, or proteins can then be isolated and identified using molecular techniques, such as PCR or metagenomics.

DNA hybridization techniques, such as fluorescence in situ hybridization (FISH), can be used to visualize specific microbial populations involved in biodegradation directly in the environment. This technique involves hybridizing fluorescently labelled nucleic acid probes to specific target sequences within the microbial cells, allowing for their visualization using fluorescence microscopy. Metabolic foot printing involves profiling the metabolites produced by microbial communities during biodegradation. This technique can provide information on the metabolic pathways involved in biodegradation and the specific microbial populations responsible for their activity. Metabolic engineering can also be used to enhance the biodegradation of specific contaminants by manipulating the metabolic pathways of microorganisms involved in biodegradation. This can be done by introducing genes encoding for specific enzymes or pathways into the microorganisms or by manipulating the environmental conditions to enhance the expression of these genes. The presence and appearance of taxonomic and operational gene markers in soil may be determined using a PCR-based quantitative investigation of soil microbial communities. Amplified PCR results are used as a starting point for direct investigation of certain molecular biomarker genes in techniques for analyzing a person's DNA.

5.15.2.3 Transcriptomics and Metatranscriptomics

Study of the transcriptome, which is the complete set of all RNA molecules, including messenger RNA (mRNA), transfer RNA (tRNA), ribosomal RNA (rRNA), and non-coding RNA (ncRNA), produced by a cell, tissue, or organism, involves the identification, quantification, and analysis of RNA transcripts present in a given sample. The set of genes that are being transcribed at a certain moment and condition is represented by the transcriptome, which serves as a vital link between the cellular phenotype, interactome, genome, and proteome. In order to adapt to environmental changes and, hence, ensure survival, the capacity to modulate gene expression is essential. Transcriptomics gives an in-depth look at this procedure across the whole human genome. An effective method for detecting the amounts of mRNA expression in transcriptomics is DNA microarray analysis (39). A transcriptomic analysis requires the isolation and enrichment of total mRNA, cDNA synthesis, and sequencing of the cDNA transcriptome. A DNA microarray can be used as a transcriptomics tool to examine and study the mRNA expression of nearly every gene in an organism. For the purpose of collecting practical knowledge about the operations of environmental microbial communities, the study of transcriptional mRNA profiles, sometimes referred to as *transcriptomics* or *metatranscriptomics*, is essential (40). Throughout the entire biodegradation process, syntrophism between bacteria and complementary metabolic pathways can be found utilizing metagenomics, genome binning, and metatranscriptomics. Researchers can utilize metatranscriptomics to examine the expression of genes (42).

5.15.2.4 Proteomics and Metabolomics

In contrast to metabolomics, which is the study of small-molecule metabolites, such as sugars, amino acids, and lipids, present in biological samples, proteomics is the large-scale study of the structure, function, and interactions of proteins within cells, tissues, or organisms. It involves the identification, quantification, and characterization of all proteins present in a biological sample, which can provide valuable insights into the molecular mechanisms underlying biological processes (28). Proteomics has been used to analyze protein abundance and compositional changes as well as to identify important proteins connected to microbes. Data from both the proteome and metabolome will be useful for cell-free bioremediation.

5.16 CONCLUSION

Global industrialization has led to environmental, freshwater, and topsoil pollution. Due to human activities, such as mining and the eventual removal of toxic metal effluents from steel mills, battery manufacturers, and energy generation, water quality has gotten worse, raising serious environmental issues. The ecology is harmed by effluents like petroleum, polythenes, and trace metals. Heavy metals are contaminants that naturally occur in the crust of the Earth and are challenging to break down. They are extracted as minerals after existing as ores in rocks. Heavy metals may be released into the environment at high exposure levels. They continue to be hazardous much longer once they are in the environment.

Bioremediation is a valuable waste management technology that may be used to remove trash from polluted regions and places. Geological characteristics of polluted sites comprising soil, pollutant type and depth, human habitation site, and performance of every bioremediation technique should be integrated in determining the most appropriate and operative bioremediation technique to successful treatment of polluted sites.

The existing time- and resource-intensive approaches of water purification have the potential to be transformed by the integration of cutting-edge technologies like IoTs, AI, machine learning, cloud computing, and 5G communication.

In conclusion, omics approach is a valuable tool in environmental pollution techniques, as it provides insights into the gene expression patterns and metabolic activities of microbial communities in contaminated environments. This information can be used to design targeted bioremediation strategies, optimize environmental conditions, monitor the progress of bioremediation processes, assess the health of microbial communities, and discover new enzymes and metabolic pathways for more effective bioremediation approaches

REFERENCES

1. Butt, H., et al. (2018). Engineering plant architecture via CRISPR/Cas9-mediated alteration of strigolactone biosynthesis. *BMC Plant Biol.* 18:174. doi: 10.1186/s12870-018-1387-1.
2. Justino, C.I.L., et al. (2017). Recent progress in biosensors for environmental monitoring: a review. *Sensors* 17:2918. doi: 10.3390/s17122918.
3. Gavrilaș, S., et al. (2022). Recent trends in biosensors for environmental quality monitoring. *Sensors (Basel)* 22(4):1513. doi: 10.3390/s22041513; PMID: 35214408; PMCID: PMC8879434.
4. Rahimi, P., et al. (2019). Enzyme-based biosensors for choline analysis: a review. *TrAC Trends Anal. Chem.* 110:367–374. doi: 10.1016/j.trac.2018.11.035.
5. Economou, A., et al. (2017). *Advances in Food Diagnostics*, 2nd ed. Hoboken, NJ, USA: John Wiley & Sons Ltd. pp. 231–250.
6. Chang, H.J., et al. (2017). Microbially derived biosensors for diagnosis, monitoring and epidemiology. *Microb. Biotechnol.* 10:1031–1035. doi: 10.1111/1751-7915.12791.
7. Huang, L., et al. (2021). Application of Aptamer-based biosensor in bisphenol A detection. *Chin. J. Anal. Chem.* 49:172–183. doi: 10.1016/S1872-2040(20)60077-9.
8. Wang, K., et al. (2014). Research and development of functionalized Aptamer based biosensor. *Chin. J. Anal. Chem.* 42:298–304. doi: 10.1016/S1872-2040(13)60712-4.
9. Xie, M.J., et al. (2022). Recent advances in aptamer-based optical and electrochemical biosensors for detection of pesticides and veterinary drugs. *Food Control.* 131:108399. doi: 10.1016/j.foodcont.2021.108399.
10. Sharma, S., et al. (2016). Antibodies and antibody-derived analytical biosensors, in *Biosensor Technologies for Detection of Biomolecules, vol. 60, Essays in Biochemistry*, ed. Estrela, P. (South Portland, ME, USA: Portland Press), 9–18.
11. Omidfar, K., et al. (2013). New analytical applications of gold nanoparticles as label in antibody based sensors. *Biosens. Bioelectron.* 43:336–347. doi: 10.1016/j.bios.2012.12.045.
12. Tschmelak, J., et al. (2005). Biosensors for unattended, cost-effective and continuous monitoring of environmental pollution: automated water analyser computer supported system (AWACSS) and river analyser (RIANA). *J. Environ. Anal. Chem.* 85:837–852.
13. Jia, M.X., et al. (2021). Recent advances on immunosensors for mycotoxins in foods and other commodities. *Trends Anal. Chem.* 136:116193. doi: 10.1016/j.trac.2021.116193.
14. Fang, L., et al. (2020). Recent progress in immunosensors for pesticides. *Biosens. Bioelectron.* 164:112255. doi: 10.1016/j.bios.2020.112255.
15. Malla, M.A., et al. (2018). Understanding and designing the strategies for the microbe-mediated remediation of environmental contaminants using omics approaches. *Front. Microbiol.* 9:1132. doi: 10.3389/fmicb.2018.01132.

16. Singh, V., et al. (2018). Recent advances in CRISPR-Cas9 genome editing technology for biological and biomedical investigations. *J. Cell. Biochem.* 119:81–94. doi: 10.1002/jcb.26165
17. Babar, M.M., et al. (2018). Omics approaches in industrial biotechnology and bioprocess engineering, in *Omics Technologies and Bio-Engineering*, eds. Barh, D. and Azevedo, V. (Cambridge, MA: Academic Press), 251–269. doi: 10.1016/B978-0-12-815870-8.00014-0.
18. Nawaz, A., et al. (2022). Microbial fuel cells: insight into simultaneous wastewater treatment and bioelectricity generation. *Process Safety Environ. Protect.* 161:357–373. doi: 10.1016/j.psep.2022.03.039; ISSN: 0957–5820.
19. Adams, B.L. (2016). The next generation of synthetic biology chassis: moving synthetic biology from the laboratory to the field. *ACS Synth. Biol.* 5(12):1328–1330. doi: 10.1021/acssynbio.6b00256; Gonzalez-Garcia, R.A., et al. (2017). Metabolic pathway and flux analysis of H2 production by an anaerobic mixed culture. *Int. J. Hydrogen Energy* 42:4069–4082. doi: 10.1186/1475-2859-13-48
20. Kale, R.D., et al. (2017). Colour removal using nanoparticles. *Text. Cloth. Sustain.* 2:4. doi: 10.1186/s40689-016-0015-4.
21. Wang, X., et al. (2010). Mass production of micro/nanostructured porous ZnO plates and their strong structurally enhanced and selective adsorption performance for environmental remediation. *J. Mater. Chem.* 20:8582–8590. doi: 10.1039/c0jm01024c.
22. Salvadori, M.R. (2019). Processing of nanoparticles by biomatrices in a green approach, in *Microbial Nanobionics*, ed. Prasad, R. (Berlin: Springer), 1–28. doi: 10.1007/978–3-030–16383–9_1.
23. Sharma, G., et al. (2015). Biological synthesis of silver nanoparticles by cell-free extract of Spirulina platensis. *J. Nanotechnol.* 2015:132675. doi: 10.1155/2015/132675.
24. Li, J., et al. (2016). Biosynthesis of gold nanoparticles by the extreme bacterium Deinococcus radiodurans and an evaluation of their antibacterial properties. *Int. J. Nanomed.* 11:5931–5944.
25. Gadipelly, C., et al. (2014). Pharmaceutical industry wastewater: review of the technologies for water treatment and reuse. *Indus. Eng. Chem. Res.* 53(29):11571–11592. doi: 10.1021/ie501210j.
26. Naim, M.M., et al. (2016). Application of silver, iron and chitosan-nanoparticles in wastewater treatment. *Int. Conf. Eur. Desalin. Soc. Desalin. Environ. Clean Water Energy* 73:268–280. doi: 10.5004/dwt.2017.20328.
27. Mostafa, M.K., et al. (2017). Application of entrapped nano zero valent iron into cellulose acetate membranes for domestic wastewater treatment, in *Proceedings of the Environmental Aspects, Applications and Implications of Nanomaterials and Nanotechnology – Topical Conference at the 2017 AIChE Annual Meeting*, Minneapolis, MN, USA, 1 November 2017.
28. Malik, S., et al. (2022). Exploring microbial-based green nanobiotechnology for waste water remediation: a sustainable strategy. *Nanomaterials* 12:4187. doi: 10.3390/nano12234187.
29. Sivaraj, R., et al. (2015). Green nanotechnology: the solution to sustainable development of environment, in *Environmental Sustainability* (Berlin, Heidelberg, Germany: Springer), 311–324.
30. Dutta, D., et al. (2021). Scope of green nanotechnology towards amalgamation of green chemistry for cleaner environment: a review on synthesis and applications of green nanoparticles. *Environ. Nanotechnol. Monit. Manag.* 15:100418.
31. Teye, G.K., et al. (2020). Photodegradation of pharmaceutical and personal care products (PPCPs) and antibacterial activity in water by transition metals. *Rev. Environ. Contam. Toxicol.* 254:131–162.
32. Mahanty, S., et al. (2020). Synergistic approach towards the sustainable management of heavy metals in wastewater using mycosynthesized iron oxide nanoparticles: biofabrication, adsorptive dynamics and chemometric modeling study. *J. Water Process Eng.* 37:101426. doi: 10.1016/j.jwpe.2020.101426.

33. Zhang, X., et al. (2017). Biochar for volatile organic compound (VOC) removal: sorption performance and governing mechanisms. *Bioresour. Technol.* 245:606–614.
34. Karaffa, L., et al. (2021). The role of metal ions in fungal organic acid accumulation. *Microorganisms* 9(6):1267. doi: 10.3390/microorganisms9061267; PMID: 34200938; PMCID: PMC8230503.
35. Lawrence, M., et al. (2013). Software for computing and an-notating genomic ranges. *PLoS Comput. Biol.* 9: e1003118. doi: 10.1371/journal.pcbi.1003118.
36. Robinson, J.R., et al. (2021). Fungal-metal interactions: a review of toxicity and homeostasis. *J. Fungi (Basel)* 7(3):225. doi: 10.3390/jof7030225; PMID: 33803838; PMCID: PMC8003315.
37. Kishore, S., et al. (2022). A comprehensive review on removal of pollutants from wastewater through microbial nanobiotechnology-based solutions. *Biotech. Gen. Eng. Rev.* 1–26.
38. Singh, D., et al. (2020). Omics (genomics, proteomics, metabolomics, etc.) tools to study the environmental microbiome and bioremediation, in *Waste to Energy: Prospects and Applications* (Singapore: Springer), 235–260.
39. Gutleben, J., et al. (2018). The multi-omics promise in context: from sequence to microbial isolate. *Crit. Rev. Microbiol.* 44:212–229. doi: 10.1080/1040841X.2017.1332003.
40. Sharma, P., et al. (2022). Omics approaches in bioremediation of environmental contaminants: an integrated approach for environmental safety and sustainability. *Environ. Res.* 211:113102. doi: 10.1016/j.envres.2022.113102; ISSN: 0013–9351.
41. Sanghvi, G., et al. (2020). Engineered bacteria for bioremediation, in *Bioremediation of Pollutants* (Amsterdam, The Netherlands: Elsevier), 359–374.
42. Guerra, A.B., et al. (2018). Metagenome enrichment approach used for selection of oil-degrading bacteria consortia for drill cutting residue bioremediation. *Environ. Pollut.* 235:869–880. doi: 10.1016/j.envpol.2018.01.014.
43. Jaiswal, S., et al. (2019). Gene editing and systems biology tools for pesticide bioremediation: a review. *Front. Microbiol.* 10:87. doi: 10.3389/fmicb.2019.00087.
44. Hakeem, K.R., et al. (2020). *Bioremediation and Biotechnology*. Cham, Switzerland: Springer.
45. Sharma, P., et al. (2022). Omics approaches in bioremediation of environmental contaminants: an integrated approach for environmental safety and sustainability. *Environ. Res.* 211:113102. doi: 10.1016/j.envres.2022.113102.
46. Gaur, V.K., et al. (2022). Sustainable strategies for combating hydrocarbon pollution: special emphasis on mobil oil bioremediation. *Sci. Total Environ.* 832:155083. doi: 10.1016/j.scitotenv.2022.155083.

6 Microbial Biopolymer from Renewable Feedstock

A Sustainable Approach for the Production of Polyhydroxyalkanoates (PHAs)

Shiva Aley Acharjee and Pranjal Bharali

6.1 INTRODUCTION

For the past 200 years, the world has observed the consumption of derivatives of fossil fuels and their impacts. This widespread use of petrochemical-derived polymers is responsible for global ecological imbalance in a number of areas, including the stability of the Earth's atmosphere, terrestrial and aquatic life, food chains, and the plant kingdom (de Paula et al. 2018). These synthetic polymers are exceedingly tenacious in nature, and an alarming buildup of such plastics has created issues with safe disposal, recycling, trash landfilling, bio- or photo-degradation, and incineration. Consequently, there is a pressing need for biodegradable polymers right now (Pradhan et al. 2020). Researchers throughout the world are more encouraged than ever to look for innovative alternatives as a result of greater awareness of the adverse impacts of synthetic plastics on humans and the environment. A replacement for synthetic plastics with biodegradable polymers made by microorganisms, particularly bacteria, has been considered as a means of reducing environmental pollution (Krishnan et al. 2021). Biodegradable "green plastics" like polyhydroxyalkanoates (PHAs) provide bioeconomy notions, curtail emission of greenhouse gases (GHGs) (by about 200%), use fewer fossil fuel energy (by about 95%), and produce less toxic waste, according to recent life cycle assessment studies and environmental footprint tools (Dietrich et al. 2017). PHAs are therefore regarded as a better alternative to polymers made from petroleum (Andler et al. 2021). PHAs are a class of naturally occurring biopolyesters made of hydroxy fatty acids produced by various microorganisms. These PHAs are grouped into three categories based on the quantity of carbon atoms they contain: short-chain (C3–C5), medium-chain (C6–C14), and long-chain

DOI: 10.1201/9781003517108-6

lengths (>C14) (Ding et al. 2022). The most well-known biopolymer among the many PHAs is polyhydroxybutyrate (PHB), which has exceptional qualities that make it suitable for application in industry. On the other hand, medium-chain PHA is excellent for biological applications since it is relatively flexible and has a broad spectrum of thermal properties (Rahim et al. 2020). PHAs are formed intracellularly by various microorganisms and stored as carbon and energy in carbonosomes when they are under nutritional stress conditions, such as limited essential nutrients or an abundance of carbon sources (Hassan et al. 2013). They can be broken down both aerobically (with water) and anaerobically (with methane and water) (Pradhan et al. 2020). Due to their biodegradability, biocompatibility, diversity in chemical structure, and ability to be produced from renewable carbon resources, PHAs have drawn substantial research and commercial interests (Amulya et al. 2016).

The production of PHAs utilizing waste biomass has been reported in many studies. PHAs were produced from *Bacillus subtilis* NG220 using wastewater of sugar industry as a carbon substrate, with PHAs content of 51.8% of its cell dry weight (CDW) (Singh et al. 2013). Utilizing spent coffee grounds hydrolysate as a carbon substrate, PHAs were biosynthesized from *Burkholderia cepacia* with PHAs content of 54.79% of its CDW (Obruca et al. 2014). Similarly, Mayeli et al. (2015) also stated the production of PHAs by *Bacillus axaraqunsis* utilizing petrochemical wastewater as a carbon source, accumulating 66% PHAs of its CDW. More recently, Muhammad et al. (2020) used *Laminaria japonica* biomass as a carbon source with strain *Cupriavidus necator* NCIMB 11599 for the production of PHAs, and it was able to incorporate 32% PHAs CDW. Table 6.1 lists various agricultural wastes, food waste, algae, wastewater, and dairy waste that are used as low-cost carbon feedstock for the biosynthesis of PHAs.

Despite the fact that petrochemical plastics' negative effects on the environment can be efficiently avoided by using PHAs, it still only has a modest market share. PHAs can only be used on a small scale due to its high production costs compared to traditional plastics. The cost-effectiveness of PHAs is greatly influenced by the substrates, which can account for up to 50% of the total cost. It is required to acquire high productivity and high overall yield from inexpensive, widely accessible carbon feedstock in order to manufacture low-cost PHAs on an industrial scale. As a result, making PHAs from wastes has emerged as a promising method for manufacturing PHAs in industries (Liu et al. 2021). For the production of PHAs at a cheap price, it may be possible to employ waste organic material that is widely accessible. Wastes produced from a variety of sources, including agricultural wastes, food waste, algae, wastewater, and dairy waste, can be used as a renewable feedstock for microbial fermentation for the manufacture of PHAs since they are plentiful and accessible all year round. These wastes are produced in large quantities all over the world, making management of them a difficult undertaking. These wastes are also high in organic and other nutrient content, which stimulates microbial growth and could pollute the environment. Researchers have recently begun to focus on the use of wastes as a source of raw materials for microbial fermentation and PHA synthesis (Bhatia et al. 2021). The application of these wastes in biological production will benefit both the management of the environment and the economic aspects of bioproducts (Ding et al. 2022). This chapter therefore intends to highlight various waste biorefineries for

PHA synthesis as a sustainable path towards circular bioeconomy, addressing solutions to global concerns through biorefinery technologies for biopolymer production that lead to a sustainable environment and economy. This biorefinery approach to the production of green plastics will not only help in the management of these wastes but also prompt income and enhance the process's economics.

6.2 METABOLIC PATHWAYS FOR PHA SYNTHESIS

From a wide range of microorganisms, several gene encoding enzymes involved in PHA synthesis and degradation have been cloned and described so far. It is now obvious that nature has developed a variety of PHA formation pathways, each tailored to the ecological niche of the PHA-producing microorganism (Reddy et al. 2003). The three main pathways responsible for the metabolism of PHAs are glycolysis, the de novo fatty acid synthesis route, and fatty acid β-oxidation biosynthesis, which is depicted in Figure 6.1 (Luengo et al. 2003). In pathway I (glycolysis), sugar is utilized as a carbon substrate for the biosynthesis of PHAs. In pathway II (de novo synthesis), microorganisms use glucose, gluconate, or acetate, whereas in pathway III (β-oxidation), fatty acids are used as a carbon substrate. Some of the crucial coding genes include PhaA (β ketothiolase), PhaB (acetoacetyl-coenzyme A), and PhaC (PHA synthase). The end by-product of glycolysis, acetyl-CoA, is converted into acetoacetyl-CoA by enzyme PhaA, and then to 3-hydroxybutyryl-CoA by enzyme PhaB, the precursor for PHA production. Finally, PHA polymerase, which is encoded by PhaC, converts 3-hydroxybutyryl-CoA into PHAs (Pradhan et al. 2020).

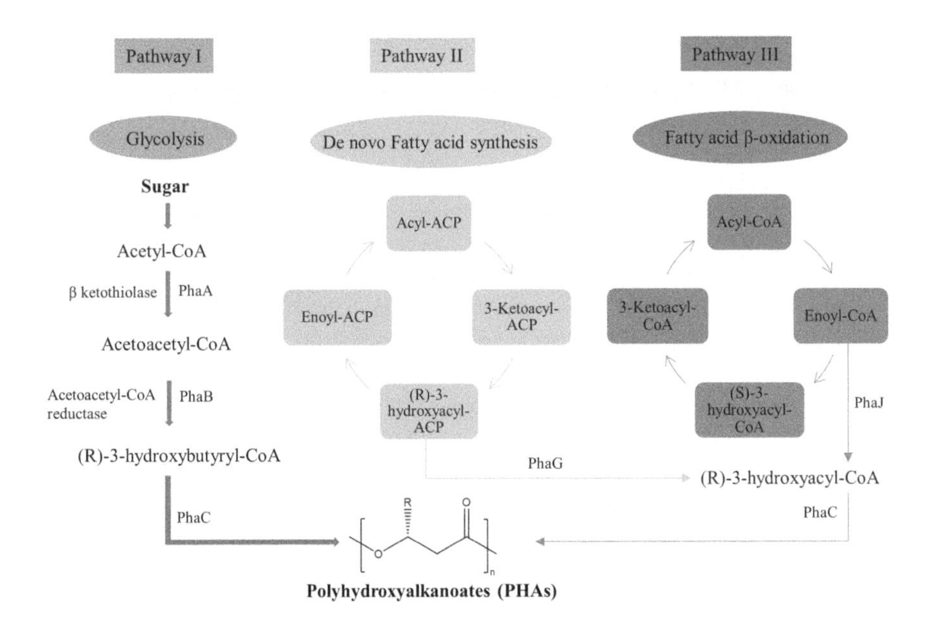

FIGURE 6.1 Metabolic pathways for the biosynthesis of PHAs.

Source: Adapted from Pakalapati et al. (2018).

Apart from the formation of PHAs from acetyl-CoA, the other pathways, namely, the de novo synthesis and β-oxidation, are the other two processes through which PHAs are produced. According to reports, the fatty acid synthesis route produces the highest PHAs (Pakalapati et al. 2018). Pathway II uses the de novo fatty acid synthesis cycle to produce R-3-hydroxyacyl-ACP for PHA production. The main enzyme in this pathway is 3-hydroxyacyl-acyl carrier protein-CoA transferase (PhaG), which transforms 3-hydroxyacyl-ACP (acyl carrier protein) into 3-hydroxyacyl-CoA (Chen et al. 2015). In pathway III, fatty acids are initially transformed to enoyl-CoA in the β-oxidation process. Enoyl-CoA is then converted into R-3-hydroxyacyl-CoA by R-3-hydroxyacyl-CoA hydratase, which is the first step in the polymerization of PHAs. Then, PHAs synthase (PhaC) catalyzes the last step of R-3-hydroxyacyl-CoA polymerization (Adeleye et al. 2020).

6.3 VARIOUS TYPES OF RENEWABLE FEEDSTOCKS FOR PHA PRODUCTION

6.3.1 Agro-Wastes

Agricultural wastes and their by-products have recently overtaken other types of waste as the leading source of pollution in the environment, with an annual production of 16 million metric tons globally. The main cause of environmental pollution in many agrarian nations is the burning of lignocellulosic wastes, including sugarcane bagasse, rice husk and straw, wheat straw, wood chips, and corn stover (Pradhan et al. 2020). A continual rise of agro-waste is witnessed every year as a result of the revolution in the agricultural industry. The agricultural waste is made up of leftovers from crop processing as well as residues from the field, which are primarily made up of stalks, uprooted roots, leaves, steams, and seedpods. These wastes are predominantly made of lignocellulosic material, which primarily consists of cellulose (40–50%), hemicellulose (25–30%), and lignin (15–20%). The lignocellulosic materials are preferred over the other feedstocks for PHA production because of their wide availability around the globe (Bhatia et al. 2021). Agricultural waste is often made up of animal waste, fertilizers, and crop wastes, all of which are essentially rich in organic components. Therefore, these organic compounds will act as the raw materials for biosynthesizing PHAs. One factor contributing to the high cost of producing PHAs is the complex nitrogen source, and production from agricultural waste will benefit since the feedstock is rich in both carbon and nitrogen (Pakalapati et al. 2018). For PHA-producing microorganisms to use the agro-wastes as a substrate, complex substances in them must be broken down by several pre-treatment processes. Agro-waste can be pre-treated in various ways to make simple sugars and fatty acids readily available for efficient microorganism utilization (Nielsen et al. 2017). Pre-treatment has numerous beneficial effects, including increasing the availability of carbon sources, reducing the concentration of organic matter, eliminating suspended solids, maintaining pH and temperature, sterilizing waste materials, and reducing any potential inhibitory effects on PHA-producing strains (Rodriguez-Perez et al. 2018). To increase PHA production, for instance, acid pre-treatment with pH 3–5 was frequently used on paper mill effluent, dairy effluent, and sugarcane molasses. The biomass of *Bacillus subtilis* and *Escherichia coli* significantly increased (by about 50%) after cane molasses was pre-treated with sulfuric acid. In the context of biological pre-treatment, raw waste materials were fermented using a mixed hydrolytic bacterial culture, which led to a nearly fourfold increase in PHA production (Li and Wilkins 2020). A study showed that about 72% of PHA's CDW was obtained using *Burkholderia sacchari* DSM 17165 and wheat straw as carbon substrate (Cesário et al. 2014). Utilizing tapioca hydrolysates as a carbon substrate, *Alcaligenes eutrophus* was able to accumulate 58% PHAs of its CDW (Sawant et al. 2016). Recently, *Halogeometricum borinquense* accumulated 44.70% CDW of PHAs from cassava waste (Salgaonkar et al. 2019). Biosynthesis of PHAs from agro-wastes is shown in Figure 6.2.

Food crops are among the most affordable and abundant bio-based raw resources, making them a potential carbon source. Due to their availability and ease of access, agricultural and fruit leftovers with high sugar content serve as alternative carbon substrate for the biosynthesis of PHAs. Fatty acids and vegetable oils, on the other hand, can reduce costs while increasing PHA output (Ding et al. 2022). *Halomonas campisalis* MCM B-1027 was investigated by Kulkarni et al. (2015) for the synthesis of PHAs using agricultural wastes such as bagasse, fruit peels, and de-oiled cakes as a carbon substrate. The best carbon substrate among all based on dry cell weight was discovered to be the aqueous extract of bagasse, which produced 47% PHAs. *Pseudomonas aeruginosa* 42A2 was examined for its ability to produce PHAs from agro-industrial oily wastes. PHA accumulation varied between 66.1% when soybean oil was utilized as a carbon source, 29.4% when waste frying oil was utilized, and 16.8% when glucose was utilized (Sathya et al. 2018). In the basal salts medium with 1.5% (w/v) of defatted oil cake at pH 7 after 48 h, the strain *Bacillus* sp. (PPECLRB-16), which was isolated from rice bran disposal area, was able to produce a significant amount (5.64 g/L) of PHB (Krishnan et al. 2021).

The composition of agro-industrial residues (AIRs) is characterized by a concentration of cellulosic and hemicellulosic polysaccharides that can be converted into monosaccharides that are highly regarded for fermentation PHA synthesis. Wheat bran, empty fruit bunches of palm oil, rice residue, and recycled paper mill waste are a few examples of AIRs that have a significant potential for upcycling into feedstocks

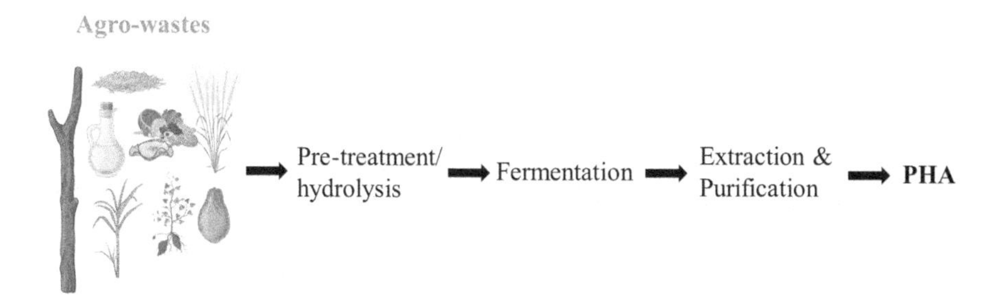

FIGURE 6.2 Biosynthesis of PHA from agro-wastes.

Source: Adapted from Vigneswari et al. (2021).

for biorefinery applications (Thomas et al. 2022). Sindhu et al. (2013) examined *Bacillus firmus* NII 830 to determine its capacity to produce PHB utilizing rice straw as a carbon source. The bacterial strain was able to accumulate 89% PHB content in the cell while growing in the hydrolysate medium without any detoxification.

Fruit residues, also known as pomaces, are often made up of fruit skins, pulp, stalks, and seeds and are regarded as useless waste. The solid portion of fruit remnants is typically discarded, utilized as animal feed, and infrequently used for low-impact applications. As fruit waste does not directly compete with the food chain, it is a sustainable substitute for fossil-based resources. Fruit residues have higher promise since extensive pre-treatments can be omitted to reduce production costs and environmental effect while still containing high levels of accessible fermentable sugars (Andler et al. 2021). Numerous food sector wastes comprising cellulose, starch, proteins, or lipids have been identified as potential feedstock for PHA production. One of the most popular beverages consumed worldwide is wine. When wine is made, a lot of solid waste is produced, 60% of which is made up mostly of grape pomace, which is a mixture of peel and seeds. Researchers are interested in grape pomace because it is an excellent fermentation substrate and contains a high concentration of glucose and fructose (Liu et al. 2021). In order to produce PHB from *Cupriavidus necator*, Kovalcik et al. (2020) employed grape winery waste as a carbon substrate that was high in oil and fermentable sugars. After 29.5 h of incubation in a 2 L bioreactor, they reported 8.3 g/L PHB content (63% DCW). According to the study of Rao et al. (2019), *Bacillus subtilis* MTCC 144 was able to accumulate 57.64, 63.27, and 77.89% of PHA CDW from mixed fruit peels, mixed vegetable peels, and green pea shells, respectively. For the first time, Rayasam et al. (2020) assessed pomegranate peels to lower the effective cost of PHAs and acquired encouraging findings by synthesizing 83% PHAs by *Bacillus halotolerans* DSM8802.

6.4 FOOD WASTE

Food waste is a widespread issue that occurs throughout the food production process, from harvesting to storage, packing, and disposal. A total of 88 MT (million tons) of food are estimated to be wasted in Europe, 57 MT of which originate from households and food services. According to the Environmental Protection Agency, over 37 MT of foods were dumped in municipal solid waste systems in the United States in 2013, accounting for 14% of all waste generated there. Due to its accessibility and low cost as a carbon source, food waste has the potential to both produce PHAs and address major waste management problems (Nielsen et al. 2017). It is possible to simultaneously minimize environmental pollution and the health risk posed by food spoilage by finding a sustainable, affordable, and effective alternative for converting these excess food wastes into a variety of value-added goods. These food wastes are a rich source of carbohydrates, proteins, minerals, fats, and oils that can be used as a substrate for the biosynthesis of PHAs (Pradhan et al. 2020). Every food source varies and has unique requirements for pre-treatments, microbial strains, culturing environments, and downstream processing. Organics frequently found in food wastes are complex substances that PHA-producing microorganisms cannot directly utilize. In these situations, it is required to use a pre-treatment or processing technique to change the complex compounds present in food waste into PHA substrates. Simple sugars like lactose or glucose, as well as fatty acids like propionic or acetic acids, are examples of substrates. Many simple food wastes are hydrolyzed to produce acceptable precursor compounds from the food waste before fermentation with selected microbial strain (Nielsen et al. 2017). Various food wastes from household and industry have been used as carbon substrate for the biosynthesis of PHAs. Biosynthesis of PHAs from food wastes is depicted in Figure 6.3. Food wastes like malt waste (Yu et al. 1998), soybean and rapeseed oil (Taniguchi et al. 2003), waste frying oil (Fernández et al. 2005), soy molasses (Solaiman et al. 2006), waste potato starch (Haas et al. 2008), wheat bran (Van-Thuoc et al. 2008), molasses (Saranya and Shenbagarathai 2011), spent wash (Chaudhry et al. 2011), kitchen waste (Omar et al. 2011), crude palm kernel oil (Chee et al. 2012), pea shells (Patel et al. 2012), sugar beet juice (Wang et al. 2013), restaurant waste (Eshtaya et al. 2013), spent coffee grounds hydrolysate (Obruca et al. 2014), cooking oil (Cruz et al. 2015), teff straw (Getachew and Woldesenbet 2016), coconut oil (Basnett et al. 2018), waste fish oil (Van Thuoc et al. 2019), cashew apple juice (Arumugam et al. 2020), waste cooking oil (Sangkharak et al. 2021), etc. were successfully tested for PHA production. Yu et al. (1998) reported the use of malt waste as sole carbon source in fermentation using microbial strain *Alcaligenus eutrophus* DSM1124 with PHAs content of 70% of its CDW. Again, Fernández et al. (2005) used waste frying oil as carbon source using strain *Pseudomonas aeruginosa* NCIB 40045, and it was able to accumulate 29% PHAs. Utilizing waste potato starch as carbon source, strain

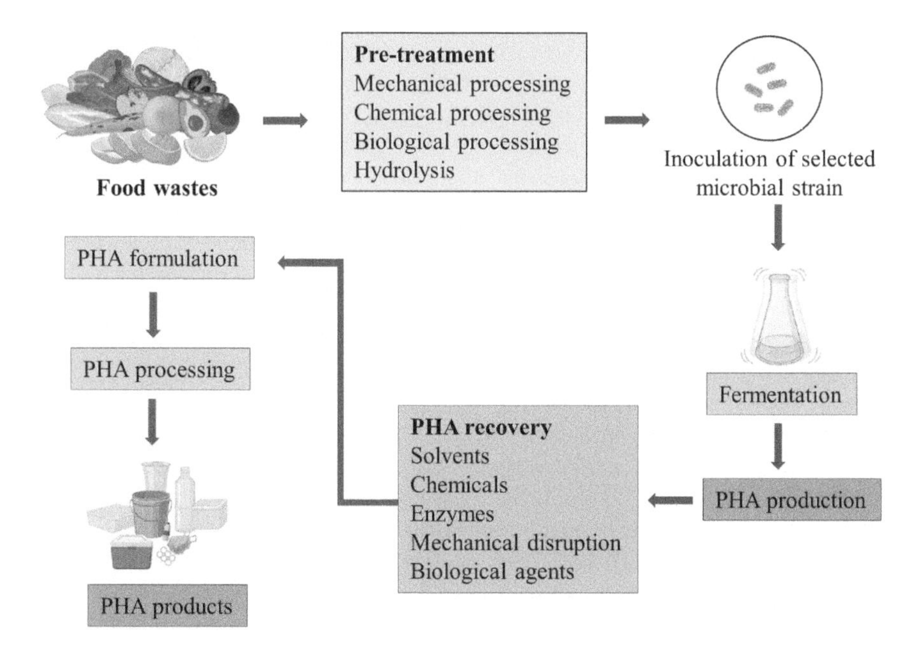

FIGURE 6.3 Biosynthesis of PHA from food wastes.

Source: Adapted from Brigham and Riedel (2018).

Ralstonia eutropha NCIMB 11599 produced 52.5% PHAs of its CDW (Haas et al. 2008). Eshtaya et al. (2013) also used restaurant waste as carbon source with strain *Escherichia coli* pnDTM2 (recombinant) and was able to accumulate 45% PHAs of its CDW. Similarly, Obruca et al. (2014) used spent coffee grounds hydrolysate to produce PHAs using *Burkholderia cepacia* and was able to accumulate 54.79% PHAs of its CDW. Recently, Arumugam et al. (2020) investigated PHA production using cashew apple juice as carbon source using *Cupriavidus necator*, and it accumulated 80.6% PHAs of its CDW. More recently, Sangkharak et al. (2021) used waste cooking oil as carbon source using *Bacillus thermoamylovorans* for PHA production and incorporated 74.4% CDW. The high cost of the carbon feedstocks for microbial culture and the downstream processing of the biopolymers are two major issues that limit PHAs' ability to compete with traditional plastics on large-scale industrial production. Therefore, using food wastes as carbon feedstock can solve the first issue as long as waste collection and transportation costs are not a barrier (Brigham and Riedel 2018).

6.5 ALGAL SOURCES

Recent studies have also focused on the utilization of algal biomass as a carbon substrate for PHA synthesis due to the high carbohydrate yield and the deficiency of lignin, which promotes the inexpensive recovery of fermentable sugars and has significant potential as feedstock biofactories (Tan et al. 2022). Algae are autotrophic or heterotrophic organisms that can be grown in a variety of soil types, even those that are not even good for agriculture. Because of their adaptable metabolism, they are more productive than terrestrial plants (Mohan et al. 2020). Algae can be split into two types: microalgae and macroalgae, depending on their size and form. Microalgae are excellent for a broad spectrum of applications owing to their numerous characteristics. In comparison to other terrestrial plants, microalgae grow more quickly (Talan et al. 2022). Moreover, algae do not compete with traditional food sources, preventing any issues with the availability of food for humans. To efficiently use macroalgae as carbon sources for PHA production, knowledge of their carbohydrate compositions is required, because different species of macroalgae have different types of carbohydrates (Ghosh et al. 2019). The production of PHAs from algae is shown in Figure 6.4.

A study employed algal biomass as carbon substrate with *Escherichia coli* for the biosynthesis of PHAs and was able to accumulate 34% PHA CDW (Sathish et al. 2014). In the following year, Bera et al. (2015) also used *Jatropha* biodiesel waste as carbon source with strain *Halomonas hydrothermalis* MTCC 5445, and it accumulated 75.8% of PHAs CDW. Similarly, Alkotaini et al. (2016) utilized *Gelidium amansii*, a red marine macroalgae, as feedstock with strain *Bacillus megaterium* KCTC 2194, and it produced 51.4% PHAs CDW. Using the same feedstock *Gelidium amansii*, Sawant et al. (2018) also produced PHAs using strain *Saccharophagus degradans*. However, the PHA yield was comparatively lower than with strain *Bacillus megaterium* KCTC 2194 (Alkotaini et al. 2016), with 17–27% PHAs CDW. Utilizing *Sargassum* sp. as carbon feedstock for PHA production with *Cupriavidus necator* PTCC 1615, it accumulated 74% PHAs CDW (Azizi et al. 2017). Recently,

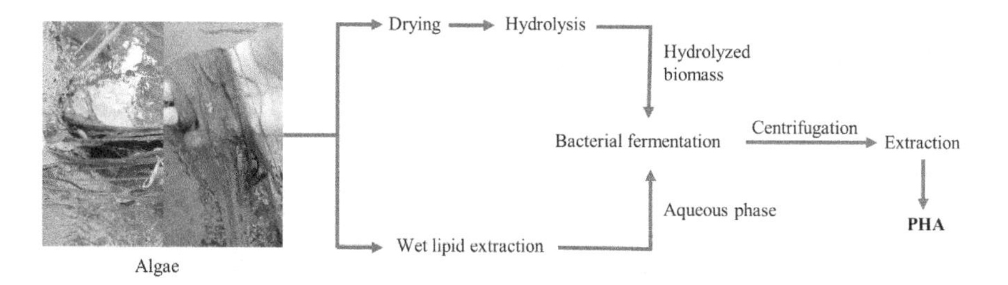

FIGURE 6.4 PHA production from algae.

Source: Adapted from Onen Cinar et al. (2020).

Khomlaem et al. (2021) examined the synthesis of PHAs using defatted *Chlorella* biomass as carbon substrate with microbial strains *Bacillus megaterium* ALA2, *Haloferax mediterranei* DSM 1411, and *Cupriavidus necator* KCTC 2649. The yield was 29.7%, 55.5%, and 75.4%, respectively, where *Cupriavidus necator* KCTC 2649 was able to accumulate the highest. Again, Abd El-malek et al. (2021) utilized *Corallina mediterranea* hydrolysates as carbon source with two bacterial strains, namely, *Halomonas salifodiane* ASL11 and *Halomonas pacifica* ASL10, for the biosynthesis of PHAs. The strain *Halomonas salifodiane* ASL11 accumulated 63%, while *Halomonas pacifica* ASL10 accumulated the highest with 67% PHAs CDW. Accordingly, using algal biomass as a carbon feedstock for PHA production may help with both the environment's carbon dioxide fixation and the cost-effective synthesis of PHAs. By lessening the demand on non-renewable resources, this strategy holds significant promise for a sustainable environment and economy.

6.6 WASTEWATER

In an effort to transition to a bio-based society, wastewater treatment facilities that were once thought of as end-of-pipe operations are now increasingly being taken into account inside biorefinery frameworks (De Vegt et al. 2012). Furthermore, the production of PHAs enhances the value of wastewater treatment by reusing carbon, directing it toward bioproducts, and lowering the production of sludge (Morgan-Sagastume et al. 2014). An effective way to deal with the issues of environmental pollution is to process wastewater to remove pollutants and organic content while simultaneously producing biopolymers (Pradhan et al. 2020). Because wastewater frequently contains a variety of organic and inorganic nutrients, treating them to produce purified effluent requires a great deal of work and is quite challenging. Therefore, it is very much beneficial to utilise the wastewater directly for production of PHAs without the need for expensive treatment (Bhuwal et al. 2013). Biosynthesis of PHAs from wastewater is shown in Figure 6.5. Khardenavis et al. (2007) investigated the potential of food processing wastewater effluent, jowar grain–based distillery spent wash, and rice grain–based distillery spent wash as carbon substrate for the biosynthesis of PHB using activated sludge consortium. It was able to produce PHB content of 39%, 42.3%, and 67%, respectively. Again, using mixed microbial culture with tomato

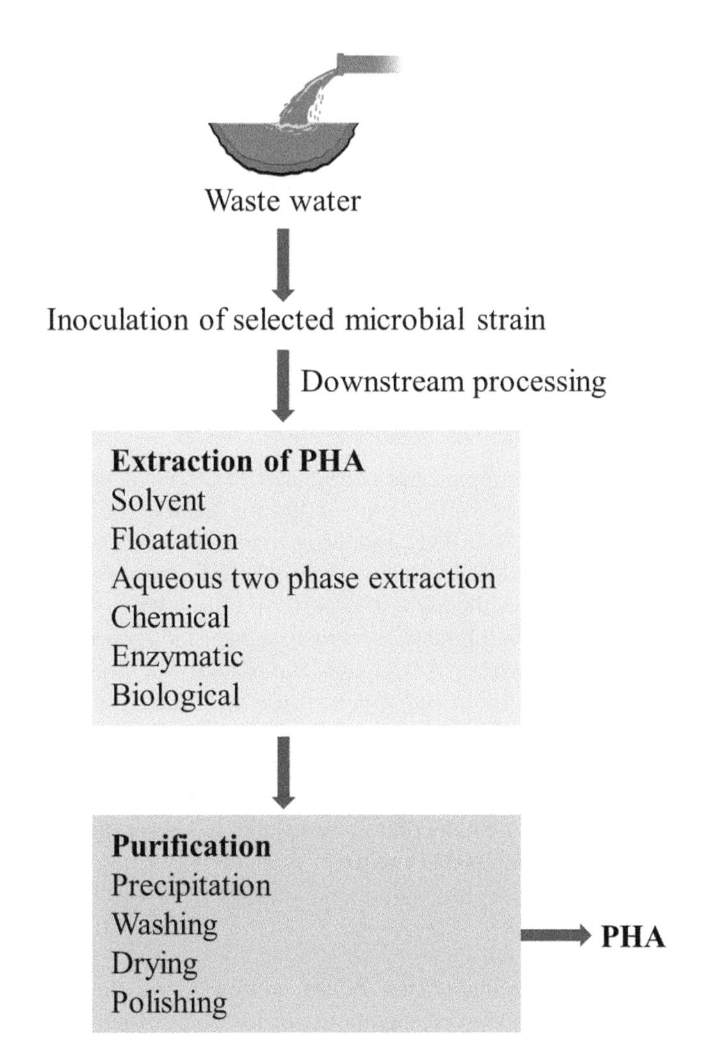

FIGURE 6.5 Biosynthesis of PHA from wastewater.

Source: Adapted from Kumar et al. (2020).

cannery wastewater as carbon ssubstrate, 20% of PHA CDW was obtained (Liu et al. 2008). Using swine wastewater as carbon substrate for the synthesis of PHAs, bacterial strain *Azotobacter vinelandii* UWD accumulated 58% of PHAs CDW (Ryu et al. 2008). A study employed wastewater of sugar industry as cheap carbon substrate using strain *Bacillus subtilis* NG220 for the production of PHAs, and it was able to accumulate 51.8% of PHAs CDW (Singh et al. 2013). Bhuwal et al. (2013) utilized wastewater of cardboard industry as carbon substrate for PHA production using bacterial strain *Enterococcus* sp. NAP11 and *Brevundimonas* sp. NAC1. The strain *Enterococcus* sp. NAP11 accumulated the highest PHAs with 79.27%, while the strain *Brevundimonas* sp. NAC1 was able to accumulate 77.63% of PHAs CDW.

Similarly, the bacterial strain *Bacillus tequilensis* MSU 112 accumulated 79.2% PHAs CDW using cassava starch wastewater as carbon substrate (Chaleomrum et al. 2014). Mayeli et al. (2015) utilized petrochemical wastewater as a carbon substrate for the biosynthesis of PHAs using *Bacillus axaraqunsis*, and it accumulated 66% PHAs CDW. Martinez et al. (2015) also reported the biosynthesis of PHAs from strain *Cupriavidus necator* DSM 545 using olive mill wastewater as a carbon substrate, and it accumulated 55% of PHAs CDW. Brewery wastewater was also used as carbon substrate for the production of PHAs using activated sludge consortium, and it successfully produced 39% PHAs of its CDW (Ben et al. 2016). Recently, Reddy et al. (2017) utilized the strain *Pseudomonas pseudoflava* for the synthesis of PHAs using synthetic wastewater as a carbon substrate, and 57% PHAs was obtained. More recently, Sabapathy et al. (2018) reported the use of *Acinetobacter junii* BP25 for the synthesis of PHAs using rice mill effluent as carbon substrate, and the strain was able to effectively produce 85.93% PHAs CDW.

6.6.1 Dairy Waste

Waste products from the processing of dairy industry are receiving more attention as a possible source of PHAs. However, the increasing production of whey is outpacing market demand, and the extra is disposed of as wastes. Due to whey's high biological oxygen demand (BOD) and chemical oxygen demand (COD), environmental concerns have been raised about its disposal. Nevertheless, due to its high protein, lipids, lactose, fat, vitamin, and mineral content, the nutrient-rich surplus whey actually has the potential to be used as a potential carbon substrate to produce PHAs (Sen and Baidurah 2021). The potential for using dairy whey as a cheap substrate for PHA synthesis is promising. However, whey has a low C–N ratio, which makes it challenging to produce PHAs; hence, it cannot be utilized directly to produce PHAs. Whey must therefore undergo pre-treatment before being used to make PHAs. In order to convert whey into bioplastics, lactose must first be hydrolyzed into monosaccharides like glucose and galactose, which can then be used to produce PHAs (Talan et al. 2022). In Asia, the dairy industry generates 113,340 MT (million tons) of milk, of which a large amount of serum is produced annually as a waste product. Out of the total amount of dairy produced in Africa and America, roughly 1 MT of whey are generated as a waste product, and 47% of this whey is dumped into drains and rivers, causing contamination and health issues (Tripathi et al. 2021). The benefits of using whey lactose for the production of PHAs are obvious because it would significantly cut PHA production costs without affecting human food supply while also resolving an environmental issue (Amaro et al. 2019). The biosynthesis of PHAs from dairy waste is depicted in Figure 6.6. Using cheese whey as carbon source, strains like *Ralstonia eutropha* produced 37% PHAs of its CDW (Marangoni et al. 2002), *Methylobacterium* sp. ZP24 produced 67% PHAs of its CDW (Nath et al. 2008), *Thermus thermophilus* HB8 produced 35% PHAs of its CDW (Pantazaki et al. 2009), *Bacillus megaterium* CCM 2037 produced 51% PHAs of its CDW (Obruca et al. 2011), and *Haloferax mediterranei* produced 66% PHAs of its CDW (Koller 2015). *Alcaligenus latus* was reported to produce 66.56% of PHAs using bean curd waste as carbon source (Kumalaningsih et al. 2011). More recently, Mozejko-Ciesielska et al. (2022)

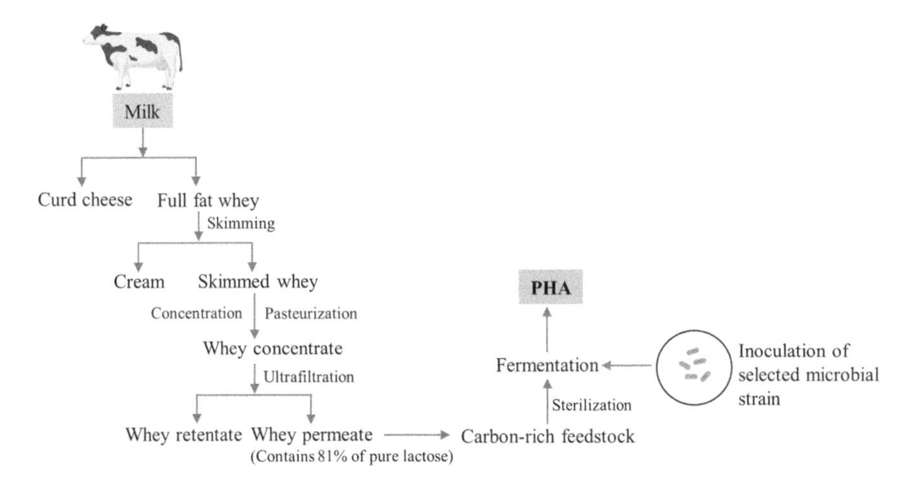

FIGURE 6.6 Biosynthesis of PHA from dairy waste.

Source: Adapted from Koller and Braunegg (2018)

TABLE 6.1
Production of PHAs by Different Microorganisms from Various Waste Feedstocks as Carbon Source

Microorganisms	Substrates	PHA Content (%)	References
PHA Production Using Agro-Wastes as Carbon Source			
Bacillus megaterium	Beet molasses	50	Omar et al. 2001
Haloferax mediterranei	Cornstarch	50.8	Chen et al. 2006
Brevindomonas vesicularis	Sawdust hydrolysates	64	Kulpreecha et al. 2009
Psuedomonas species	Corn oil	35.63	Chaudhry et al. 2011
Ralstonia eutropha	Jambul seed	41.77	Preethi et al. 2012
Bacillus firmus NII 0830	Rice straw	89	Sindhu et al. 2013
Bacillus thuringiensis	Mango peel	51.3	Gowda and Shivakumar 2014
Burkholderia sacchari DSM 17165	Wheat straw	72	Cesário et al. 2014
Burkholderia sp	Sugarcane bagasse	48	Lopes et al. 2014
Halomonas campilasis MCM B-1027	Bagasse	47	Kulkarni et al. 2015
Alcaligenes eutrophus	Tapioca hydrolysates	58	Sawant et al. 2016
Halogeometricum borinquense	Cassava waste	44.70	Salgaonkar et al. 2019
Bacillus subtilis MTCC 144	Mixed fruit peels, mixed vegetable peels, green pea shells	57.64, 63.27, 77.89	Rao et al. 2019

TABLE 6.1 *(Continued)*
Production of PHAs by Different Microorganisms from Various Waste Feedstocks as Carbon Source

Microorganisms	Substrates	PHA Content (%)	References
Cupriavidus necator	Grape winery waste	63	Kovalcik et al. 2020
Bacillus halotolerans DSM8802	Pomegranate peel	83	Rayasam et al. 2020
Haloferax mediterranei	Date palm waste	25	Alsafadi et al. 2020
Klebsiella pneumoniae	Watermelon peels, papaya peels	32.90, 27.47	Valdez-Calderón et al. 2020
PHA Production Using Food Wastes as Carbon Source			
Alcaligenus eutrophus DSM1124	Malt waste	70	Yu et al. 1998
Ralstonia eutropha	Soybean and rapeseed oil	57	Taniguchi et al. 2003
Pseudomonas aeruginosa NCIB 40045	Waste frying oil	29	Fernández et al. 2005
Pseudomonas corrugata	Soy molasses	5–17	Solaiman et al. 2006
Ralstonia eutropha NCIMB 11599	Waste potato starch	52.5	Haas et al. 2008
Halomonas boliviensis LC1	Wheat bran	34	Van-Thuoc et al. 2008
Escherichia coli (recombinant)	Molasses	75.5	Saranya and Shenbagarathai 2011
Psuedomonas sp.	Spent wash	25.46	Chaudhry et al. 2011
Cupriavidus necator CCGUG 52238	Kitchen waste	52.7	Omar et al. 2011
Burkholderia sp. USM (JCM 15050) (recombinant)	Crude palm kernel oil	66	Chee et al. 2012
Bacillus cereus Strain EGU3	Pea shells	71	Patel et al. 2012
Azohydromonas lata	Sugar beet juice	65.60	Wang et al. 2013
Escherichia coli pnDTM2 (recombinant)	Restaurant waste	45	Eshtaya et al. 2013
Burkholderia cepacia	Spent coffee grounds hydrolysate	54.79	Obruca et al. 2014
Cupriavidus necator	Cooking oil	63	Cruz et al. 2015
Bacillus sp.	Teff straw	37.4	Getachew and Woldesenbet 2016
Pseudomonas mendocina CH50	Coconut oil	58	Basnett et al. 2018
Salinivibrio species M318	Waste fish oil	52	Van Thuoc et al. 2019
Cupriavidus necator	Cashew apple juice	80.6	Arumugam et al. 2020
Bacillus thermoamylovorans	Waste cooking oil	74.4	Sangkharak et al. 2021

(*Continued*)

TABLE 6.1 *(Continued)*
Production of PHAs by Different Microorganisms from Various Waste Feedstocks as Carbon Source

Microorganisms	Substrates	PHA Content (%)	References
PHA Production Using Algal Carbon Source			
Escherichia coli	Algal biomass	34	Sathish et al. 2014
Halomonas hydrothermalis MTCC 5445	*Jatropha* biodiesel waste	75.8	Bera et al. 2015
Bacillus megaterium KCTC 2194	*Gelidium amansii*	51.4	Alkotaini et al. 2016
Cupriavidus necator PTCC 1615	*Sargassum* sp.	74	Azizi et al. 2017
Saccharophagus degradans	*Gelidium amansii*	17–27	Sawant et al. 2018
Haloferax mediterranei	*Ulva* sp.	48.15	Ghosh et al. 2019
Cupriavidus necator NCIMB 11599	*Laminaria japonica* biomass	32	Muhammad et al. 2020
Halomonas ventosae	Algal biodiesel waste	–	Dubey and Mishra 2021
Halomonas daqingensis	Algal biodiesel waste	–	Dubey and Mishra 2021
Bacillus megaterium ALA2	Defatted *Chlorella* biomass	29.7	Khomlaem et al. 2021
Cupriavidus necator KCTC 2649	Defatted *Chlorella* biomass	75.4	Khomlaem et al. 2021
Haloferax mediterranei DSM 1411	Defatted *Chlorella* biomass	55.5	Khomlaem et al. 2021
Halomonas pacifica ASL10	*Corallina mediterranea* hydrolysates	67	Abd El-malek et al. 2021
Halomonas salifodiane ASL11	*Corallina mediterranea* hydrolysates	63	Abd El-malek et al. 2021
PHA Production Using Wastewater as Carbon Source			
Activated sludge consortium	Food processing wastewater effluent, jowar grain–based distillery spent wash, rice grain–based distillery spent wash	39, 42.3, 67	Khardenavis et al. 2007
Azotobacter vinelandii UWD	Swine wastewater	58	Ryu et al. 2008
Mixed microbial culture	Tomato cannery wastewater	20	Liu et al. 2008
Bacillus subtilis NG220	Sugar industry wastewater	51.8	Singh et al. 2013
Enterococcus sp. NAP11	Cardboard industry wastewater	79.27	Bhuwal et al. 2013

TABLE 6.1 *(Continued)*
Production of PHAs by Different Microorganisms from Various Waste Feedstocks as Carbon Source

Microorganisms	Substrates	PHA Content (%)	References
Brevundimonas sp. NAC1	Cardboard industry wastewater	77.63	Bhuwal et al. 2013
Bacillus tequilensis MSU 112	Cassava starch wastewater	79.2	Chaleomrum et al. 2014
Bacillus axaraqunsis	Petrochemical wastewater	66	Mayeli et al. 2015
Cupriavidus necator DSM 545	Olive mill wastewater	55	Martinez et al. 2015
Activated sludge consortium	Brewery wastewater	39	Ben et al. 2016
Pseudomonas pseudoflava	Synthetic wastewater	57	Reddy et al. 2017
Wastewater microbes	Olive oil mill wastewater	11.3	Campanari et al. 2017
Acinetobacter junii BP25	Rice mill effluent	85.93	Sabapathy et al. 2018
PHA Production Using Dairy Waste as Carbon Source			
Ralstonia eutropha	Cheese whey	37	Marangoni et al. 2002
Methylobacterium sp. ZP24	Cheese whey	67	Nath et al. 2008
Thermus thermophilus HB8	Cheese whey	35	Pantazaki et al. 2009
Bacillus megaterium CCM 2037	Cheese whey	51	Obruca et al. 2011
Alcaligenus latus	Bean curd waste	66.56	Kumalaningsih et al. 2011
Haloferax mediterranei	Cheese whey	66	Koller 2015

studied the generation of PHAs by *Paracoccus homiensis* utilizing cheese whey mother liquor (CWML) and cheese whey (CW). The maximum biopolymer concentration was found with CWML (1.09 g/L) and CW (0.80 g/L).

6.7 CONCLUSION AND FUTURE OUTLOOK

Microbially derived PHAs have attracted a great deal of interest in scientific study as well as commercial applications as a replacement for synthetic polymers because of its bio-original, non-toxic, biodegradable, and biocompatible nature. The production of PHAs from a variety of sources has been reported in recent years. However, the high cost of PHA production on an industrial scale is a concern. Using cheaper and more environmentally friendly renewable feedstocks (such as agricultural wastes, food waste, algae, wastewater, and dairy waste) as substrate can reduce production costs. In addition to being less expensive, organic waste is always readily available. Utilizing organic wastes is a sustainable approach that complies with zero-waste policies and lowers manufacturing costs by eliminating transportation expenditures

(Tripathi et al. 2021). PHA production attempts to decouple economic growth from resource depletion and environmental damage, making them extremely important today. The potential to produce PHAs from organic waste has the added benefit of using waste to produce another valuable product. As a result, less organic waste is dumped in landfills, which reduces the amount of methane produced there. Methane is a potent GHG that is partially responsible for climate change and global warming (Stavroula et al. 2020). In order to achieve a "green future," commitment to the development of integrated biorefinery technologies will aid in reducing the global burden of synthetic plastics and GHG emissions. In addition, the government should intensify the implementation of bio-based policies and encourage current society to reduce the use of synthetic plastics by promoting the use of bioplastics in current applications. In light of this, standardization, certification, and labelling of biodegradable, compostable, and recyclable plastics are essential for consumer comprehension of appropriate waste disposal. Although biodegradable and compostable plastics are technically recyclable, only 1% of bioplastics are now recycled because there is no European standard for appropriately labelling bioplastics in various environments (Raj et al. 2022). The sustainable and renewable characteristics of PHAs deliver a friendly option for the next generation of environmentally benign biopolymers as efforts to find a solution to the worldwide environmental pollution catastrophe escalates. Additionally, global prediction indicates that PHA production will increase in the upcoming years as a result of continuous attempts to secure a large supply of them. Therefore, the continued development of PHAs from renewable feedstock will make it a lot easier to attain a sustainable and green environment in the near future (Adeleye et al. 2020). For the sake of the environment, and to enhance quality of life, the scientific community and the industrial sector should pay much more attention to PHA generation from organic waste and also use a waste biorefinery circular bioeconomy strategy to address global issues. Therefore, this study might motivate and direct future research aimed at achieving commercially feasible PHA production from renewable feedstock.

REFERENCES

Abd El-malek, F., Rofeal, M., Farag, A., Omar, S., & Khairy, H. (2021). Polyhydroxyalkanoate nanoparticles produced by marine bacteria cultivated on cost effective Mediterranean algal hydrolysate media. *Journal of Biotechnology*, *328*, 95–105. https://doi.org/10.1016/j.jbiotec.2021.01.008

Adeleye, A. T., Odoh, C. K., Enudi, O. C., Banjoko, O. O., Osiboye, O. O., Odediran, E. T., & Louis, H. (2020). Sustainable synthesis and applications of polyhydroxyalkanoates (PHAs) from biomass. *Process Biochemistry*, *96*, 174–193. https://doi.org/10.1016/j.procbio.2020.05.032

Alkotaini, B., Koo, H., & Kim, B. S. (2016). Production of polyhydroxyalkanoates by batch and fed-batch cultivations of *Bacillus megaterium* from acid-treated red algae. *Korean Journal of Chemical Engineering*, *33*, 1669–1673. https://doi.org/10.1007/s11814-015-0293-6

Alsafadi, D., Ibrahim, M. I., Alamry, K. A., Hussein, M. A., & Mansour, A. (2020). Utilizing the crop waste of date palm fruit to biosynthesize polyhydroxyalkanoate bioplastics with favorable properties. *Science of the Total Environment*, *737*, 139716. https://doi.org/10.1016/j.scitotenv.2020.139716

Amaro, T. M., Rosa, D., Comi, G., & Iacumin, L. (2019). Prospects for the use of whey for polyhydroxyalkanoate (PHA) production. *Frontiers in Microbiology*, *10*, 992. https://doi.org/10.3389/fmicb.2019.00992

Amulya, K., Reddy, M. V., Rohit, M. V., & Mohan, S. V. (2016). Wastewater as renewable feedstock for bioplastics production: understanding the role of reactor microenvironment and system pH. *Journal of Cleaner Production*, *112*, 4618–4627. https://doi.org/10.1016/j.jclepro.2015.08.009

Andler, R., Valdés, C., Urtuvia, V., Andreeßen, C., & Díaz-Barrera, A. (2021). Fruit residues as a sustainable feedstock for the production of bacterial polyhydroxyalkanoates. *Journal of Cleaner Production*, *307*, 127236. https://doi.org/10.1016/j.jclepro.2021.127236

Arumugam, A., Anudakshaini, T. S., Shruthi, R., Jeyavishnu, K., Sundarra Harini, S., & Sharad, J. S. (2020). Low-cost production of PHA using cashew apple (*Anacardium occidentale* L.) juice as potential substrate: optimization and characterization. *Biomass Conversion and Biorefinery*, *10*(4), 1167–1178. https://doi.org/10.1007/s13399-019-00502-5

Azizi, N., Najafpour, G., & Younesi, H. (2017). Acid pretreatment and enzymatic saccharification of brown seaweed for polyhydroxybutyrate (PHB) production using *Cupriavidus necator*. *International Journal of Biological Macromolecules*, *101*, 1029–1040. https://doi.org/10.1016/j.ijbiomac.2017.03.184

Basnett, P., Marcello, E., Lukasiewicz, B., Panchal, B., Nigmatullin, R., Knowles, J. C., & Roy, I. (2018). Biosynthesis and characterization of a novel, biocompatible medium chain length polyhydroxyalkanoate by *Pseudomonas mendocina* CH50 using coconut oil as the carbon source. *Journal of Materials Science: Materials in Medicine*, *29*, 1–11. https://doi.org/10.1007/s10856-018-6183-9

Ben, M., Kennes, C., & Veiga, M. C. (2016). Optimization of polyhydroxyalkanoate storage using mixed cultures and brewery wastewater. *Journal of Chemical Technology & Biotechnology*, *91*(11), 2817–2826. https://doi.org/10.1002/jctb.4891

Bera, A., Dubey, S., Bhayani, K., Mondal, D., Mishra, S., & Ghosh, P. K. (2015). Microbial synthesis of polyhydroxyalkanoate using seaweed-derived crude levulinic acid as co-nutrient. *International Journal of Biological Macromolecules*, *72*, 487–494. https://doi.org/10.1016/j.ijbiomac.2014.08.037

Bhatia, S. K., Otari, S. V., Jeon, J. M., Gurav, R., Choi, Y. K., Bhatia, R. K., . . . Yang, Y. H. (2021). Biowaste-to-bioplastic (polyhydroxyalkanoates): conversion technologies, strategies, challenges, and perspective. *Bioresource Technology*, *326*, 124733. https://doi.org/10.1016/j.biortech.2021.124733

Bhuwal, A. K., Singh, G., Aggarwal, N. K., Goyal, V., & Yadav, A. (2013). Isolation and screening of polyhydroxyalkanoates producing bacteria from pulp, paper, and cardboard industry wastes. *International Journal of Biomaterials*, *2013*. https://doi.org/10.1155/2013/752821

Brigham, C. J., & Riedel, S. L. (2018). The potential of polyhydroxyalkanoate production from food wastes. *Applied Food Biotechnology*, *6*(1), 7–18. https://doi.org/10.22037/afb.v6i1.22542

Campanari, S., Augelletti, F., Rossetti, S., Sciubba, F., Villano, M., & Majone, M. (2017). Enhancing a multi-stage process for olive oil mill wastewater valorization towards polyhydroxyalkanoates and biogas production. *Chemical Engineering Journal*, *317*, 280–289. https://doi.org/10.1016/j.cej.2017.02.094

Cesário, M. T., Raposo, R. S., de Almeida, M. C. M., van Keulen, F., Ferreira, B. S., & da Fonseca, M. M. R. (2014). Enhanced bioproduction of poly-3-hydroxybutyrate from wheat straw lignocellulosic hydrolysates. *New Biotechnology*, *31*(1), 104–113. https://doi.org/10.1016/j.nbt.2013.10.004

Chaleomrum, N., Chookietwattana, K., & Dararat, S. (2014). Production of PHA from cassava starch wastewater in sequencing batch reactor treatment system. *APCBEE Procedia*, *8*, 167–172. https://doi.org/10.1016/j.apcbee.2014.03.021

Chaudhry, W. N., Jamil, N., Ali, I., Ayaz, M. H., & Hasnain, S. (2011). Screening for polyhydroxyalkanoate (PHA)-producing bacterial strains and comparison of PHA production from various inexpensive carbon sources. *Annals of Microbiology*, *61*(3), 623–629. https://doi.org/10.1007/s13213-010-0181-6

Chee, J. Y., Lau, N. S., Samian, M. R., Tsuge, T., & Sudesh, K. (2012). Expression of *Aeromonas caviae* polyhydroxyalkanoate synthase gene in *Burkholderia* sp. USM (JCM15050) enables the biosynthesis of SCL-MCL PHA from palm oil products. *Journal of Applied Microbiology*, *112*(1), 45–54. https://doi.org/10.1111/j.1365-2672.2011.05189.x

Chen, C. W., Don, T. M., & Yen, H. F. (2006). Enzymatic extruded starch as a carbon source for the production of poly (3-hydroxybutyrate-co-3-hydroxyvalerate) by *Haloferax mediterranei. Process Biochemistry*, *41*(11), 2289–2296. https://doi.org/10.1016/j.procbio.2006.05.026

Chen, G. Q., Hajnal, I., Wu, H., Lv, L., & Ye, J. (2015). Engineering biosynthesis mechanisms for diversifying polyhydroxyalkanoates. *Trends in Biotechnology*, *33*(10), 565–574. https://doi.org/10.1016/j.tibtech.2015.07.007

Cruz, M. V., Sarraguça, M. C., Freitas, F., Lopes, J. A., & Reis, M. A. (2015). Online monitoring of P(3HB) produced from used cooking oil with near-infrared spectroscopy. *Journal of Biotechnology*, *194*, 1–9. https://doi.org/10.1016/j.jbiotec.2014.11.022

de Paula, F. C., de Paula, C. B., & Contiero, J. (2018). Prospective biodegradable plastics from biomass conversion processes. *Biofuels-State of Development*, 245–272. http://dx.doi.org/10.5772/intechopen.75111

De Vegt, O., Werker, A., Fetter, B., Hopman, R., Krins, B., & Winters, R. (2012). PHA from waste water. Transformation of residual materials and waste water into valuable bioplastics. *Bioplastics Magazine*, *7*(4), 26–28.

Dietrich, K., Dumont, M. J., Del Rio, L. F., & Orsat, V. (2017). Producing PHAs in the bioeconomy – towards a sustainable bioplastic. *Sustainable Production and Consumption*, *9*, 58–70. https://doi.org/10.1016/j.spc.2016.09.001

Ding, Z., Kumar, V., Sar, T., Harirchi, S., Dregulo, A. M., Sirohi, R., . . . Awasthi, M. K. (2022). Agro waste as a potential carbon feedstock for poly-3-hydroxy alkanoates production: commercialization potential and technical hurdles. *Bioresource Technology*, 128058. https://doi.org/10.1016/j.biortech.2022.128058

Dubey, S., & Mishra, S. (2021). Efficient production of polyhydroxyalkanoate through halophilic bacteria utilizing algal biodiesel waste residue. *Frontiers in Bioengineering and Biotechnology*, *9*, 624859. https://doi.org/10.3389/fbioe.2021.624859

Eshtaya, M. K., Nor 'Aini, A. R., & Hassan, M. A. (2013). Bioconversion of restaurant waste into Polyhydroxybutyrate (PHB) by recombinant *E. coli* through anaerobic digestion. *International Journal of Environment and Waste Management*, *11*(1), 27–37. https://doi.org/10.1504/IJEWM.2013.050521

Fernández, D., Rodríguez, E., Bassas, M., Viñas, M., Solanas, A. M., Llorens, J., . . . Manresa, A. (2005). Agro-industrial oily wastes as substrates for PHA production by the new strain Pseudomonas aeruginosa NCIB 40045: effect of culture conditions. *Biochemical Engineering Journal*, *26*(2–3), 159–167. https://doi.org/10.1016/j.bej.2005.04.022

Getachew, A., & Woldesenbet, F. (2016). Production of biodegradable plastic by polyhydroxybutyrate (PHB) accumulating bacteria using low cost agricultural waste material. *BMC Research Notes*, *9*(1), 1–9. https://doi.org/10.1186/s13104-016-2321-y

Ghosh, S., Gnaim, R., Greiserman, S., Fadeev, L., Gozin, M., & Golberg, A. (2019). Macroalgal biomass subcritical hydrolysates for the production of polyhydroxyalkanoate (PHA) by *Haloferax mediterranei. Bioresource Technology*, *271*, 166–173. https://doi.org/10.1016/j.biortech.2018.09.108

Gowda, V., & Shivakumar, S. (2014). Agrowaste-based polyhydroxyalkanoate (PHA) production using hydrolytic potential of *Bacillus thuringiensis* IAM 12077. *Brazilian Archives of Biology and Technology*, *57*, 55–61. https://doi.org/10.1590/S1516-89132014000100009

Haas, R., Jin, B., & Zepf, F. T. (2008). Production of poly (3-hydroxybutyrate) from waste potato starch. *Bioscience, Biotechnology, and Biochemistry*, *72*(1), 253–256.

Hassan, M. A., Yee, L. N., Yee, P. L., Ariffin, H., Raha, A. R., Shirai, Y., & Sudesh, K. (2013). Sustainable production of polyhydroxyalkanoates from renewable oil-palm biomass. *Biomass and Bioenergy*, *50*, 1–9. https://doi.org/10.1016/j.biombioe.2012.10.014

Khardenavis, A. A., Kumar, M. S., Mudliar, S. N., & Chakrabarti, T. (2007). Biotechnological conversion of agro-industrial wastewaters into biodegradable plastic, poly β-hydroxybutyrate. *Bioresource Technology*, *98*(18), 3579–3584. https://doi.org/10.1016/j.biortech.2006.11.024

Khomlaem, C., Aloui, H., & Kim, B. S. (2021). Biosynthesis of polyhydroxyalkanoates from defatted *Chlorella* biomass as an inexpensive substrate. *Applied Sciences*, *11*(3), 1094. https://doi.org/10.3390/app11031094

Koller, M. (2015). Recycling of waste streams of the biotechnological poly (hydroxyalkanoate) production by *Haloferax mediterranei* on whey. *International Journal of Polymer Science*, *2015*. https://doi.org/10.1155/2015/370164

Koller, M., & Braunegg, G. (2018). Advanced approaches to produce polyhydroxyalkanoate (PHA) biopolyesters in a sustainable and economic fashion. *EuroBiotech Journal*, *2*(2), 89–103. https://doi.org/10.2478/ebtj-2018–0013

Kovalcik, A., Pernicova, I., Obruca, S., Szotkowski, M., Enev, V., Kalina, M., & Marova, I. (2020). Grape winery waste as a promising feedstock for the production of polyhydroxyalkanoates and other value-added products. *Food and Bioproducts Processing*, *124*, 1–10. https://doi.org/10.1016/j.fbp.2020.08.003

Krishnan, S., Chinnadurai, G. S., Ravishankar, K., Raghavachari, D., & Perumal, P. (2021). Valorization of agro-wastes for the biosynthesis and characterization of polyhydroxybutyrate by *Bacillus* sp. isolated from rice bran dumping yard. *3 Biotech*, *11*(4), 1–14. https://doi.org/10.1007/s13205-021-02722-x

Kulkarni, S. O., Kanekar, P. P., Jog, J. P., Sarnaik, S. S., & Nilegaonkar, S. S. (2015). Production of copolymer, poly (hydroxybutyrate-co-hydroxyvalerate) by *Halomonas campisalis* MCM B-1027 using agro-wastes. *International Journal of Biological Macromolecules*, *72*, 784–789. https://doi.org/10.1016/j.ijbiomac.2014.09.028

Kulpreecha, S., Boonruangthavorn, A., Meksiriporn, B., & Thongchul, N. (2009). Inexpensive fed-batch cultivation for high poly (3-hydroxybutyrate) production by a new isolate of *Bacillus megaterium*. *Journal of Bioscience and Bioengineering*, *107*(3), 240–245. https://doi.org/10.1016/j.jbiosc.2008.10.006

Kumalaningsih, S., Hidayat, N., & Aini, N. (2011). Optimization of polyhydroxyalkanoates (PHA) production from liquid bean curd waste by *Alcaligenes latus* bacteria. *Journal of Agriculture and Food Technology*, *1*, 63–67.

Kumar, M., Rathour, R., Singh, R., Sun, Y., Pandey, A., Gnansounou, E., . . . Thakur, I. S. (2020). Bacterial polyhydroxyalkanoates: opportunities, challenges, and prospects. *Journal of Cleaner Production*, *263*, 121500. https://doi.org/10.1016/j.jclepro.2020.121500

Li, M., & Wilkins, M. R. (2020). Recent advances in polyhydroxyalkanoate production: feedstocks, strains and process developments. *International Journal of Biological Macromolecules*, *156*, 691–703. https://doi.org/10.1016/j.ijbiomac.2020.04.082

Liu, H. Y., Hall, P. V., Darby, J. L., Coats, E. R., Green, P. G., Thompson, D. E., & Loge, F. J. (2008). Production of polyhydroxyalkanoate during treatment of tomato cannery wastewater. *Water Environment Research*, *80*(4), 367–372. https://doi.org/10.2175/106143007X221535

Liu, H. Y., Kumar, V., Jia, L., Sarsaiya, S., Kumar, D., Juneja, A., . . . Awasthi, M. K. (2021). Biopolymer poly-hydroxyalkanoates (PHA) production from apple industrial waste residues: a review. *Chemosphere*, *284*, 131427. https://doi.org/10.1016/j.chemosphere.2021.131427

Lopes, M. S. G., Gomez, J. G. C., Taciro, M. K., Mendonça, T. T., & Silva, L. F. (2014). Polyhydroxyalkanoate biosynthesis and simultaneous remotion of organic inhibitors from sugarcane bagasse hydrolysate by *Burkholderia* sp. *Journal of Industrial Microbiology and Biotechnology*, *41*(9), 1353–1363. https://doi.org/10.1007/s10295-014-1485-5

Luengo, J. M., Garcıa, B., Sandoval, A., Naharro, G., & Olivera, E. R. (2003). Bioplastics from microorganisms. *Current Opinion in Microbiology*, *6*(3), 251–260. https://doi.org/10.1016/S1369-5274(03)00040-7

Marangoni, C., Furigo Jr, A., & de Aragão, G. M. (2002). Production of poly (3-hydroxybu tyrate-co-3-hydroxyvalerate) by *Ralstonia eutropha* in whey and inverted sugar with propionic acid feeding. *Process Biochemistry*, *38*(2), 137–141. https://doi.org/10.1016/S0032-9592(01)00313-2

Martinez, G. A., Bertin, L., Scoma, A., Rebecchi, S., Braunegg, G., & Fava, F. (2015). Production of polyhydroxyalkanoates from dephenolised and fermented olive mill wastewaters by employing a pure culture of *Cupriavidus necator*. *Biochemical Engineering Journal*, *97*, 92–100. https://doi.org/10.1016/j.bej.2015.02.015

Mayeli, N., Motamedi, H., & Heidarizadeh, F. (2015). Production of polyhydroxybutyrate by *Bacillus axaraqunsis* BIPC01 using petrochemical wastewater as carbon source. *Brazilian Archives of Biology and Technology*, *58*, 643–650. https://doi.org/10.1590/S1516-8913201500048

Mohan, S. V., Hemalatha, M., Chakraborty, D., Chatterjee, S., Ranadheer, P., & Kona, R. (2020). Algal biorefinery models with self-sustainable closed loop approach: trends and prospective for blue-bioeconomy. *Bioresource Technology*, *295*, 122128. https://doi.org/10.1016/j.biortech.2019.122128

Morgan-Sagastume, F., Valentino, F., Hjort, M., Cirne, D., Karabegovic, L., Gerardin, F., . . . Werker, A. (2014). Polyhydroxyalkanoate (PHA) production from sludge and municipal wastewater treatment. *Water Science and Technology*, *69*(1), 177–184. https://doi.org/10.2166/wst.2013.643

Mozejko-Ciesielska, J., Marciniak, P., Moraczewski, K., Rytlewski, P., Czaplicki, S., & Zadernowska, A. (2022). Cheese whey mother liquor as dairy waste with potential value for polyhydroxyalkanoate production by extremophilic *Paracoccus homiensis*. *Sustainable Materials and Technologies*, *33*, e00449. https://doi.org/10.1016/j.susmat.2022.e00449

Muhammad, M., Aloui, H., Khomlaem, C., Hou, C. T., & Kim, B. S. (2020). Production of polyhydroxyalkanoates and carotenoids through cultivation of different bacterial strains using brown algae hydrolysate as a carbon source. *Biocatalysis and Agricultural Biotechnology*, *30*, 101852. https://doi.org/10.1016/j.bcab.2020.101852

Nath, A., Dixit, M., Bandiya, A., Chavda, S., & Desai, A. J. (2008). Enhanced PHB production and scale up studies using cheese whey in fed batch culture of *Methylobacterium* sp. ZP24. *Bioresource Technology*, *99*(13), 5749–5755. https://doi.org/10.1016/j.biortech.2007.10.017

Nielsen, C., Rahman, A., Rehman, A. U., Walsh, M. K., & Miller, C. D. (2017). Food waste conversion to microbial polyhydroxyalkanoates. *Microbial Biotechnology*, *10*(6), 1338–1352. https://doi.org/10.1111/1751-7915.12776

Obruca, S., Benesova, P., Petrik, S., Oborna, J., Prikryl, R., & Marova, I. (2014). Production of polyhydroxyalkanoates using hydrolysate of spent coffee grounds. *Process Biochemistry*, *49*(9), 1409–1414. https://doi.org/10.1016/j.procbio.2014.05.013

Obruca, S., Marova, I., Melusova, S., & Mravcova, L. (2011). Production of polyhydroxyalkanoates from cheese whey employing *Bacillus megaterium* CCM 2037. *Annals of Microbiology*, *61*(4), 947–953. https://doi.org/10.1007/s13213-011-0218-5

Omar, F. N., Rahman, N. A. A., Hafid, H. S., Mumtaz, T., Yee, P. L., & Hassan, M. A. (2011). Utilization of kitchen waste for the production of green thermoplastic polyhydroxybutyrate (PHB) by *Cupriavidus necator* CCGUG 52238. *African Journal of Microbiology Research*, *5*(19), 2873–2879. https://doi.org/10.5897/AJMR11.156

Omar, S., Rayes, A., Eqaab, A., Voß, I., & Steinbüchel, A. (2001). Optimization of cell growth and poly (3-hydroxybutyrate) accumulation on date syrup by a *Bacillus megaterium* strain. *Biotechnology Letters*, *23*(14), 1119–1123. https://doi.org/10.1023/A:1010559800535

Onen Cinar, S., Chong, Z. K., Kucuker, M. A., Wieczorek, N., Cengiz, U., & Kuchta, K. (2020). Bioplastic production from microalgae: a review. *International Journal of Environmental Research and Public Health*, *17*(11), 3842. https://doi.org/10.3390/ijerph17113842

Pakalapati, H., Chang, C. K., Show, P. L., Arumugasamy, S. K., & Lan, J. C. W. (2018). Development of polyhydroxyalkanoates production from waste feedstocks and applications. *Journal of Bioscience and Bioengineering*, *126*(3), 282–292. https://doi.org/10.1016/j.jbiosc.2018.03.016

Pantazaki, A. A., Papaneophytou, C. P., Pritsa, A. G., Liakopoulou-Kyriakides, M., & Kyriakidis, D. A. (2009). Production of polyhydroxyalkanoates from whey by *Thermus thermophilus* HB8. *Process Biochemistry*, *44*(8), 847–853. https://doi.org/10.1016/j.procbio.2009.04.002

Patel, S. K., Singh, M., Kumar, P., Purohit, H. J., & Kalia, V. C. (2012). Exploitation of defined bacterial cultures for production of hydrogen and polyhydroxybutyrate from pea-shells. *Biomass and Bioenergy*, *36*, 218–225. https://doi.org/10.1016/j.biombioe.2011.10.027

Pradhan, S., Dikshit, P. K., & Moholkar, V. S. (2020). Production, characterization, and applications of biodegradable polymer: polyhydroxyalkanoates. In *Advances in Sustainable Polymers* (pp. 51–94). Singapore: Springer. https://doi.org/10.1007/978-981-15-1251-3_4

Preethi, R., Sasikala, P., & Aravind, J. (2012). Microbial production of polyhydroxyalkanoate (PHA) utilizing fruit waste as a substrate. *Research in Biotechnology*, *3*(1), 61–69.

Rahim, A. H. A., Man, Z., Sarwono, A., Muhammad, N., Khan, A. S., Hamzah, W. S. W., . . . Elsheikh, Y. A. (2020). Probe sonication assisted ionic liquid treatment for rapid dissolution of lignocellulosic biomass. *Cellulose*, *27*, 2135–2148. https://doi.org/10.1007/s10570-019-02914-y

Raj, T., Chandrasekhar, K., Kumar, A. N., & Kim, S. H. (2022). Lignocellulosic biomass as renewable feedstock for biodegradable and recyclable plastics production: a sustainable approach. *Renewable and Sustainable Energy Reviews*, *158*, 112130. https://doi.org/10.1016/j.rser.2022.112130

Rao, A., Haque, S., El-Enshasy, H. A., Singh, V., & Mishra, B. N. (2019). RSM–GA based optimization of bacterial PHA production and in silico modulation of citrate synthase for enhancing PHA production. *Biomolecules*, *9*(12), 872. https://doi.org/10.3390/biom9120872

Rayasam, V., Chavan, P., & Kumar, T. (2020). Polyhydroxyalkanoate synthesis by bacteria isolated from landfill and ETP with pomegranate peels as carbon source. *Archives of Microbiology*, *202*(10), 2799–2808. https://doi.org/10.1007/s00203-020-01995-9

Reddy, C. S. K., Ghai, R., & Kalia, V. (2003). Polyhydroxyalkanoates: an overview. *Bioresource Technology*, *87*(2), 137–146. https://doi.org/10.1016/S0960-8524(02)00212-2

Reddy, M. V., Mawatari, Y., Onodera, R., Nakamura, Y., Yajima, Y., & Chang, Y. C. (2017). Polyhydroxyalkanoates (PHA) production from synthetic waste using *Pseudomonas pseudoflava*: PHA synthase enzyme activity analysis from *P. pseudoflava* and *P. palleronii*. *Bioresource Technology*, *234*, 99–105. https://doi.org/10.1016/j.biortech.2017.03.008

Rodriguez-Perez, S., Serrano, A., Pantión, A. A., & Alonso-Fariñas, B. (2018). Challenges of scaling-up PHA production from waste streams. A review. *Journal of Environmental Management*, *205*, 215–230. https://doi.org/10.1016/j.jenvman.2017.09.083

Ryu, H. W., Cho, K. S., Goodrich, P. R., & Park, C. H. (2008). Production of polyhydroxyalkanoates by *Azotobacter vinelandii* UWD using swine wastewater: effect of supplementing glucose, yeast extract, and inorganic salts. *Biotechnology and Bioprocess Engineering*, *13*, 651–658. https://doi.org/10.1007/s12257-008-0072-x

Sabapathy, P. C., Devaraj, S., Parthiban, A., & Kathirvel, P. (2018). Bioprocess optimization of PHB homopolymer and copolymer P3 (HB-co-HV) by *Acinetobacter junii* BP25 utilizing rice mill effluent as sustainable substrate. *Environmental Technology*, *39*(11), 1430–1441. https://doi.org/10.1080/09593330.2017.1330902

Salgaonkar, B. B., Mani, K., & Bragança, J. M. (2019). Sustainable bioconversion of cassava waste to poly (3-hydroxybutyrate-co-3-hydroxyvalerate) by *Halogeometricum borinquense* strain E3. *Journal of Polymers and the Environment*, *27*(2), 299–308. https://doi.org/10.1007/s10924-018-1346-9

Sangkharak, K., Khaithongkaeo, P., Chuaikhunupakarn, T., Choonut, A., & Prasertsan, P. (2021). The production of polyhydroxyalkanoate from waste cooking oil and its application in biofuel production. *Biomass Conversion and Biorefinery*, *11*, 1651–1664. https://doi.org/10.1007/s13399-020-00657-6

Saranya, V., & Shenbagarathai, R. (2011). Production and characterization of PHA from recombinant *E. coli* harbouring phaC1 gene of indigenous *Pseudomonas* sp. LDC-5 using molasses. *Brazilian Journal of Microbiology*, *42*, 1109–1118. https://doi.org/10.1590/S1517-83822011000300032

Sathish, A., Glaittli, K., Sims, R. C., & Miller, C. D. (2014). Algae biomass based media for poly (3-hydroxybutyrate)(PHB) production by *Escherichia coli*. *Journal of Polymers and the Environment*, *22*, 272–277. https://doi.org/10.1007/s10924-014-0647-x

Sathya, A. B., Sivasubramanian, V., Santhiagu, A., Sebastian, C., & Sivashankar, R. (2018). Production of polyhydroxyalkanoates from renewable sources using bacteria. *Journal of Polymers and the Environment*, *26*(9), 3995–4012. https://doi.org/10.1007/s10924-018-1259-7

Sawant, S. S., Salunke, B. K., & Kim, B. S. (2018). Consolidated bioprocessing for production of polyhydroxyalkanotes from red algae *Gelidium amansii*. *International Journal of Biological Macromolecules*, *109*, 1012–1018. https://doi.org/10.1016/j.ijbiomac.2017.11.084

Sawant, S. S., Salunke, B. K., Tran, T. K., & Kim, B. S. (2016). Lignocellulosic and marine biomass as resource for production of polyhydroxyalkanoates. *Korean Journal of Chemical Engineering*, *33*(5), 1505–1513. https://doi.org/10.1007/s11814–016–0019–4

Sen, K. Y., & Baidurah, S. (2021). Renewable biomass feedstocks for production of sustainable biodegradable polymer. *Current Opinion in Green and Sustainable Chemistry*, *27*, 100412. https://doi.org/10.1016/j.cogsc.2020.100412

Sindhu, R., Silviya, N., Binod, P., & Pandey, A. (2013). Pentose-rich hydrolysate from acid pretreated rice straw as a carbon source for the production of poly-3-hydroxybutyrate. *Biochemical Engineering Journal*, *78*, 67–72. https://doi.org/10.1016/j.bej.2012.12.015

Singh, G., Kumari, A., Mittal, A., Yadav, A., & Aggarwal, N. K. (2013). Poly β-hydroxybutyrate production by *Bacillus subtilis* NG220 using sugar industry waste water. *BioMed Research International*, *2013,* 1–10. https://doi.org/10.1155/2013/952641

Solaiman, D. K., Ashby, R. D., Hotchkiss, A. T., & Foglia, T. A. (2006). Biosynthesis of medium-chain-length poly (hydroxyalkanoates) from soy molasses. *Biotechnology Letters*, *28*, 157–162. https://doi.org/10.1007/s10529-005-5329-2

Stavroula, K., Simos, M., & Katherine-Joanne, H. (2020). Polyhydroxyalkanoates (PHAs) from household food waste: research over the last decade. *International Journal of Biotechnology and Bioeng*ineering, *6*, 26–36.

Talan, A., Pokhrel, S., Tyagi, R. D., & Drogui, P. (2022). Biorefinery strategies for microbial bioplastics production: sustainable pathway towards circular bioeconomy. *Bioresource Technology Reports*, *17*, 100875. https://doi.org/10.1016/j.biteb.2021.100875

Tan, F. H. P., Nadir, N., & Sudesh, K. (2022). Microalgal biomass as feedstock for bacterial production of PHA: advances and future prospects. *Frontiers in Bioengineering and Biotechnology*, *10*. https://doi.org/10.3389/fbioe.2022.879476

Taniguchi, I., Kagotani, K., & Kimura, Y. (2003). Microbial production of poly (hydroxyalkanoate) s from waste edible oils. *Green Chemistry*, *5*(5), 545–548. https://doi.org/10.1039/B304800B

Thomas, C. M., Kumar, D., Scheel, R. A., Ramarao, B., & Nomura, C. T. (2022). Production of medium chain length polyhydroxyalkanoate copolymers from agro-industrial waste streams. *Biocatalysis and Agricultural Biotechnology*, 102385. https://doi.org/10.1016/j.bcab.2022.102385

Tripathi, A. D., Paul, V., Agarwal, A., Sharma, R., Hashempour-Baltork, F., Rashidi, L., & Darani, K. K. (2021). Production of polyhydroxyalkanoates using dairy processing waste–a review. *Bioresource Technology*, *326*, 124735. https://doi.org/10.1016/j.biortech.2021.124735

Valdez-Calderón, A., Barraza-Salas, M., Quezada-Cruz, M., Islas-Ponce, M. A., Angeles-Padilla, A. F., Carrillo-Ibarra, S., . . . Rivas-Castillo, A. M. (2020). Production of polyhydroxybutyrate (PHB) by a novel *Klebsiella pneumoniae* strain using low-cost media from fruit peel residues. *Biomass Conversion and Biorefinery*, 1–14. https://doi.org/10.1007/s13399-020-01147-5

Van Thuoc, D., My, D. N., Loan, T. T., & Sudesh, K. (2019). Utilization of waste fish oil and glycerol as carbon sources for polyhydroxyalkanoate production by *Salinivibrio* sp. M318. *International Journal of Biological Macromolecules*, *141*, 885–892. https://doi.org/10.1016/j.ijbiomac.2019.09.063

Van-Thuoc, D., Quillaguaman, J., Mamo, G., & Mattiasson, B. (2008). Utilization of agricultural residues for poly (3-hydroxybutyrate) production by *Halomonas boliviensis* LC1. *Journal of Applied Microbiology*, *104*(2), 420–428. https://doi.org/10.1111/j.1365-2672.2007.03553.x

Vigneswari, S., Noor, M. S. M., Amelia, T. S. M., Balakrishnan, K., Adnan, A., Bhubalan, K., . . . Ramakrishna, S. (2021). Recent advances in the biosynthesis of polyhydroxyalkanoates from lignocellulosic feedstocks. *Life*, *11*(8), 807. https://doi.org/10.3390/life11080807

Wang, B., Sharma-Shivappa, R. R., Olson, J. W., & Khan, S. A. (2013). Production of polyhydroxybutyrate (PHB) by *Alcaligenes latus* using sugarbeet juice. *Industrial Crops and Products*, *43*, 802–811. https://doi.org/10.1016/j.indcrop.2012.08.011

Yu, P. H., Chua, H., Huang, A. L., Lo, W., & Chen, G. Q. (1998). Conversion of food industrial wastes into bioplastics. *Applied Biochemistry and Biotechnology*, *70*(1), 603–614. https://doi.org/10.1007/BF02920172

7 A Sanitary Engineering Assessment of a Local Community Wastewater Plan for Nitrogen Removal

A Case Study Highlighting Decision-Making Tactics

J. T. Tanacredi, R. Reynolds, R. Nuzzi, and R. C. Tollefsen

7.1 ABBREVIATIONS

SWP	subwatershed wastewater plan
DGEIS	draft generic environmental impact statement
FGEIS	final generic environmental impact statement
NRE	nitrogen removal efficiency
NFG	nitrogen focus group
STE	septic tank effluent
STU	soil treatment unit
STP	sewage treatment plant
OWTS	on-site wastewater treatment system
I/A OWTS	innovative and advanced on-site wastewater treatment system
SWIS	subsurface wastewater infiltration system

7.2 OVERVIEW AND BACKGROUND

Coastal and near-shore ecosystems ecologically identified as ecotones or transition zones between unique ecological entities such as barrier island estuarine systems and the open ocean are extremely complex ecologically (Tanacredi, 2019). Human populations have dramatically increased along the coastline, with over 40% of the US population (approximately 130 million people) living on only 10% of the total land of the

 DOI: 10.1201/9781003517108-7

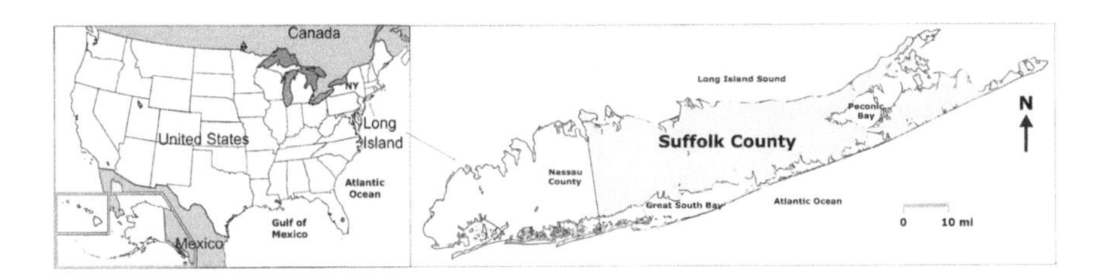

FIGURE 7.1 Map locating Suffolk County on Long Island, New York, USA.

contiguous United States. Commensurate with this increase has been a concomitant increase in built infrastructure, such as commercial, industrial, and residential development, roads, and associated infrastructure, such as wastewater treatment facilities (Tanacredi et al., 2016; Hantzsche et al., 1992). These urbanized areas have been subject to expanding urban sprawl boundaries for decades (Grimm et al., 2008). Soils associated with urban/suburban ecosystems represent one of the most important components providing essential ecosystem services, such as stormwater interception and purifications (Taka et al., 2017), groundwater recharge, wastewater nutrient cycling, and carbon sequestration (Valiela et al., 1997; Edmondson et al., 2014), which are associated with soil microbes directly linked to key soil processes, such as denitrification, and human health (Valiela et al., 1997; Li et al., 2009). Bacterial diversity and composition have been shown to be affected by soil pH, moisture, and the carbon–nitrogen ratio (Wang et al., 2018).

Compounding the urbanization of the rural fringe areas are changes related to natural processes, such as weather, sea level fluctuations, alterations of coastal morphology, and wave energy dynamics, all of which are persistent and yet incredibly difficult to forecast to any degree of certainty. Coastal and nearshore environments have unique characteristics which can only be monitored using best guesses in data acquisition. Decisions made with the aid of computer models having erroneous or insufficient data can result in unintended consequences, including undesirable ecological effects.

Approximately 25% of US homes depend on conventional gravity flow septic systems to dispose of household wastewater (USEPA, 2002); however, in Suffolk County, approximately 70% of the homes (and commercial developments) depend on conventional gravity flow septic systems. The central concept of this case study is that the final management plan targeting these septic systems in Suffolk County did not adequately analyze or recognize the physical, ecological, and social dimensions of their proposed long-term water quality improvement plan, resulting in considerable expense and little, if any, environmental benefit.

7.3 DEVELOPMENT OF THE WASTEWATER TREATMENT PLAN

In 2016, Suffolk County began developing a wastewater treatment plan known as the subwatershed wastewater plan (SWP). Based on an assumptive premise that nitrogen is the primary cause of drinking water aquifer contamination, fish kills, wetlands destruction, harmful algae blooms, and loss of the shellfish industry, Suffolk County attempted to determine nitrogen loads entering ground and surface waters.

Following the public review of a draft generic environmental impact statement (DGEIS) and issuance of a final generic environmental impact statement (FGEIS), the SWP was finalized on March 17, 2020. The plan's focus was to reduce nitrogen from wastewater to protect the environment and public health by providing an overall plan for the disposal of wastewater (sewage) generated by residences and commercial developments in the county. The SWP, relying on a series of assumptions, utilized nitrogen flow computer models to predict nitrogen loading from the 360,000 existing non–point source septic systems. While there has been much said about the 360,000 existing septic systems in Suffolk County, there is no database to indicate their status.

In the early development stages of the SWP, the county assembled a *nitrogen focus group* (NFG) with the critical task of developing the final nitrogen loading rates and efficiency factors that would be used for computer modeling, addressing nitrogen loading. The NFG was tasked to use the findings of Valiela et al. (1997) as a baseline for its model assumptions.

The NFG selected the final nitrogen "attenuation factors" significantly different from the modelling assumptions used in Valiela et al. (1997), as presented in Table 7.1. After release of the DGEIS and findings of the NFG, Reynolds et al. (2019) challenged the assumptions used in the SWP. This case study builds on that investigation focusing on the significant differences between the nitrogen attenuation factors chosen by the NFG and those originally in Valiela et al. (1997).

7.4 CASE STUDY REVIEW AND METHODOLOGY

This case study is designed to present an in-depth, multi-faceted understanding of a complex sustainability issue in a real-life context. We investigated the changes made to the nitrogen loading assumptions for non-point source on-site septic systems in the SWP and whether corrective changes were warranted. We clarified definitions for terms relating to nitrogen loading and on-site septic systems, consistent with peer-reviewed scientific engineering literature. We reviewed the process by which the changes to the nitrogen loading assumptions came about, focusing on interactions between the county, its consultants, and the SWP nitrogen focus group (NFG), who decided on the nitrogen removal efficiencies (NRE) to be used in models. We then identified and researched the various subsurface nitrogen removal zones related to on-site wastewater treatment and presented the results of our investigation, including a detailed literature search of previously documented NREs and how they compared to those chosen by the county in the SWP. Based on this investigation, we then presented conclusions and recommendations relating to NREs and on-site septic systems. This case study reveals details associated with nitrogen removal efficiencies in traditional wastewater treatment processes, which was information untapped by the county wastewater plan. The factors utilized to determine the decisions reached in the SWP were concluded to be unsustainable and ineffectual in the removal of unsubstantiated "abundant nitrogen" to groundwater and surface waters by traditional septic systems.

7.5 DEFINING NITROGEN REMOVAL TERMS FOR CONSISTENCY AND UNDERSTANDING

Wastewater environmental engineers understand that the full process of nitrogen removal and reduction in wastewater discharges from septic systems can be divided into specific subsurface zones of varying removal efficiencies, identified in Figure 7.2. Any coherent discussion of these processes requires the use of consistent and defined terms. The SWP used a myriad of terms relating to nitrogen removal and septic systems, which caused confusion and misinterpretations. These non-standardized terms included "attenuation factors," "biologically mediated natural attenuation of nitrogen," "hyporheic zone attenuation adjustment," "nitrogen reduction via OSDS," "overall attenuation from OWTS," and "percent nitrogen removal." The following definitions were identified for accuracy in this case study:

- *Septic system* (on-site wastewater treatment system [OWTS], on-site septic system, on-site sewage disposal system [OSDS], sanitary system, or conventional septic system) refers to subsurface structures consisting of a septic tank and soil treatment unit (STU), designed to receive household wastewater, provide primary and secondary treatment, and disperse effluent into the ground.
- *Soil treatment unit* (STU) refers to a subsurface wastewater treatment unit utilizing imported or native substrate (e.g., soils, sand, gravel) to treat septic tank effluent and facilitate infiltration to the surrounding subsoil, also known as leach field, leaching lines, subsurface wastewater infiltration or absorption system, leaching pool, SWIS, cluster system, NRB, or drain field (Seitzinger, 1988; Lusk et al., 2017; Reynolds et al., 2021; USEPA, 2016; Siegrist, 2014).
- *Nitrogen removal efficiency* (NRE) or percent nitrogen attenuation is the percentage of nitrogen removed from wastewater or groundwater by naturally occurring or engineered processes, which may include denitrification, DE ammonification, volatilization, cation exchange, and decomposition.
- *Overall NRE for wastewater from a septic system* (overall NRE for septic system wastewater) is the total percentage of nitrogen removed from wastewater by subsurface treatment zones, identified in Figure 7.2, including the septic tank, the STU, the vadose zone, the aquifers, the riparian zone, and the hyporheic zone, expressed as a compounded percentage.

7.6 SUBSURFACE NITROGEN TREATMENT ZONES

Wastewater disposal on land by traditional gravity flow septic systems decreases coastal nitrogen loading when compared with direct treated sewage discharge to surface waters, as significant quantities of nitrogen can be lost during transport through the soil (Christensen et al., 1990; Groffman and Tiedje, 1989; Parkin, 1987), unsaturated zone (Valiela et al., 1997; DeSimone and Howes, 1998), and aquifer (Korom, 1992; Pabich, 2001; Pabich et al., 2001; Valiela et al., 1992, 2000; Westgate et al., 2000) before being discharged to coastal surface estuarine waters (Colman et al., 2004).

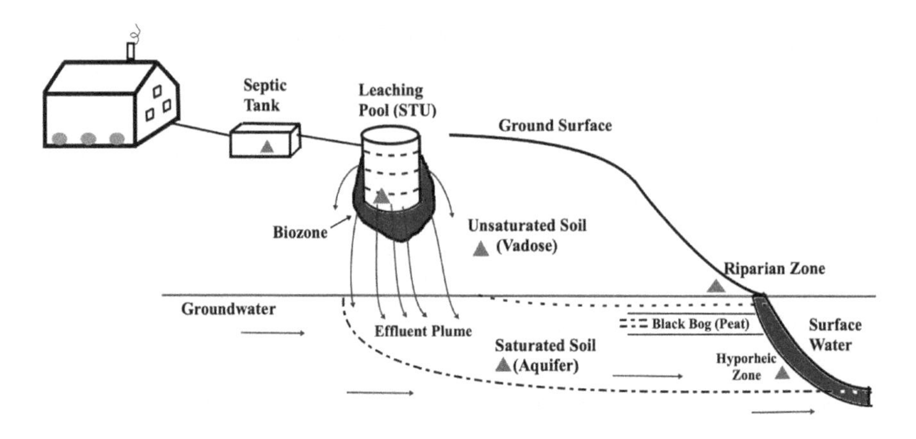

FIGURE 7.2 Subsurface septic system and naturally occurring nitrogen removal zones remove nitrogen from wastewater.

Reduction or "attenuation" of nitrogen is known to occur as wastewater effluent passes through several man-made and naturally occurring subsurface treatment zones, as exhibited in Figure 7.2.

Since the zones are in series and removal of the nitrogen has a compounding effect, each subsequent zone removes a percentage of the remaining nitrogen from the previous zone. The cumulative, subsurface removal of nitrogen from septic system wastewater can be calculated using the following equations, where *NL* is the nitrogen load of wastewater in grams, *NRE* is the nitrogen removal efficiency (expressed as a %), *s* is the septic system NRE/100, *r* is the riparian zone NRE/100, *v* is the vadose zone NRE/100, *h* is the hyporheic zone NRE/100, and *a* is the aquifer NRE/100.

$$\text{NL}\,(1-\text{s})\,(1-\text{v})\,(1-\text{a})\,(1-\text{r})\,(1-\text{h}) = \text{NL remaining} \quad (1)$$

$$\text{NL} - \text{NL remaining} = \text{NL removed} \quad (2)$$

$$(\text{NL removed}/\text{NL}) \times 100 = \textit{overall NRE for septic system wastewater} \quad (3)$$

7.7 CONVENTIONAL GRAVITY FLOW SEPTIC SYSTEMS IN SUFFOLK COUNTY, NEW YORK

A conventional septic system used in Suffolk County is simple in design and offers a cost-effective way to treat wastewater in small-town, rural, and suburban areas lacking centralized large-scale wastewater treatment plants. These systems have two main components: (1) a septic tank and (2) a soil treatment unit (STU), as depicted in Figure 7.2. Residential systems are designed based upon the number of bedrooms, soil conditions, topography, and depth to groundwater. The approximate residence time of wastewater in a septic tank is 2–3 days; this allows primary treatment with anaerobic digestion and separation of solids, grease, and other floatable solids that could cause hydraulic failure or clogging of the infiltrative surface of the STU. An STU receives and treats the effluent from a septic tank; the primary type of STUs used in Suffolk County are precast leaching pools (Suffolk County, 2015). These systems are maintained easily and do not require power to operate. Properly maintained septic systems, coupled with naturally occurring subsurface zones, can remove 75–96% of the nitrogen load of any private home on Long Island, as presented in Table 7.2.

Wastewater passing through an on-site septic system into the subsoil can move through multiple zones that potentially contribute to nitrogen removal, as depicted in Figure 7.2. While it is difficult to directly measure the nitrogen removal efficiency (NRE) of each of the 360,000 septic units in Suffolk County, there is considerable literature documenting the NRE of individual nitrogen removal zones that collectively can be utilized to predict overall NRE and nitrogen loading from septic systems.

7.8 DISCUSSION OF NITROGEN REMOVAL EFFICIENCIES (NRE)

7.8.1 Septic System NRE

A *septic system* is designed to receive household wastewater, provide primary and secondary treatment, and disperse the effluent into the ground. The term *NRE for a septic system* refers to the percentages of nitrogen removed from wastewater in the septic tank and the soil treatment unit (STU) and is expressed as a compounded percentage.

In agreement with the SWP, research has revealed that, on average, 6% nitrogen removal occurs in the septic tank, with estimates ranging from 1 to 12% (Costa et al., 2002; Darby and Leverenz, 2004; Christensen et al., 1990; Valiela et al., 1997; Stony Brook University, 2016).

The SWP did not identify any qualified studies applicable to determining nitrogen removal by leaching pools (STU). Of the 360,000 existing septic systems utilizing leaching pools in Suffolk County, none had been effectively analyzed for nitrogen removal efficiency. Assumptions were made in the model that disregarded nitrogen removal by leaching pools, resulting in a 0% NRE for a leaching pool (Suffolk County, 2020a).

Both leaching pools and leaching lines (STUs) provide nitrogen removal by discharging septic tank effluent (STE) into surrounding soil through a soil interface and biomat (biozone), and therefore, treatment by leaching lines and leaching pools can be equated as STUs. Soil pH, temperature, texture, water content, microbial community, and availability of organic C associated with biomats (Lusk et al., 2017; Reynolds et al., 2021; Siegrist, 2014; USEPA, 2002) that have been consistently observed by sanitary engineers around leaching pools were not accounted for in the SWP. There appears, therefore, to be no justification for the SWP reducing the NRE for a leaching pool to 0%.

Valiela et al. (1997) estimated the NRE basic standard for a septic system to be in the range of 38–40% (7% NRE for septic tank + 35% NRE for STU). Colman et al. (2004) estimated a 35% NRE for a septic system. Katz et al. (2010) found NREs as high as 40% for a septic system, utilizing a chloride tracer in the STU to assess nitrogen attenuation at various depths below a leaching field. They reported that dilution with groundwater accounted for about 10–25% of N losses from the septic tank effluent (STE) with depth. It was concluded by Lusk et al. (2017) that the "STU was able to remove 25–40% of the STE-borne nitrogen through de-nitrification, NH4 sorption, and NH_3 volatilization."

The incorrect assumption by the SWP that a leaching pool (STU) provides no (0%) attenuation of nitrogen was due to at least three causes:

1. The SWP lacked empirical treatment data for Suffolk County leaching pools.
2. The SWP did not investigate the treatment mechanisms of leaching pools.
3. The SWP did not consider leaching fields or leaching pools in the treatment processes.

The planning result includes neither septic system utilizing leaching fields nor leaching pools being considered as options in the SWP regional wastewater disposal strategy. The options considered are conventional sewage treatment plants (STP) that treat wastewater from multiple residential and commercial establishments and discharge the effluent into the ground, STPs that discharge directly into surface waters, and innovative and advanced on-site wastewater treatment systems (I/A OWTS). Conventional OWTS (septic systems) are not included in the strategy and are deemed unacceptable in the SWP, based on deficient nitrogen removal capability.

The FGEIS indicated that the NFG (the panel of experts) had decided that the NRE for a septic system in Suffolk County was 15 to 28%, depending on the underlying soils. This conflicted with the Valiela et al. (1997) finding of 38 to 40%. More importantly, it begs the question as to why the SWP uses a 6% NRE (septic tank NRE of 6% and the leaching pool NRE of 0%) (Suffolk County, 2020a).

The NFG misinterpreted the NRE for a septic system by including the NRE for the vadose and the aquifer and disregarding the NRE of the STU. The septic system NRE only includes the septic tank and NRE of the STU and does not include the vadose or aquifers, which are separate nitrogen removal zones (Costa et al., 2002; USEPA, 2016).

The FGEIS indicated that the NFG determination of 15 to 28% nitrogen removal in a septic system is consistent with the 25% NRE used in the Massachusetts Estuary Project (Massachusetts Department of Environmental Protection, 2012). The 25% nitrogen removal was based on a study by Costa et al. (2002) that appropriately only included nitrogen removal in the septic tank and STU to determine its finding of 25% NRE. The NFG incorrectly included (added) the nitrogen removal for the vadose and the aquifers to determine the 15 and 28% NRE for the septic system. Based on the NFG determination of 0% NRE for the leaching pool and 6% NRE for the septic tank, the NRE for the *septic system* would be 6%, not 15 or 28%. This 6% is in sharp contrast and a critically misdiagnosed value when compared to Valiela et al. (1997) and others who found the NRE for a septic system ranges from 25 to 40% (Costa et al., 2002; Katz et al., 2010; Colman et al., 2004; Andreoli et al., 1979; Valiela et al., 1997; USEPA, 2002), suggesting that an NRE of 35% for any wastewater model is appropriate for the septic system.

The NFG used the findings of DeSimone, L. A., and Howes, B. (1998), to support their conclusions about nitrogen removal by septic systems "in a similar hydrogeological environment" (Suffolk County, 2020a). *In* fact, DeSimone and Howe had previously studied the effluent from a "facility that treats septage pumped from commercial and residential septic systems" (a scavenger plant), *not a septic system.*

Effluent from a scavenger plant is not equivalent to that of a conventional septic system. It was identified that "the liquid component of the septage receives HCl and phosphoric acid addition, biological treatment to reduce oxygen demand and ultraviolet disinfection" (DeSimone and Howes, 1998). The treatment processes in a scavenger plant result in an effluent quite different from the residential wastewater typically found in a septic tank, exhibiting chloride concentrations approximately ten times higher than the average (i.e., 950 mg/L vs. 65 mg/l) (DeSimone and Howes, 1998; Xu, 2007) and nitrogen characteristics different from those found in a typical septic tank and leaching pool system. Nitrogen found in a septic tank is primarily in the form of ammonium (NH4), and when it flows into the leaching pools, it averages 60 to 75mg/l-N (Lusk et al., 2017; Xu, 2007). However, the nitrogen in the effluent from the scavenger plant was a mix of ammonium and nitrate, 27 mg/l-N and 9.4 mg/l-N, respectively, with a total nitrogen concentration of 46 mg/l-N.

The scavenger plant also used rapid infiltration beds (STU) to discharge the effluent into the underlying soil. It is important to the denitrification process that a biomat (biozone) be established at the soil interface and maintained to maximize treatment efficiency. Conventional septic systems are designed to produce this biomat (biozone) and provide sufficient time for treatment, while rapid infiltration systems are not. It is understood by professional sanitary engineers that the naturally occurring denitrification depends on a balanced wastewater and soil environment, which includes pH, bacteria, temperature, and a food source. Because the effluent from a scavenger plant is not representative of the effluent from a conventional septic system, comparisons of NREs are not valid.

7.8.2 Vadose NRE

The vadose zone is the unsaturated sediment through which wastewater effluent (leachate) travels as it passes from the STU into the aquifer, where it disperses, forming a septic plume. Valiela et al. (1997) indicated there would be a 34% reduction of nitrogen in the *septic plume* as it passed through the vadose zone and into the aquifer. The NFG did not specifically include NRE for the septic plume in the modelling assumptions but did include 10% NRE for the vadose zone, based in part on the findings of DeSimone and Howes (1998). The NFG assigned a 10% NRE for the vadose, concluding that "minimal biologically mediated natural attenuation of nitrogen occurs in Suffolk County's highly aerobic upper glacial aquifer which is consistent with DeSimone, L.A., and Howes, B., 1998" (Suffolk County, 2020a). Since nitrogen removal for a septic plume is a function of interactions in the vadose and aquifer, our assessment did not assign a specific NRE to the septic plume but considered such nitrogen removals to be incorporated in the vadose and aquifer zones.

7.8.3 Aquifer NRE

7.8.3.1 Aquifer NRE Determined by Soil Type in SWP

For the purposes of this discussion, we use the term *NRE of the aquifer* or *aquifer NRE* and define it as the percentage of nitrogen removed from groundwater in the zone of saturated sediments (aquifer). The Valiela et al. (1997) baseline NRE for the aquifer was 35%.

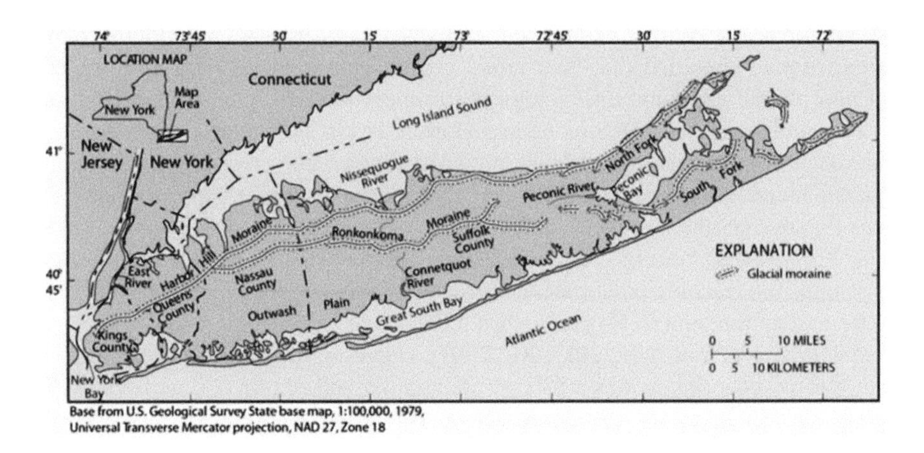

FIGURE 7.3 The figure shows glacial moraine in Suffolk County, Long Island, New York.

Depending on the location in Suffolk County, the NFG indicated two different NREs for the glacial aquifer. The NFG used the types of soils in an area ("glacial till" versus "glacial sand") to differentiate the glacial aquifer and assumed levels of nitrogen removal. The NFG assigned a higher NRE for the aquifer north of the "glacial moraine" (containing glacial till) than the aquifer south of the moraine (containing glacial sands). The general location of the glacial moraine on Long Island is depicted in Figure 7.3. The NFG assigned a 15% NRE for the aquifer north of the *glacial moraine* (generally the north shore) and assigned a 0% NRE for the aquifer south of glacial moraine (generally the south shore, known as the *glacial outwash*). Portions of the area north of the glacial moraine are underlain by "*glacial sands*," not "*glacial till*," as is the glacial outwash area (Young et al., 2013). The SWP did not provide support for differentiating aquifer NREs based upon soil types and lowering the aquifer NREs from the Valiela baseline of 35% to 0% and 15%.

7.8.3.2 Glacial Outwash Plain Assumed as 0% NRE

In general, the glacial aquifer south of the moraine was assigned a 0% NRE, and the area north of the moraine was assigned a 15% NRE, based on soil differences. The FGEIS indicated that the glacial outwash plain provided no (0%) nitrogen removal for septic system wastewater. The FGEIS stated "no denitrification through the coastal plain sediments was included" (Suffolk County, 2020a). This contradicts evidence that has consistently shown sediments, like those in Long Island's glacial outwash plain and coastal area, provide a great potential for natural denitrification (Reynolds et al., 2019; Parkin, 1987; Reynolds et al., 2021; USEPA, 2002, 2016; Bachman et al., 2000). *Denitrification* is a bacterial process that converts nitrates and nitrites to nitrogen gas. Other mechanisms for nitrogen removal (attenuation) in the north shore aquifer also affect NRE determinations for that aquifer, including cation exchange of ammonium onto negatively charged soil particles in the unsaturated or saturated zone (Hanson and Schoonen, 1999), anaerobic ammonium oxidation (Bohlke et al., 2006), and direct denitrification to nitrogen gas in the unsaturated or saturated zone (Abit et al., 2008;

Groffman et al., 2006; Hanson and Schoonen, 1999; Seitzinger et al., 2006). The glacial outwash plain of Long Island is comprised of soils like those described by the USGS (2000), which promote naturally occurring denitrification in the coastal areas.

7.8.3.3 Glacial Moraine Aquifer Assumed as 15% NRE

The NFG assumptions, lowering the NRE of the aquifers, are primarily based upon a study by Young et al. (2013) which used the ratio of nitrogen gas to argon gas in groundwater to predict 15% denitrification in the upper glacial aquifer underlying the hamlet of Northport in Suffolk County (Suffolk County, 2020a, 2020c). One of the goals of the Young et al. study "was to examine the extent of denitrification in Long Island's north shore aquifers and to determine if previously modeled 35% nitrogen attenuation is accurate in this setting" (Young et al., 2013). The study concluded, "Results show an average 15% of total nitrogen in the system was *denitrified*, significantly lower than model predictions of 35% *denitrification*." As Young et al. (2013) only addressed *denitrification*, it clearly reveals that the NFG incorrectly interpreted the 15% reduction of nitrogen to include all methods of *nitrogen attenuation* and did not evaluate other mechanisms for nitrogen removal in this zone to account for the pre-established 35% NRE. A 15% NRE for the north shore glacial aquifer should be considered representative of only a portion of the nitrogen removal that occurs there.

Further, the county depended on the findings of DeSimone and Howes (1998), determining that lowering the pre-established 35% NRE of the aquifer to 0% and 15% was *consistent* with the study's findings. The study investigated a wastewater plume located in an aquifer consisting of course-grain sediments like that of glacial outwash sediments on Long Island. Review of DeSimone and Howes (1998) reveals that this study found an 18% removal of nitrogen in the aquifer, not the 0% reported in the FGEIS. Reportedly, there was an 18% removal of nitrogen in the aquifer, primarily due to ammonium sorption. If we accept the findings as a guideline for determining NRE of the aquifers on Long Island, then the NRE for the glacial outwash aquifer would have been at least 18%, not 0%, and the NRE for the north shore aquifer would be higher than 18% (due to its reported increased clay and silt content). The NFG assigned a 0% NRE to the glacial outwash aquifer and a 15% NRE to the north shore glacial aquifer, which are not compatible with the findings of DeSimone and Howes (1998). The NFG incorrectly estimated the NRE of the aquifer north of the glacial moraine to be 15% and inappropriately predicted an increase in nitrogen loading to surface waters. The rationale for reducing the NRE for the aquifer south of the glacial moraine to 0% is not verifiable.

7.8.3.4 Aquifer NRE Reduced Based on Dilution Theory

The NFG recommended lowering the glacial aquifer NRE from 35% to as low as 0% based upon the assumption "that the 35 percent loss term (*NRE*) based on the observed decline in nitrogen concentrations moving down gradient from the OSDS (*on-site sewage disposal system*) in Valiela et al. (1997) most likely resulted from dilution, rather than nitrogen loss in the shallow aquifer" (Suffolk County, 2020a). A "most likely" finding is dramatically insufficient to justify the NFG-recommended reduction of the aquifer NRE. Valiela et al. (1997) accounted for dilution and concluded that despite dilution, nitrogen removal occurred, and their 35% NRE of the aquifer appears to have included nitrogen losses due to sediment interactions, riparian and hyporheic losses, and for the aquifer is reasonable (Valiela et al., 1997; Kinney and Valiela, 2011; Robertson et al., 1991; Colman et al., 2004).

7.8.4 Riparian and Hyporheic Zone NRE

Riparian zones are vegetated regions adjacent to streams and wetlands that can be effective at intercepting and controlling nitrogen loads entering water bodies. Wilhelm et al. (1994) noted that "[t]he most complete treatment of wastewater occurs when a natural anaerobic setting, which is capable of denitrification follows the aerobic treatment zone." The *hyporheic zone* is the area of saturated sediments within a bed of a surface water body where discharge water mixes with the surface water. As presented in Table 7.1, none of the models specified an NRE for the riparian zone, and only the SWP specified an NRE for the hyporheic zone. This lack of riparian zone NRE inclusion was most disturbing, considering the preponderance of literature reporting that it may be as high as 50–80% (Mayer et al., 2005; Clausen et al., 2000; Hanson et al., 1994; Vellidis et al., 2003).

The Suffolk County Department of Health Service's records document that the riparian zone has extensive layers of subsurface black peat (bog) throughout the outwash plain along the south shore of Long Island. Such peat layers, as depicted in Figure 7.2, provide the necessary dissolved organic carbon source that, when coupled with low dissolved oxygen (DO) levels and increased detention times (traversing peat), provide an opportunity for natural de-nitrification in coastal areas, especially in shallow groundwater areas near the shoreline, which include the riparian zones (Cooper, 1990; Mayer et al., 2005; Clausen et al., 2000; Anderson, 1998; Young, 2010; Xu, 2007; Schubert, 2010).

Mayer et al. (2005) noted that:

> Riparian buffers of various types are effective at reducing nitrogen in riparian zones, especially nitrogen flowing in the subsurface. Buffers will be most effective at controlling nitrogen through de-nitrification when 1) water flow (overland and subsurface) is evenly distributed and soil infiltration rates are high, 2) anaerobic (saturated) conditions persist in the subsurface, and 3) sufficient organic carbon is present. Soil type, subsurface hydrology (e.g., soil saturation, groundwater flow paths), and subsurface biogeochemistry (organic carbon supply, nitrate inputs) are significantly crucial factors governing nitrogen removal in buffers.

The nitrogen removal efficiency of wetland buffer zones has been estimated to be from 12 to 80% (Yates and Sheridan, 1983; Brüsch and Nilsson, 1993; Velinsky et al., 2017), and greater than 95% of nitrate can be removed within 1 m (Burns and Nguyen, 2002). Cooper (1990) found that subsurface nitrate removal from highly organic, saturated soils was ~90%. Clausen et al. (2000) observed a 49–58% reduction in subsurface nitrate concentrations (95% of all nitrate loss) across a 5 m poorly to very poorly drained alluvium wetland. Vellidis et al. (2003) observed a similar reduction in nitrate, 78%, from sandy, forested wetlands (38 m wide) (Mayer et al.,

2005). The denitrification process in shallow groundwater areas can be rapid and substantial. Increased travel distance and detention time for nitrogen in groundwater can be beneficial, but it is not always necessary (Burns and Nguyen, 2002; Cooper, 1990; Colman et al., 2004; Anderson, 1998; Young and Liu, 2010; Robertson et al., 1991). Considerable investigations (Bowen et al., 2007; Harvey et al., 1990; O'Shaughnessy, 2016; Ranalli and Macalady, 2010; Rivett et al., 2008; Slater and Capone, 1987; Zarnetske et al., 2011; USEPA, 2016) provide corroborating evidence that natural denitrification occurs in unconfined sandy aquifers as exist on Long Island, especially the south shore, as a geologically denoted paleo-glacial remnant outwash plain (Anderson, 1998; Xu, 2007; Young, 2010; Starr and Gillham, 1993; Robertson et al., 1991).

Durand (2014) reported a nitrogen reduction of 24% to 34% in a hyporheic zone on the south shore of Long Island; the SWP used 10 to 15% in its modelling.

Kinney and Valiela (2011) reported that "in all cases, estimated nitrogen loads to the subwatersheds (in the eastern United States) were considerably larger than nitrogen loads to the receiving waters." These differences ranged between 70% and 90% and point to considerable retention of nitrogen within subwatersheds. Retention of this magnitude is well within the range of other watersheds in the eastern United States and reiterates the argument that coastal watersheds, despite significant human inputs, nevertheless still manage to provide a significant water quality subsidy protecting receiving waters. In addition, "the watershed retention of nitrogen comes largely via accretion in forests and soils and via denitrification and absorption within soils and aquifers" (Bowen et al., 2007). Identification of 70% to 90% overall NRE is counterintuitive to the SWP projection of 28–39% NRE for septic system wastewater in Suffolk County.

7.9 IMPACT OF MISDIAGNOSED NRE ON WASTEWATER STRATEGY

As previously discussed, based on misdiagnosed NREs, neither septic system utilizing leaching fields nor leaching pools were considered as options in the SWP regional wastewater disposal strategy, thereby committing the county to the replacement of 360,000 conventional septic systems, based upon the unsubstantiated premise that such an effort will result in a positive environmental end point. The decision to eliminate conventional septic systems ignores the previous county strategy of limiting the number of septic systems allowed per acre of land (i.e., limiting wastewater nitrogen loading per acre) and the positive environmental advantages provided by utilization of conventional septic systems. Suffolk County (2015) found that the residential areas in Suffolk that complied with the septic system restriction limiting the number of septic systems per acre met target nitrate concentrations in underlying groundwater wells. Disregarding this septic system strategy was not supported by the SWP. Conventional systems are water conservation devices by design and, as this case study shows, have a significant potential for the sustainable treatment of wastewater, especially when coupled with the naturally occurring subsurface treatment zones. The unilateral elimination of cost-effective conventional septic systems results in reliance on expensive, high-maintenance treatment alternatives with subsurface electric power needs replacing gravity flow systems not requiring any power to operate (i.e., STP and I/A OWTS), which can create problems, economically, socially, and environmentally, where power is lost due to extreme weather events (Suffolk County 2020a, 2020c; LICAP, 2017).

7.10 RESULTS

TABLE 7.1
A Comparison of Nitrogen Removal Efficiencies (NRE) for Selected Reports and Studies

Zones of Nitrogen Removal and Treatment (a)	2016 USEPA Chesapeake Bay NRE (f)	2016 Eastern Bays Report NRE		2016 SWP Baseline Valiela (1997) NRE (e)		2020 SWP Final N Focus Group NRE	
Septic tank	(i)	7%		**6%**		**6%**	
Soil treatment unit (STU)	20% (OWTS)	35%	40% (OWTS)	**34%**	38% (OWTS)	**0%** (excluded)	6% (OWTS)
Vadose		Unspecified		Unspecified		**10%**	
Septic plume	60%	35%		**35%**		**0%** (excluded)	
Aquifer		15%		**35%**		**0%** outwash area **15%** moraine area (b)	
Riparian zone (g)		Unspecified		Unspecified		**0%** (excluded)	
Hyporheic zone (h)		Unspecified		Unspecified		**10–15%** (used 15%) (d)	
Overall NRE for septic system wastewater (c)	**68%**	**66%**		**Valiela 74%**		**SWP 28–39%** (outwash vs. moraine)	

Notes:

a. Zones used are derived from SBU (2016; page 43) and Valiela et al. (1997).
b. SWP specifies 0% NRE in the glacial outwash aquifer (generally south shore) and 15% in aquifer north of glacial moraine (generally north shore) (Suffolk County, 2020a, 2020c).
c. N reductions have been compounded as per Suffolk County correspondence.
d. In lieu of empirical data for hyporheic denitrification, SWP used estimates from a wetlands study (Hamersley, 2001) to estimate the denitrification by hyporheic zones in Suffolk (Suffolk County, 2020c; Section 2.1.5.1.5). We used the 15% estimate for the purposes of calculations in this comparison.
e. Based on Figure 3 in CDM Smith QAPP report, June 19, 2017, Final SWP Appendices (Suffolk County, 2020c), and Valiela et al. (1997).
f. Based on comments by USEPA to nitrogen focus group, August 24, 2016 (Suffolk County, 2020c; Final Appendices), and USEPA (2016).
g. NRE estimates for the riparian zone were not specified or were excluded.
h. NRE estimates for the hyporheic zone were not specified, except in SWP.
i. NRE for OWTS (on-site wastewater treatment system or septic system) includes compounded NRE for septic tank and STU.

TABLE 7.2
A Comparison of Nitrogen Removal Efficiencies (NRE) Using Findings from Peer-Reviewed Literature and the 2020 SWP

Zones of Nitrogen Removal and Treatment (a)	Peer-Reviewed NRE Range (g)	2020 SWP NRE
Septic system (septic tank + STU)	20–40% (h)	6%
Vadose	Not determined	10%
Aquifer (i)	15–75%	0% for outwash area, 15% for moraine area (b)
Riparian zone (e)	49–78%	Not included
Hyporheic zone (f)	24–50%	10–15% (d)
Overall NRE for septic system wastewater (compounded) (c)	**74–98%**	**28–39%** (outwash vs. moraine)

a. Zones used are derived from SBU (2016) (page 43) and Valiela et al. (1997), with riparian and hyporheic zones added.
b. SWP specifies 0% NRE in the glacial outwash aquifer (generally south shore) and 15% in aquifer north of glacial moraine aquifer (generally north shore) (Suffolk County, 2020a, 2020c).
c. Reductions have been compounded as per Suffolk County correspondence.
d. In lieu of empirical data for hyporheic denitrification, SWP used estimates from a wetlands study (Hamersley, 2001) to estimate the denitrification by hyporheic zones to be 10–15% in Suffolk County (Suffolk County, 2020c, Section 2.1.5.1.5). We used the higher 15% estimate for the purposes of calculations in this comparison.
e. The riparian range is based upon discussions in this report, including references to Mayer et al. (2005), Clausen et al. (2000), Hanson et al. (1994), and Vellidis et al. (2003).
f. The hyporheic range is based upon discussions in this report, including references to Durand J. (2014), Hamersley (2001), Seitzinger (1988), and Smith (1999)
g. The septic system range is based upon discussions in this report, including references to Katz et al. (2010), Costa et al. (2002), Colman et al. (2004), Andreoli et al. (1979), Valiela et al. (1997), USEPA (2002), and USEPA (2016).
h. NREs for septic tanks and STU have been compounded and presented as septic system NRE (%).
i. DeSimone and Howes, 1998; Young et al., 2013; Valiela et al., 1997; Colman et al., 2004; USEPA NFG Comments, Suffolk County, 2020c; Kinney and Valiela (2011); Robertson et al., 1991; Suffolk County, 2020a, 2020c; USEPA, 2016.

7.11 CONCLUSION

This case study provides a critical assessment and detailed review of the SWP and FGEIS used as a paradigm of the large-scale infrastructure planning, attempting to initiate a local government replacement plan of 360,000 traditional gravity flow septic systems under the unsubstantiated premise that such an effort will result in a positive environmental end point. It suggests that the final plan underestimated the nitrogen-reduction ability of traditional gravity flow septic systems, resulting in overestimated projections of nitrogen loading from those systems. If so, it may result in unnecessary management actions that are environmentally unsound and can result in potentially negative long-term socioeconomic impacts.

In summary, the following facts suggest that the pollution attributed to septic discharges from traditional existing residential gravity flow septic systems is unjustified.

1. The SWP lowered the NRE for a septic system to 15% without any experimental data substantiating such action. No research on existing septic systems that utilize leaching pools (STU) was performed.
2. The changes to the nitrogen removal efficiencies in the SWP, including the use of a 15% *NRE for septic systems* and a 27% *overall NRE for septic systems wastewater*, resulted in a doubling of the projected nitrogen load to surface waters. No substantive scientific evidence exists to support the use of such low nitrogen removal efficiencies.
3. The final NREs used in the SWP were based upon "*discussions and deliberations*" by a nitrogen focus group (NFG), which did not get adequately substantiated in the SWP, the NRE's used in planning models, and public documents.
4. DeSimone and Howe (1998), a study of a "scavenger waste treatment plant," is inappropriate for the substantiation of the ability of septic systems to treat wastewater or determine the NRE of an STU or the underlying vadose zone.
5. No reason was provided for changing the original 35% *NRE for the aquifer* presented in Valiela et al. (1997), which was supported and promoted by The Nature Conservancy (2014), to 0% *NRE for the aquifer* in the "outwash plain area." Contrary to a 0% NRE, studies have shown that sediments, like those in Long Island's outwash plain and coastal area, provide for natural nitrogen removal. The original 35% *NRE for the aquifer* would be substantially higher if the riparian zone is included as part of the *NRE for the aquifer.*

Tables 7.1 and 7.2 exhibit that the SWP significantly underestimated NRE while simultaneously overestimating the nitrogen loading to Suffolk County surface waters from traditional gravity flow septic systems.

The most effective corrective action is the re-evaluation of all nitrogen removal efficiencies used in the SWP. The models used for the SWP require more accurate nitrogen loading assumptions.

Plans to upgrade existing residential septic systems must be based on scientific investigation into problems associated with waste disposal. At present, the education of homeowners as to the proper care of traditional gravity flow systems, replacement or upgrade of failing systems, and continuing research would appear to be the most prudent path.

On-site septic systems utilizing leaching fields or leaching pools should have been considered as options in the Suffolk County SWP regional wastewater disposal strategy, based upon the information presented in this case study.

Improving water quality must always be determined by accurate and appropriately applied science and engineering (Tanacredi et al., 2016).

7.11.1 Compliance with Ethical Standards

Conflict of Interest: All authors in this chapter have no conflict of interest.

Ethical Approval: This chapter does not contain any studies with human participants or animals performed by any of the authors.

7.11.2 Conflict of Interest Statement

On behalf of all authors, the corresponding author states that there is no conflict of interest.

REFERENCES

Abit, S.M., Amoozegar, A., Vepraskas, M.J., and Niewoehner, C.P. 2008. Fate of nitrate in the capillary fringe and shallow groundwater in a drained sandy soil. *Geoderma* 146:209–215. doi:10.1016/j.geoderma.2008.05.015

Anderson, D. 1998, March 8–10. Natural denitrification in groundwater impacted by onsite wastewater treatment systems, in *Eighth National Symposium on Individual and Small Community Sewage Systems*, Orlando, FL. https://eurekamag.com/research/003/209/003209783.php

Andreoli, A., Bartilucci, N., Forgione, R., and Reynolds, R. 1979. Nitrogen removal in a subsurface disposal system. *Journal Water Pollution Control Federation* 51(4):841–854. https://drive.google.com/file/d/1PzC8jNJ3nZhvuaEKrWcmBdWwHQNxO3ZR/view?usp=sharing

Bachman, L., and Krantz, D. 2000, April. *The Potential for Denitrification of Groundwater by Coastal Plain Sediments in the Patuxent River Basin, Maryland*, USGS Fact Sheet FS-053–00. https://pubs.usgs.gov/fs/fs05300/

Bohlke, J.K., Smith, R.L., and Miller, D.N. 2006. Ammonium transport and reaction in contaminated groundwater: application of isotope tracers and isotope fractionation studies. *Water Resources Research* 42. https://agupubs.onlinelibrary.wiley.com/doi/full/10.1029/2005WR004349

Bowen, J.L., Ramstack, J.M., Mazzilli, S., and Valiela, I. 2007. NLOAD: an interactive, web-based modeling tool for nitrogen management in estuaries. *Ecological Applications* 17:S17–S30. https://doi.org/10.1890/05-1460.1

Brüsch, W., and Nilsson, B. 1993. Nitrate transformation and water movement in a wetland area. *Hydrobiologia* 251:103–111. https://link.springer.com/article/10.1007/BF00007170

Burns, D.A., and Nguyen, L. 2002. Nitrate movement and removal along a shallow groundwater flow path in a riparian wetland within a sheep-grazed pastoral catchment: results of a tracer study. *New Zealand Journal of Marine and Freshwater Research* 36:371–385. https://www.tandfonline.com/doi/abs/10.1080/00288330.2002.9517094

Christensen, S., Simkins, S., and Tiedje, J.M. 1990. Spatial variation in denitrification: dependency of activity centers on the soil environment. *Journal of the Soil Science Society of America* 54:1608–1613. https://doi.org/10.2136/sssaj1990.03615995005400060016x

Clausen, J.C., Guillard, K., Sigmund, C.M., and Martin Dors, K. 2000. Water quality changes from riparian buffer restoration in Connecticut. *Journal of Environmental Quality* 29:1751–1761. https://doi.org/10.2134/jeq2000.00472425002900060004x

Colman, J., Masterson, J., Pabich, W., and Walter, D. 2004. Effects of aquifer travel time on nitrogen transport to a coastal embayment. *Groundwater Oceans Issue* 42(7):1069–1078. https://doi.org/10.1111/j.1745-6584.2004.tb02644.x

Cooper, A.B. 1990. Nitrate depletion in the riparian zone and stream channel of a small headwater catchment. *Hydrobiologia* 202:13–26. https://link.springer.com/article/10.1007/BF02208124

Costa, J.E., Heufelder, G., Foss, S., Milham, N., and Howes, B. 2002. Nitrogen removal efficiencies of three alternative septic system technologies and a conventional septic system. *Environment Cape Cod* 5(1). https://www.semanticscholar.org/paper/Nitrogen-Removal-Efficiencies-of-Three-Alternative-Costa-Heufelder/1bdec00a5e3be2d200a60a2cd9333f56860d32c1

Darby, J.L., and Leverenz, H. 2004. *Virus, Phosphorus, and Nitrogen Removal in Onsite Wastewater Treatment Processes*. UC Water Resources Center Technical Completion Report, Project No. W-953. https://escholarship.org/uc/item/9rc0j398

DeSimone, L., and Howes, B. 1998. Nitrogen transport and transformations in a shallow aquifer receiving wastewater discharge: a mass balance approach. *Water Resources Research* 34(2):271–285. https://doi.org/10.1029/97WR03040

Durand, J. 2014. Characterization of the spatial and temporal variations of submarine groundwater discharge using electrical resistivity and seepage measurements. *Environmental Science, Geology*. https://www.semanticscholar.org/paper/Characterization-of-the-Spatial-and-Temporal-of-and-Durand/a43e15c92d3289a99f1ac333560366ebcfee65e9

Edmondson, J.L., Dories, Z.G., McCormack, S.A., Gaston, K.J., and Leake, J.R. 2014. Land-cover effects on soil organic carbon stocks in European City. *Science of the Total Environment* 472:444–453. https://doi.org/10.1016/j.scitotenv.2013.11.025

Grimm, N.B., Faeth, S.H., Golubiewski, N.E., Redman, C.L., Wu, J., Bai, X., and Briggs, J.M. 2008. Global change and the ecology of cities. *Science* 319(5864):756–760. https://doi.org/10.1126/science.1150195

Groffman, P.M., Altabet, M.A., Bohlke, J.K., Butterbach-Bahl, K., David, M.B., Firestone, M.K., Giblin, A.E., Kana, T.M., Nielsen, L.P., and Voytek, M.A. 2006. Methods for measuring denitrification: diverse approaches to a difficult problem. *Ecological Applications* 16:2091–2122. https://doi.org/10.1890/1051-0761(2006)016[2091:mfmdda]2.0.co;2

Groffman, P.M., and Tiedje, J.M. 1989. Denitrification in north temperate forest soils: Spatial and temporal patterns at the landscape and seasonal scales. *Soil Biology Biochemistry* 21:613–620. https://doi.org/10.1016/0038-0717(89)90053-9

Hamersley, M.R. 2001. *The Role of Denitrification in the Nitrogen Cycle of New England Salt Marshes*, PhD dissertation. https://apps.dtic.mil/sti/pdfs/ADA405876.pdf

Hanson, G., and Schoonen, M. 1999. *A Geochemical Study of the Effects of Land Use on Nitrate Contamination in the Long Island Aquifer System*. http://pbisotopes.ess.sunysb.edu/reports/nitrate/ (accessed July 2011).

Hanson, G.C., Groffman, P.M., and Gold, A.J. 1994. Symptoms of nitrogen saturation in a riparian wetland. *Ecological Applications* 4:750–756. https://digitalcommons.uri.edu/cgi/viewcontent.cgi?article=1067&context=nrs_facpubs

Hantzsche, N., and Finnemore, E. 1992. Predicting ground-water nitrate-nitrogen impacts. *Groundwater* 30(4):490–499. https://doi.org/10.1111/j.1745-6584.1992.tb01524.x

Harvey, J.W., and Odum, W.E. 1990. The influence of tidal marshes on upland groundwater discharge to estuaries. *Biogeochemistry* 10:217–236. https://doi.org/10.1007/bf00003145

Katz, B.G., Griffin, D.W., McMahon, P.B., Harden, H.S., Wade, E., Hicks, R.W, and Chanton, J.P. 2010. Fate of effluent-borne contaminants beneath septic tank drain fields overlying a Karst Aquifer. *Journal of Environmental Quality* 39(4):1181–1195. http://dx.doi.org/10.2134/jeq2009.0244

Kinney, E., and Valiela, I. 2011. Nitrogen loading to Great South Bay: Land use, sources, retention, and transport from land to bay. *Journal of Coastal Research* 672–686. https://doi.org/10.2112/JCOASTRES-D-09-00098.1

Korom, S.F. 1992. Natural denitrification in the saturated zone: a review. *Water Resources Research* 28:1657–1668. https://doi.org/10.1029/92WR00252

Li, Y.T., Rowland, C., Benedetti, M., Li, F., Pando, A., Lavell, P., and Dai, J. 2009. Microbial biomass, enzyme and mineralization activity in relation to soil organic C, N and P turnover influenced by acid metal stress. *Soil Biology and Biochemistry* 41(5):969–977. https://doi.org/10.1016/j.soilbio.2009.01.021

Long Island Commission for Aquifer Protection. 2017. *Groundwater Resources Protection Plan.* LICAP. https://drive.google.com/file/d/1PwwMpnwUTRMHILxIIETGwHluj-FiRUUEEM/view?usp=sharing

Lusk, M.G., Toor, G.S., Yanga, Y., Mechtensimera, S., Dea, M., and Obrezac, T.A. 2017. A review of the fate and transport of nitrogen, phosphorus, pathogens, and trace organic chemicals in septic systems. *Critical Reviews in Environmental Science and Technology* 47(7):455–541. https://doi.org/10.1080/10643389.2017.1327787

Massachusetts Department of Environmental Protection. 2012. *Massachusetts Estuaries Project, Linked Watershed-Embayment Approach to Determine Critical Nitrogen Loading Thresholds for the Waquoit Bay and Eel Pond Embayment System Towns of Falmouth and Mashpee, Massachusetts.* https://www.falmouthma.gov/DocumentCenter/View/1128/Full-Waquoit-Bay-MEP-Report-PDF

Mayer, P.M., Reynolds, S., Canfield, T., and McCutchen, M. 2005, October. *Riparian Buffer Width, Vegetative Cover, and Nitrogen Removal Effectiveness: A Review of Current Science and Regulations*, EPA/600/R-05/118. https://cfpub.epa.gov/si/si_public_record_report.cfm?Lab=NRMRL&dirEntryId=140503

The Nature Conservancy. 2014. *Nitrogen Load Modeling to Forty-Three Sub Watersheds of the Peconic Estuary.* PEP. https://www.peconicestuary.org/wp-content/uploads/2017/06/Nitrogenloadmodelingtoforty-thr.pdf

O'Shaughnessy, M. 2016. *Mainstream DE Ammonification – WERF Report (INFR 6R11).* Water Environment Research Foundation, Alexandria, VA. https://cfpub.epa.gov/si/si_public_record_report.cfm?Lab=NRMRL&dirEntryId=314210

Pabich, W. 2001. *Denitrification of Anthropogenic Nitrogen in Groundwater: Measurement and Modeling Using Stable Isotopic and Mass Balance Approaches*, Ph.D. thesis. Massachusetts Institute of Technology, Department of Civil and Environmental Engineering, Cambridge, MA. https://www.semanticscholar.org/paper/Denitrification-of-anthropogenic-nitrogen-in-%3A-and-Pabich/f0ddaac904658814d60a8c49b378d6d1ec3f4694

Pabich, W.J., Hemond, H.F., and Valiela, I. 2001. Denitrification rates in groundwater, Cape Cod, USA: control by nitrate and DOC concentrations. *Biogeochemistry* 55:247–268. https://link.springer.com/article/10.1023/A:1011842918260

Parkin, T.B. 1987. Soil microsites as a source of denitrification variability. *Soil Science Society of America Journal* 51:1194–1199. https://doi.org/10.2136/sssaj1987.03615995005100050019x

Ranalli, A., and Macalady, D. 2010. The importance of the Riparian zone and in-stream processes in nitrate attenuation in undisturbed and aquacultural watersheds – a- review. *Journal of Hydrology* 389:406–415. https://doi.org/10.1016/j.jhydrol.2010.05.045

Reynolds, R., Akras, P., Costa, S., Mattice, J., Nuzzi, R., Tanacredi, J., Tollefsen, R., and Trent, M. 2019, September. *A Review of the Draft Generic Environmental Impact Statement for the Suffolk County Subwatersheds Wastewater Plan, SWP Appendix D.* https://suffolkcountyny.gov/Portals/0/formsdocs/planning/CEQ/2020/Appendix%20D_SC%20SWP_DGEIS_%20Public%20Written%20Comments_022020.pdf

Reynolds, R., Nuzzi, R., and Tanacredi, J. 2021. Discussion of microbial communities in partially and fully treated effluent of the nitrogen-removing biofilters by Langlois et al. *Journal of Sustainable Water in the Built Environment.* https://doi.org/10.1061/JSWBAY.0000; https://ascelibrary.org/doi/abs/10.1061/JSWBAY.0000947

Rivett, M., Buss, S.R., Morgan, P., Smith, J.W., and Bemment, C.D. 2008. Nitrate attenuation in groundwater: a review of biogeochemical controlling processes. *Water Research* 42:4215–4232. https://doi.org/10.1016/j.watres.2008.07.020

Robertson, W.D., Cherry, J.A., and Sudicky, E.A. 1991. Groundwater contamination from two small septic systems on sand aquifers. *Groundwater* 29(1). https://www.semanticscholar.org/paper/Ground%E2%80%90Water-Contamination-from-Two-Small-Septic-on-Robertson-Cherry/59280faad950585ecc340a3f2deb5c6d1667214f

Schubert, C.E. 2010. *Analysis of the Shallow Groundwater Flow System at Fire Island National Seashore, Suffolk County, New York*. U.S. Geological Survey Scientific Investigations Report 2009–5259, p. 106. https://pubs.usgs.gov/sir/2009/5259/

Seitzinger, S. 1988. Denitrification in freshwater and coastal marine ecosystems: ecological and geochemical significance. *American Society of Limnology and Oceanography* https://doi.org/10.4319/lo.1988.33.4part2.0702

Seitzinger, S., Harrison, J.A., Bohlke, J.K., Bouwman, A.F., Lowrance, R., Peterson, B., Tobias, C., and Van Drecht, G. 2006. Denitrification across landscapes and waterscapes: a synthesis. *Ecological Applications* 16:2064–2090. http://dx.doi.org/10.1890/1051-0761(2006)016[2064:DALAWA]2.0.CO;2

Siegrist, R.L. 2014, April 7–8. Engineering design of a modern soil treatment unit, in *Soil Science Society of America Conference*, Albuquerque, NM. pp. 13–26.

Slater, J., and Capone, D. 1987. Denitrification in aquifer soil and nearshore marine sediments influenced by groundwater nitrate. *Applied and Environmental Microbiology*. http://dx.doi.org/10.1128/AEM.53.6.1292-1297.1987

Smith, K.N. 1999. *The Role of Denitrification in the Nitrogen Cycle of New England Salt Marshes*. Mashapaquit Marsh, West Falmouth, MA. https://doi.org/10.1575/1912/2748; https://hdl.handle.net/1912/2748

Starr, R., and Gillham, R. 1993. Denitrification and organic carbon availability in two aquifers. *Ground Water* 31(6). https://doi.org/10.1111/j.1745-6584.1993.tb00867.x

Stony Brook University. 2016. *Long Island South Shore Estuary Reserve Eastern Bays Project: Nitrogen Loading, Sources and Management Options*. SBU, School of Marine and Atmospheric Sciences. https://dos.ny.gov/system/files/documents/2020/04/finalreporteasternbaysnitrogenloadingsourcesandmgmtoptions.pdf

Suffolk County. 2015. *Suffolk County Comprehensive Water Resources Management Plan (SCCWRMP)*. https://www.suffolkcountyny.gov/Departments/Health-Services/Environmental-Quality/Water-Resources/Comprehensive-Water-Resources-Management-Plan

Suffolk County. 2020a. *Final Generic Environmental Impact Statement (FGEIS). 2020 Suffolk County Sub Watershed Wastewater Plan*. https://www.suffolkcountyny.gov/Portals/0/formsdocs/planning/CEQ/2020/Suffolk%20County_SWP_Final%20GEIS_02262020.pdf

Suffolk County. 2020c. *Final Sub Watershed Wastewater Plan*. Suffolk County Department of Health Services. https://www.suffolkcountyny.gov/Portals/0/formsdocs/planning/CEQ/2020/RevisedComplete%20SWP2-21-20.pdf

Taka, M., Kokkonen, T., Kuoppamäki, K., Niemi, T., Sillanpää, N., Valtanen, M., Warsta, L., and Setälä, H. 2017, April 15. Spatio-temporal patterns of major ions in urban stormwater under cold climate. *Hydrological Processes* 31(8):1564–1577. https://doi.org/10.1002/hyp.11126

Tanacredi, J.T. (2019). *The Redesigned Earth: An Introduction to Ecology for Engineers*. Springer Nature Switzerland AG. ISBN 978-3-030-31237-4 (eBook) and ISBN 978-3-030-31235-0; p. 259.

Tanacredi, J.T., Schreibman, M.P., and McDonnell, K. 2016. Questioning ecosystem assessment and restoration practices in a major urban estuary: perpetuating myths of degradation despite the facts. *Journal of Environmental Science and Engineering B* 5:78–108. David Publishing. https://digitalcommons.molloy.edu/cgi/viewcontent.cgi?referer=&httpsredir=1&article=1000&context=cercom_fac

USEPA. 2002, February. *Onsite Wastewater Treatment Design Manual.* EPA/625/R-00/008/. https://cfpub.epa.gov/si/si_public_record_Report.cfm?Lab=NRMRL&dirEntryID=55133

USEPA. 2016. *Nutrient Attenuation in Chesapeake Bay Watershed Onsite Wastewater Treatment Systems – Final Report.* Tetra Tech. https://docs.google.com/document/d/1UzForoSrM624tJanvnnAkJ1R19sM2LhYbxviulrMkm8/edit

USGS. 2000. *Potential for Denitrification of Groundwater by Coastal Plain Sediments in the Patuxent River Basin, Maryland.* https://pubs.usgs.gov/fs/fs05300/

Valiela, I., Collins, G., Kremer, J., Lajtha, K., Geist, M., Seely, B., Brawley, J., and Sham, C.H. 1997. Nitrogen loading from coastal watersheds to receiving estuaries: New method and application. *Ecological Applications* 7(2):358–380. Ecological Society of America. https://doi.org/10.1890/1051-0761(1997)007[0358:NLFCWT]2.0.CO;2

Valiela, I., Foreman, K., LaMontagne, M., Hersh, D., Costa, J., Peckol, P., DeMeo-Anderson, B., D'Avanzo, C., Babione, M., Sham, C.H., Brawley, J., and Lajtha, K. 1992. Couplings of watersheds and coastal waters: sources and consequences of nutrient enrichment in Waquoit Bay, Massachusetts. *Estuaries* 15:443–457. http://www.edc.uri.edu/nrs/classes/nrs555/assets/readings_2017/Valiela_etal_1992_nutrients_Estuaries.pdf

Valiela, I., Geist, M., McClelland, J., and Tomasky, G. 2000. Nitrogen loading from watersheds to estuaries: verification of the Waquoit Bay nitrogen loading model. *Biogeochemistry* 49:277–293. https://www.jstor.org/stable/1469621

Velinsky, D.J., Paudel, B., Quirk, T., Piehler, M., and Smyth, A. 2017. Salt marsh denitrification provides a significant nitrogen sink in Barnegat Bay, New Jersey. *Journal of Coastal Res*earch 78(10078):70–78. https://doi.org/10.2112/SI78-007.1

Vellidis, G., Lowrance, R., Gay, P., and Hubbard, R.K. 2003. Nutrient transport in a restored riparian wetland. *Journal of Environmental Quality* 32:711–726. https://doi.org/10.2134/jeq2003.7110

Wang, X., Wu, J., and Krimari, D. 2018. Composition and functional genes analysis of bacterial communities from urban parks of Shanghai, China their role in ecosystem functionality. *Landscaping and Urban Planning* 177(5):83–91. https://doi.org/10.1016/j.landurbplan.2018.05.003

Westgate, E., Kroeger, K.D., Pabich, W.J., and Valiela, I. 2000. Fate of anthropogenic nitrogen in a nearshore Cape Cod aquifer. *The Biological Bulletin* 199:221–223. [Westgate].

Wilhelm, S.R., Schiff, S.L., and Cherry, J.A. 1994. Biogeochemical evolution of domestic wastewater in septic systems: 1 conceptual model. *Groundwater* 32(6). https://doi.org/10.1111/j.1745-6584.1994.tb00930.x

Xu, X. 2007. *Chemical Signature of a Sewage Plume for a Cesspool, Long Island, New York*, SBU dissertation. https://www.semanticscholar.org/paper/Chemical-Signature-of-a-Sewage-Plume-From-a-Long-Xu/b0b6636c8d9515267fe7118043db8c2af3beb55f

Yates, P., and Sheridan, J.M. 1983. Estimating the effectiveness of vegetated floodplains/wetlands as nitrate-nitrite and orthophosphoric filters. *Agriculture, Ecosystems and Environment* 9:303–314.

Young, C. 2010. *Extent of Denitrification in Northport Groundwater*, SBU thesis. https://www.geo.sunysb.edu/reports/young-ms.pdf

Young, C., Kroeger, K., and Hanson, G. 2013. Limited denitrification in glacial deposit aquifers having thick unsaturated zones (Long Island, USA). *Hydrology Journal* 21:1773–1786. [Young Report]. https://link.springer.com/article/10.1007%2Fs10040-013-1038-4
Young, C., and Liu, F. 2010. *Nitrogen Loss in a Barrier Island Shallow Aquifer System.* SBU. https://pbisotopes.ess.sunysb.edu/lig/Conferences/abstracts11/young-liu.pdf
Zarnetske, J.P., Haggerty, R., Wondzell, S.M., and Baker, M.A. 2011. Labile dissolved organic carbon supply limits hyporheic denitrification. *Journal of Geophysical Research* 116:G04036. https://doi.org/10.1029/2011JG001730

8 Bioremediation by Using Green Nanotechnologies

A Novel Biological Approach for Environmental Cleanup

Maheswari Behera, Sunanda Mishra, Prateek Ranjan Behera, Biswajita Pradhan, Debasis Dash, and Lakshmi Singh

8.1 INTRODUCTION

In the 21st century, all human beings are more or less familiar with terms like *nano-science*, *nanotechnology*, *nanomaterials*, or *nano-chemistry*. Among the rapidly developing fields of technology, nanotechnology is one. It requires fabricating and utilizing nanoparticles, which significantly modify metal or alloy properties (Kavitha et al., 2013). The word *nano* means "dwarf" or "extremely little" (10^{-9}). Richard Feynman, the eminent physicist and Nobel laureate, delivered a talk in 1959 that laid the conceptual foundations of nanotechnology. Nanotechnology is generally concerned with pharmacology, atomic physics, metallurgy, biology, and engineering (Amudha & Shanmugasundaram, 2014; Chen et al., 2004).

Nowadays, it has seen a rise in enthusiasm in the chemical, electronic, and biological sciences about green procedures to produce various nanoparticles, which has led to progress in the domain of nanotechnology (Singh & Walker, 2006). There may be several advantages for the planet that nanotechnology can provide. There are three distinct approaches to this problem: pollution detection and sensing, its treatment, and its prevention. Site cleanup and water purification are the main topics of these nanotech discussions. Several nanotechnologies are now being developed for air remediation in addition to their uses for soil, groundwater, and wastewater treatment. The reduction in particle size makes it possible to create more compact sensors, which, in turn, facilitates their deployment in difficult-to-access areas. Progress in nanotechnology's ability to decrease pollution is underway, and it can spur the most stringent improvements in the environmental profession to date (Zhang, 2003; Singh, 2009). Among the many uses for nanotechnology, wastewater treatment utilizing nanoparticles is a particular area of focus (Kowshik et al., 2002). One of the speedily growing

DOI: 10.1201/9781003517108-8

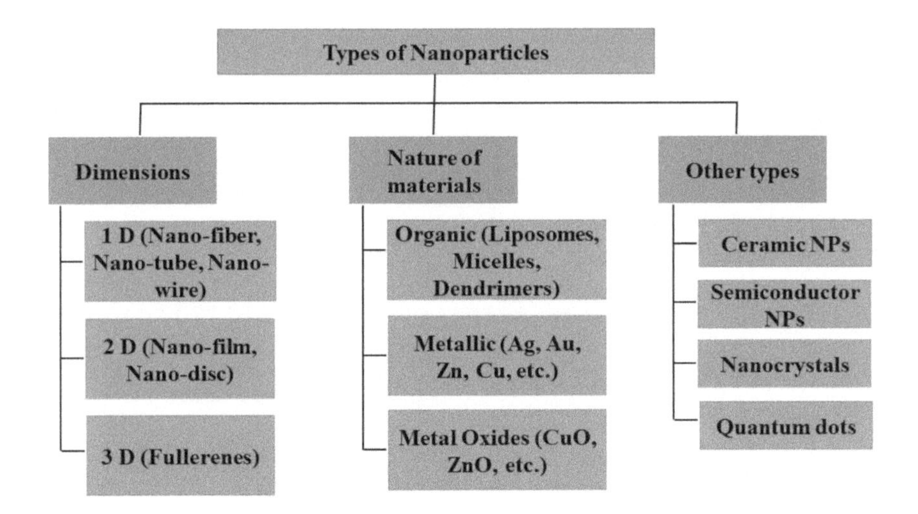

FIGURE 8.1 Illustrative diagrams depicting the various types of nanoparticles according to their size, shape, and chemical characteristics.

subfields in green nanotechnology is the study of how microorganisms can be used in the manufacture of nanoparticles. Every day, novel biological agents are employed to produce nanoparticles, offering a reliable alternative to conventional chemical and physical techniques. By fine-tuning the procedures, nanoparticles of the required morphology and size can be produced in a brief period with minimal contamination. In today's global economy, nanotechnologies have become ubiquitous as a means of addressing pressing problems. Improved comprehension of innovations based on nanotechnology requires the creation of new ways to evaluate progress (Balaji et al., 2009; Wu et al., 2019; Chatterjee et al., 2020; Jadoun et al., 2021).

Hence, this chapter examines the possibilities of nanotechnology in the bioremediation sector. A variety of green nano-material classes and their potential implementations in heavy metal withdrawal and wastewater treatment were also described.

8.2 NANOPARTICLES AND THEIR CLASSIFICATIONS

Nanoparticles are the foundational constituent of nanotechnology. These are the particles of size 1–100 nm. Based on their size, chemical characteristics, and shape, nanoparticles are typically divided into various categories. Some of the classes of NPs are multidimensional, inorganic, organic nanoparticles, etc. (Shin et al., 2016). However, metal-based nanoparticles such as silver, aluminum, cobalt, zinc, cadmium, gold, copper, iron, and lead are the most extensively employed because of their specific chemical, physical, and optical properties (Sigmund et al., 2006; Sun et al., 2008). Figure 8.1 depicts the different types of nanoparticles classified based on their dimensions and the nature of the materials.

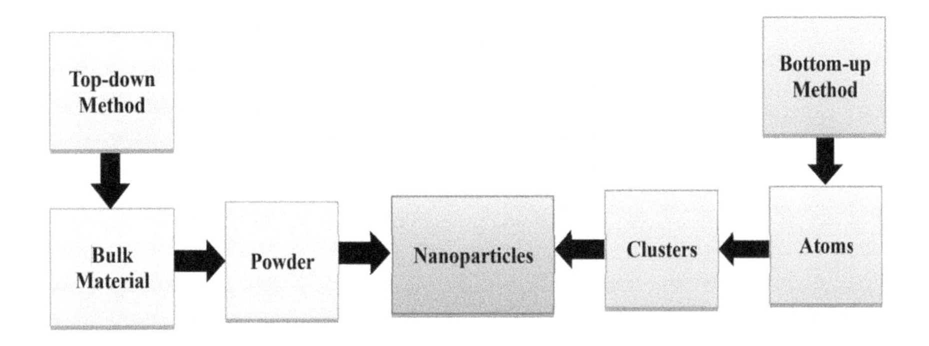

FIGURE 8.2 Synthesis process of nanoparticles.

8.3 SYNTHESIS OF NANOPARTICLES

Nanoparticles are generated using a variety of processes, including chemical, biological, enzymatic, and physical. From ancient times, two ways of nanoparticles preparation have been known: (1) bottom-to-up approach and (2) top-to-down approach. The techniques are classified into subcategories, depending on the type of reaction, operation, and procedures employed. Figure 8.2 illustrates a diagram that displays the procedures for synthesizing nanoparticles.

8.3.1 Top-to-down Approach

It is a destructive technique that involves reducing the size of bulk materials to obtain particles at the nanoscale level (Iravani, 2011). Some commonly used methods for synthesizing nanoparticles include mechanical milling, laser ablation, thermal decomposition, nanolithography, and sputtering.

8.3.2 Bottom-to-Up Approach

This method involves a sequential process where individual atoms are first transformed into nuclei, and then into nanoparticles. Several techniques are commonly used for this approach, including sol-gel, chemical vapor deposition (CVD), spinning, biosynthesis, and pyrolysis. These methods have been extensively studied and utilized for nanoparticle synthesis, as reported in research by Kammler et al. in 2001 and Kavitha et al. in 2013.

8.4 BIOLOGICAL PRODUCTION OF NANOPARTICLES

Different drawbacks of nonconventional procedures employed to produce nanoparticles have shifted the attention of the researcher's community to develop an alternative method which is known as biosynthesis or green synthesis method. This approach for the production of nanoparticles is non-poisonous, biodegradable, and

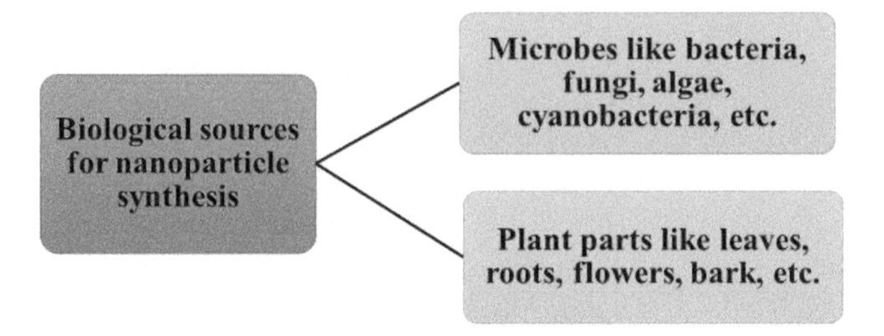

FIGURE 8.3 Sources of biological synthesis of nanoparticles.

environment-friendly in nature, involving different microorganisms, namely, bacteria, yeast, fungi, etc., and plant tissues, namely, fruit, leaves, peel, latex, flower, stem, root, etc. Phytochemicals such as polyphenols, lipids, proteins, carboxylic acids, saponins, polysaccharides, amino acids, enzymes, etc. present in plants are employed as lessen and capping agents. Biosynthesized nanoparticles have distinct and increased features that make them suitable for different applications in various sectors (Kavitha et al., 2013; Latha & Gowri, 2014; Mohanty & Jena, 2017). Figure 8.3 depicts different biological sources used for the synthesis of nanoparticles.

8.4.1 Bacteria as a Source of Nanoparticle Production

There has been widespread usage of bacterial species in commercial biotechnological applications, viz., genetic engineering, bioremediation, and bioleaching. Bacteria can play a salient character in the production of nanoparticles because of their capacity to decrease metal ions. Nanoparticles of metals and metal oxides have been widely synthesized using prokaryotic bacteria and actinomycetes. Since microorganisms may be easily manipulated, their use for the creation of nanoparticles has become a popular method. *Bacillus cereus*, *Escherichia coli*, *Lactobacillus casei*, *Enterobacter cloacae*, *Plectonema boryanum UTEX 485*, *Pseudomonas proteolytica*, *Geobacter* spp., *Arthrobacter gangotriensis*, *Shewanella oneidensis*, *Rhodopseudomonas capsulate*, *Corynebacterium* sp. SH09, etc. are a few exemplars of bacteria that are employed to generate various nanoparticles with certain sizes and shapes (Sunkar & Nachiyar, 2012; Iravani, 2014; Tsekhmistrenko et al., 2020; Salem & Fouda, 2021).

8.4.2 Fungi as a Source of Nanoparticle Production

Monodisperse metal/metal oxide nanoparticles with clear-cut morphologies can also be synthesized via fungi. Because of their broad range of enzymes, fungi can be utilized as organic agents in the manufacturing of metal and metal oxide nanoparticles. Fungi have the capacity to synthesize a greater number of nanoparticles than bacteria. Additionally, fungi possess several advantages over other organisms as they have proteins, enzymes, and reducing components present on the exterior of their cells. Metal nanoparticle creation is thought to originate from enzymatic reduction within the fungal cell wall, occurring in an intracellular manner. Fungi of various species are employed in the synthesis of metal/metal oxide nanoparticles. This is evidenced by numerous studies, including those conducted by Vahabi et al. (2011), Narayanan et al. (2013), Peiris et al. (2018), Chatterjee et al. (2020), Faramarzi et al. (2020), and Priyadarshini et al. (2021).

8.4.3 Algae as a Source of Nanoparticle Production

There has been much recent attention given to algae as a potential nanoparticle production source. Both microalgae and macroalgae have a character in the synthesis of nanoparticles. Algal cells contain cations that serve as reducing agents and are primarily accountable for NP synthesis. One instance is the research conducted by El-Sheekh et al. (2022) which involved the production of Ag- and AuNPs from *Oscillatoria* sp. and *Spirulina platensis*, respectively. The team discovered that the synthesized nanoparticles demonstrated in vitro antiviral activity against herpes simplex (HSV-1) virus. Another example is the work of Fatima et al. (2020), who utilized *Portieria hornemannii*, a red algal species with potent antioxidant properties, to produce AgNPs. The resulting nanoparticles exhibited antibacterial activity against a *Vibrio* species. One category of algae, macroalgae, includes seaweeds. Organisms like these can be found adhering to the surface of rocks or, less commonly, in the water or on coasts. Certain classes of proteins, polysaccharides, and phenolic chemicals make up their makeup. They are employed as medicinal and pharmacological agents due to their abundance of bioactive chemicals. For example, *Turbinaria ornata* was used to create AgNPs, and its anti-tumor activity was tested in retinoblastoma Y79 cells by Remya et al. (2018).

8.4.4 Plant as a Source of Nanoparticle Production

Biosynthetic techniques utilizing plant extracts have gained attention as a practical and cost-effective alternative to traditional nanoparticle synthesis methods due to their simplicity and versatility. The "one-pot" production technique allows for the use of multiple plant parts to balance the metallic nanoparticles. The green synthesis of metal/metal oxide nanoparticles from plant leaf extracts is one such method that has captured the interest of researchers for its wide range of potential applications. Several scientists have utilized this method, including Adil et al. (2015), Vijayaraghavan and AshokKumar (2017), Kumar and Kumar (2018).

Various plants such as alfalfa, aloe vera, lemon, oat, tulsi, neem, mustard, coriander, and lemongrass have been utilized to synthesize a range of nanoparticles (Singh et al., 2010; Christensen et al., 2011; Nath & Banerjee, 2013; Fazlzadeh et al., 2017; Singh et al., 2019).

8.5 APPLICATIONS OF GREEN NANOPARTICLES

Because of their nanoscale dimensions, the materials and systems studied by nanotechnology boast vastly enhanced physical, biological, and chemical capabilities that allow for the exploitation of previously untapped phenomena and processes. Nanoscale particles have been studied and used in numerous fields due to their exceptional magnetic, chemical, electrical, optical, and other capabilities. These fields include biotechnology, microelectronics, catalysis, data storage, energy storage, and many more. The potential for technologically enabled material modification has become an ingredient in the production of high-performance materials. This technology is not just about materials and applications; it also has applications in the life sciences. One of the most important ways nanoparticles are used is in nano-bioremediation (Komarneni, 2003; Wu et al., 2019; Banu et al., 2021).

8.5.1 Nano-Bioremediation

Many factors, including cost-effectiveness, complexity, risk, resource availability, and time commitment, are considered when determining the most appropriate strategy for the remediation of polluted sites. It has been noted that selecting only one technology to remove impurities may not be the best option. As a result, it is crucial to blend applications of numerous technologies to mitigate the challenges associated with the use of one approach. Nano-bioremediation is one such technique, and it is currently the subject of extensive research in a wide range of polluted settings due to its potential efficacy. This cutting-edge method is quickly becoming the standard for cleaning polluted areas. Nanomaterials are used in this method to diminish the toxicity of the pollutants to a level more amenable to biodegradation. Nanoparticles biologically generated from phytoextracts or microorganisms are used for the cleanup of contaminated land and water locations in nano-bioremediation (Rathoure, 2018; Phenrat et al., 2020). Nano-bioremediation relies on three key features to be effective: (1) the use of environmentally friendly nanomaterials, (2) the provision of a solution for the elimination of contaminants, and (3) the deployment of sensors for environmental agents. Because of its increased surface area, tiny size, and appealing chemical characteristics, nanotechnology has gained popularity for application in remediation. This method relies on a nanocatalyst's ability to penetrate deeply into a pollutant, before using a combination of microbes and phytoextracts to break down the waste in a way that does not affect the environment. By mineralizing organic pollutants to products like carbon dioxide and water, microorganisms and phytoextracts can transform metalloids and heavy metals into the non-toxic structure. Nanoparticles, due to their tiny size, can pierce the contaminated site, yielding superior results compared to more traditional bioremediation strategies (Tratnyek & Johnson, 2006; Tosco et al., 2014; Pete et al., 2021).

8.5.1.1 Techniques of Nano-Bioremediation

Biogenic nanoparticles primarily use adsorption and reduction mechanisms to interact with and eliminate metals and metalloids. To remediate heavy metals, physical or chemical adsorption methods can be utilized. Porous structures are sufficient for physical adsorption, while chemical adsorption requires the presence of functional groups on the surface of the adsorbent. Adsorption using chemicals is preferred since it is more effective at removing pollutants from a site. There are two main ways that heavy metals can be reduced: (1) the heavy metals are reduced directly by the nanoparticles; (2) the heavy metals are adsorbed onto the nanoparticle surface and subsequently further reduced to lower valances. Upon reaching non-toxic

TABLE 8.1
Examples of the Different Processes Involved in Nano-Bioremediation

Name of the Process	Target Compounds	Nanomaterials Used	References
Adsorption	Metals, organic compounds, phosphate, DDT, dioxin	Iron oxides, carbon-based nanomaterials	Stafiej & Pyrzynska, 2007; Pan et al., 2010; Bhaumik et al., 2012
Photocatalysis	Organic pollutants and Dyes, PAHs	TiO_2, ZnO, Fe_2O_3, Fe_3O_4	Khedr et al., 2009;Yadav et al., 2017
Redox reactions	PCB, PAH	Nanoscale calcium peroxide, nanoscale zero-valent iron	Zhang 2003; Tratnyek & Johnson, 2006; Varanasi et al., 2007
Disinfection	Organic compounds	Ag/TiO_2 nanocomposites and CNTs	Hu et al., 2005;
Membranes	Chlorinated compounds, polyethylene	NanoAg/TiO_2/zeolites/magnetite and CNTs	Yadav et al., 2017

concentrations, these nanoparticles are easily biodegradable due to an enhanced rate of biodegradation (Wu et al., 2019; Latif et al., 2020). Table 8.1 demonstrates different processes used in nano-bioremediation.

8.5.1.2 Present Status of Nano-Bioremediation

As the optical and catalytic properties of AgNPs are generally high and the catalytic and electrical properties of biologically produced AgNPs are considerably higher, the most common application of these catalytic characteristics is in the purification of wastewater. The azo dyes in wastewater can be degraded by using biologically produced AgNPs made from *Konjac glucomannan* extract (Chen et al., 2021). The wastewater treatment industry can benefit from this catalytic activity for pollutant degradation. Because of this, the release of potentially deleterious by-products from chemical wastewater treatment will be more efficiently controlled. *Coscinium fenestratum* stem extract was utilized to produce silver nanoparticles (AgNPs) for wastewater treatment, and these AgNPs were then used in the spectroscopic examination of wastewater to detect the fungicide thiram. These AgNPs have also been shown to be very effective at neutralizing acidic pollutants in wastewater (Ragam & Mathew, 2020).

Qin et al. (2013) described a simple, environmentally friendly way to make photoluminescent nanosized carbon dots (C-dots) by taking crumble as a key ingredient. When used to detect Hg^{2+} ions, the synthesized carbon dots demonstrated discriminating selectivity toward Hg^{2+}. Because of their ultra-fine size (1–4 nm), Pb ions were also sensitive to changes in the environment and were selectively absorbed by carbon dots.

An optical Au nanosensor was developed by Bindhu and Umadevi (2014) for the environmentally friendly detection of Fe^{3+} ions. In order to create the AuNPs, they used a reducing agent derived from the leaves of the cannabinoid plant *Hibiscus cannabinus*. Incredibly selective and discriminating behaviors towards tiny concentrations of Fe^{3+} were seen in the produced nanostructures, which were measured to be around 22 nm in size. For a long time, gold nanoparticles have been used as a high-end metal detector.

Tripathi et al. (2014) reported an eco-friendly protocol for detecting Hg^{2+} ions in water samples employing gold nanoparticles that were coated with a fungal extract obtained from *Trichoderma harzianum*.

Graphene quantum dots (GQDs) are used for perceiving Pb^{2+} ions and were synthesized by Zhou et al. (2014) by using L-ascorbic acid and sodium polystyrene sulfonate in an environmentally friendly nature. Additionally, green-produced AgNPs developed by using leaf extract of *Gongura hibiscus sabdariffa* were able to detect Hg^{2+}, Cd^{2+}, and Pb^{2+} at the ppm level (Kumar et al., 2014). Dar et al. (2014) described a procedure for detecting As^{3+} in aqueous solutions using AgNPs graphene oxide composites and stabilizing and reducing agents such as ascorbic acid and beta-cyclodextrin.

Fluorescent carbon nanoparticles (FCNPs) have been synthesized in a green manner by Singh and Mishra (2016), who were inspired by the excellent sensing capabilities of fluorosensors to develop a method for detecting Fe^{3+} ions. Having a small detection range of between 18 ppm and 56 ppb for Fe^{3+} ions, fluorescent carbon NPs (FCNPs) were produced from D-glucose.

Ha et al. (2016) utilized biologically synthesized palladium nanoparticles to eliminate hexavalent chromium from polluted water. Kimber et al. (2018) reported that silver nanoparticles produced by *Aspergillus niger* had the ability to decolorize the organic dye within two days. The biosynthesis of nanoparticles is feasible, but it is a challenging process with certain drawbacks, including low yields, the possibility of contamination of living cells, and challenges in separation. Therefore, there is a significant need for new and durable microbes to generate nanoparticles (Jiang et al., 2018). *Aspergillus tubingensis*, for example, was able to produce iron oxide nanoparticles that efficiently removed nickel, lead, copper, and zinc from wastewater. Additionally, a study on the reusability of these nanoparticles demonstrated that they had excellent regenerative capacity, with the ability to undergo up to five cycles of desorption and adsorption (Mahanty et al., 2020). Producing nanoparticles from biological sources and controlling their shape, size, and distribution are interesting topics that need more research.

Chatterjee and colleagues (2020) used a fungus, *Aspergillus niger BSC-1*, which is commonly found in mangroves, to produce iron oxide nanoparticles that can be used for the removal of metals from wastewater. These nanoparticles had a nanoflake shape and displayed effective adsorption properties for the removal of chromium. Meanwhile, San Keskin and co-workers (2018) developed nanofibers coated with *Lysinibacillus* sp. and cyclodextrin that showed promise in the removal of hexavalent chromium, nickel, and dye. Subramaniyam et al. (2015) utilized *Chlorococcum* sp. MM11 to produce iron nanoparticles that were effective in remediating hexavalent chromium to its trivalent form, with a reduction rate of 92%.

Mukherjee et al. (2016) developed a plant-based biogenic nanoparticle composition utilizing *Aloe vera* leaves. Both the adsorption and extraction of arsenic from polluted water were successful when employing this environmentally friendly technique. In another experiment, Al-Qahtani (2017) demonstrated that silver nanoparticles generated from *Ficus benjamina* leaf extract were able to efficiently remove cadmium. Venkateswarlu et al. (2015) created biogenic iron oxide nanoparticles that were modified using 3-mercaptopropionic acid. These nanoparticles were then used as an adsorbent to eliminate nickel from water.

Wang et al. (2018) employed the bacteria *Citrobacter freundii Y9* to produce selenium nanoparticles that reacted with mercury to produce Hg-Se. These nanoparticles were utilized to eliminate mercury from groundwater effectively. On the other hand, Choudhury et al. (2017) developed an efficient method for removing toxic metals from the environment. They immobilized *Saccharomyces cerevisiae* on Titania nanopowder to create a nanocomposite. The effectiveness of this composition in purifying water contaminated with Cr (VI) was determined to be 99.92%.

NCs synthesized from plant-derived materials (husks, wood shavings, straw, etc.) help with imperishable development (Nakbanpote et al., 2007). To eliminate arsenic ions from water, polyaniline/rice husk nanocomposites were produced through a batch adsorption technique (Lashkenari et al., 2011).

Siddiqui and Chaudhry (2017) reported that the use of binary metal oxides such as iron-manganese oxide, titanium-iron oxide, iron-zirconium oxide, and cerium-iron oxide in carbon frameworks to form nanocomposites has resulted in significant advancements in water purification due to their excellent adsorption properties. In this regard, $MnFe_2O_4$/biochar and *Nigella sativa* seed-based nanocomposites have been developed as effective adsorbents for water treatment, leading to the production of safe and cost-effective drinking water (Siddiqui & Chaudhry, 2018). Therefore, employing new nanotechnology-based methods is crucial for the successful remediation of polluted water. Table 8.2 summarizes different bionanoparticles used in the remediation process.

8.6 CHALLENGES AND FUTURE PERSPECTIVES

Environmentally friendly nanoparticles may have multiple applications in water remediation processes, including photocatalytic/chemically catalyzed degradation, adsorption of contaminants, pollutant sensing, and detection. More study is needed, though, to learn how the matrix of the composite and the nanofiller interact with the polluted water. Green synthetic nanoparticle synthesis on a big scale and additional applications are also possible. Using biogenic nanoparticles in water remediation on a large scale has shown that they may effectively remove metals and organic pollutants from contaminated water. The water remediation and cost-effectiveness fields can still be improved by increasing selectivity in material removal, long-term stability, and resistance to changes in pH and chemical concentrations in polluted water. Therefore, it is important to conduct further investigations into the long-term effectiveness of biogenic nanoparticles in the future.

8.7 CONCLUSIONS

There are significant technological and economic impediments to the broad range of take-up for commercialization of green nanoparticles, despite their extensive use for

TABLE 8.2
Various Biogenic Nanoparticles Used for Remediation Purposes

Biological Agent	Class of Biological Agent	Type of Nanoparticles	Shape and Size	Targeted Pollutant	References
E. coli	Bacteria	Zinc oxide	Spherical; 10 nm	Lead and cadmium	Somu & Paul, 2018
Fusarium oxysporum	Fungus	Iron oxide	Spherical; 26.78 nm	Arsenic	Balakrishnan et al., 2020
Shewanella putrefaciens 200	Fungus	Mercury-selenium nanocomposite	–	Mercury	Jiang et al., 2012
Azadirachta indica and *Mentha longifolia*	Plant	Nanoscale zero-valent iron	–	Nickel and lead	Francy et al., 2020
Deinococcusradiodurans R1	Bacteria	Iron	Spherical; 141.8–164.2 nm	Arsenic	Kim et al., 2019
Enterococcus faecalis	Bacteria	Iron oxide	Cubical, hexagonal, brick, and irregular; 15.4 nm	Hexavalent chromium	Samuel et al., 2021
Catharanthus roseus	Plant	Silver	Spherical; 100 nm	Chromium and cadmium	Verma & Bharadvaja, 2022
Euphorbia heterophylla	Plant	Ag/Ti_{O2} nanocomposite	–	Organic dyes	Atarod et al., 2016
Thymbra spicata	Plant	Silver	–	Organic dyes	Veisi et al., 2018
Azadirachta indica	Plant	Silver	Spherical; 11–35 nm	Congo red	Shaikh & Sukalyan, 2018
Anabaena variabilis and *Spirulina platensis*	Cyanobacteria	Silver	Spherical–oval; 17.9–26.4 nm	Malachite green	Ismail et al., 2021
Camellia sinensis	Plant	Silver-gold nanocomposites	Spherical; 20–200 nm	Congo red, 4-nitrophenol	Kang & Kolya, 2021
Vaccinium corymbosum	Plant	Iron, nanoscale zero-valent iron	Spherical; 61.1–106 nm	Arsenic	Manquián-Cerda et al., 2017
Syzygiumaromaticum	Plant	Iron-Aluminum nanocomposites	Spherical; 458.9 nm	Fluoride	Mondal & Purkait, 2019
Chlorella vulgaris	Algae	TiO_2/Ag nanocomposites	Spherical; 36.7 nm	Hexavalent chromium	Wang et al., 2017
Sargassum bovinum	Algae	Palladium	Octahedral; 5–10 nm	Hydrogen peroxide	Momeni & Nabipour, 2015

a number of critical applications in many industrial domains. Because of this, new methods of synthesis are needed to create materials with novel characteristics and improve their performance in industrial settings. For this reason, we need to come up with techniques that are cheaper, more environmentally amiable, and scalable to keep up with the expanding needs brought on by scientific and technological progress. The use of various biological entities is an alternative to researching the shape-controlled and broad-size distributions of the NPs using physiochemical methods of synthesis. The use of nanoparticles has been demonstrated in a variety of contexts, including the purification of water, the removal of toxic metals and chemicals, the detoxification of industrial waste, and the detoxification of substances, including drug residues, dyes, and pesticides. Wastewater treatment can greatly benefit from the use of green nanoparticles, which are ecologically safe, biocompatible, and efficient in water remediation primarily through adsorption and photocatalytic processes.

REFERENCES

Adil, S. F., Assal, M. E., Khan, M., Al-Warthan, A., Siddiqui, M. R. H., & Liz-Marzán, L. M. (2015). Biogenic synthesis of metallic nanoparticles and prospects toward green chemistry. *Dalton Transactions*, *44*(21), 9709–9717.

Al-Qahtani, K. M. (2017). Cadmium removal from aqueous solution by green synthesis zero valent silver nanoparticles with Benjamina leaves extract. *The Egyptian Journal of Aquatic Research*, *43*(4), 269–274.

Amudha, M., & Shanmugasundaram, K. K. (2014). Biosynthesis and characterization of silver nanoparticles using the aqueous extract of Vitex negundo Linn. *World Journal of Pharmacy and Pharmaceutical Sciences (WJPPS)*, *3*(8), 1385–1393.

Atarod, M., Nasrollahzadeh, M., & Sajadi, S. M. (2016). Euphorbia heterophylla leaf extract mediated green synthesis of Ag/TiO2 nanocomposite and investigation of its excellent catalytic activity for reduction of variety of dyes in water. *Journal of Colloid and Interface Science*, *462*, 272–279.

Balaji, D. S., Basavaraja, S., Deshpande, R., Mahesh, D. B., Prabhakar, B. K., & Venkataraman, A. (2009). Extracellular biosynthesis of functionalized silver nanoparticles by strains of Cladosporium cladosporioides fungus. *Colloids and Surfaces B: Biointerfaces*, *68*(1), 88–92.

Balakrishnan, G. S., Rajendran, K., & Kalirajan, J. (2020). Microbial synthesis of magnetite nanoparticles for arsenic removal. *Journal of Applied Biology and Biotechnology*, *8*(3), 70–75.

Banu, A. N., Kudesia, N., Raut, A. M., Pakrudheen, I., & Wahengbam, J. (2021). Toxicity, bioaccumulation, and transformation of silver nanoparticles in aqua biota: A review. *Environmental Chemistry Letters*, *19*(6), 4275–4296.

Bhaumik, M., Maity, A., Srinivasu, V. V., & Onyango, M. S. (2012). Removal of hexavalent chromium from aqueous solution using polypyrrole-polyaniline nanofibers. *Chemical Engineering Journal*, *181*, 323–333.

Bindhu, M. R., & Umadevi, M. (2014). Green synthesized gold nanoparticles as a probe for the detection of Fe 3+ ions in water. *Journal of Cluster Science*, *25*, 969–978.

Chatterjee, S., Mahanty, S., Das, P., Chaudhuri, P., & Das, S. (2020). Biofabrication of iron oxide nanoparticles using manglicolous fungus Aspergillus niger BSC-1 and removal of Cr (VI) from aqueous solution. *Chemical Engineering Journal*, *385*, 123790.

Chen, J., Liu, Y., Xiong, Y., Wei, D., Peng, J., Mahmud, S., & Liu, H. (2021). Konjac glucomannan reduced-stabilized silver nanoparticles for mono-azo and di-azo contained wastewater treatment. *Inorganica Chimica Acta*, *515*, 120058.

Chen, R. J., Choi, H. C., Bangsaruntip, S., Yenilmez, E., Tang, X., Wang, Q., . . . Dai, H. (2004). An investigation of the mechanisms of electronic sensing of protein adsorption on carbon nanotube devices. *Journal of the American Chemical Society*, *126*(5), 1563–1568.

Choudhury, P. R., Bhattacharya, P., Ghosh, S., Majumdar, S., Saha, S., & Sahoo, G. C. (2017). Removal of Cr (VI) by synthesized titania embedded dead yeast nanocomposite: Optimization and modeling by response surface methodology. *Journal of Environmental Chemical Engineering*, *5*(1), 214–221.

Christensen, L., Vivekanandhan, S., Misra, M., & Kumar Mohanty, A. (2011). Biosynthesis of silver nanoparticles using Murraya koenigii (curry leaf): An investigation on the effect of broth concentration in reduction mechanism and particle size. *Advanced Materials Letters*, *2*(6), 429–434.

Dar, R. A., Khare, N. G., Cole, D. P., Karna, S. P., & Srivastava, A. K. (2014). Green synthesis of a silver nanoparticle–graphene oxide composite and its application for As (III) detection. *RSC Advances*, *4*(28), 14432–14440.

El-Sheekh, M. M., Shabaan, M. T., Hassan, L., & Morsi, H. H. (2022). Antiviral activity of algae biosynthesized silver and gold nanoparticles against Herps Simplex (HSV-1) virus in vitro using cell-line culture technique. *International Journal of Environmental Health Research*, *32*(3), 616–627.

Faramarzi, S., Anzabi, Y., & Jafarizadeh-Malmiri, H. (2020). Nanobiotechnology approach in intracellular selenium nanoparticle synthesis using Saccharomyces cerevisiae – fabrication and characterization. *Archives of Microbiology*, *202*(5), 1203–1209.

Fatima, R., Priya, M., Indurthi, L., Radhakrishnan, V., & Sudhakaran, R. (2020). Biosynthesis of silver nanoparticles using red algae Portieria hornemannii and its antibacterial activity against fish pathogens. *Microbial Pathogenesis*, *138*, 103780.

Fazlzadeh, M., Rahmani, K., Zarei, A., Abdoallahzadeh, H., Nasiri, F., & Khosravi, R. (2017). A novel green synthesis of zero valent iron nanoparticles (NZVI) using three plant extracts and their efficient application for removal of Cr (VI) from aqueous solutions. *Advanced Powder Technology*, *28*(1), 122–130.

Francy, N., Shanthakumar, S., Chiampo, F., & Sekhar, Y. R. (2020). Remediation of lead and nickel contaminated soil using nanoscale zero-valent iron (nZVI) particles synthesized using green leaves: First results. *Processes*, *8*(11), 1453.

Ha, C., Zhu, N., Shang, R., Shi, C., Cui, J., Sohoo, I., . . . Cao, Y. (2016). Biorecovery of palladium as nanoparticles by Enterococcus faecalis and its catalysis for chromate reduction. *Chemical Engineering Journal*, *288*, 246–254.

Hu, J., Chen, G., & Lo, I. M. (2005). Removal and recovery of Cr (VI) from wastewater by maghemite nanoparticles. *Water Research*, *39*(18), 4528–4536.

Iravani, S. (2011). Green synthesis of metal nanoparticles using plants. *Green Chemistry*, *13*(10), 2638–2650.

Iravani, S. (2014). Bacteria in nanoparticle synthesis: Current status and future prospects. *International Scholarly Research Notices*, *2014*, Article ID 359316.

Ismail, G. A., Allam, N. G., El-Gemizy, W. M., & Salem, M. A. (2021). The role of silver nanoparticles biosynthesized by Anabaena variabilis and Spirulina platensis cyanobacteria for malachite green removal from wastewater. *Environmental Technology*, *42*(28), 4475–4489.

Jadoun, S., Arif, R., Jangid, N. K., & Meena, R. K. (2021). Green synthesis of nanoparticles using plant extracts: A review. *Environmental Chemistry Letters*, *19*, 355–374.

Jiang, S., Ho, C. T., Lee, J. H., Van Duong, H., Han, S., & Hur, H. G. (2012). Mercury capture into biogenic amorphous selenium nanospheres produced by mercury resistant Shewanella putrefaciens 200. *Chemosphere*, *87*(6), 621–624.

Jiang, Z., Zhang, S., Klausen, L. H., Song, J., Li, Q., Wang, Z., . . . Dong, M. (2018). In vitro single-cell dissection revealing the interior structure of cable bacteria. *Proceedings of the National Academy of Sciences*, *115*(34), 8517–8522.

Kammler, H. K., Mädler, L., & Pratsinis, S. E. (2001). Flame synthesis of nanoparticles. *Chemical Engineering & Technology: Industrial Chemistry-Plant Equipment-Process Engineering-Biotechnology*, *24*(6), 583–596.
Kang, C. W., & Kolya, H. (2021). Green synthesis of Ag-Au bimetallic nanocomposites using waste tea leaves extract for degradation Congo red and 4-nitrophenol. *Sustainability*, *13*(6), 3318.
Kavitha, K. S., Baker, S., Rakshith, D., Kavitha, H. U., Yashwantha Rao, H. C., Harini, B. P., & Satish, S. (2013). Plants as green source towards synthesis of nanoparticles. *International Journal of Scientific Research in Biological Sciences*, *2*(6), 66–76.
Khedr, M. H., Halim, K. A., & Soliman, N. K. (2009). Synthesis and photocatalytic activity of nano-sized iron oxides. *Materials Letters*, *63*(6–7), 598–601.
Kim, H. K., Jeong, S. W., Yang, J. E., & Choi, Y. J. (2019). Highly efficient and stable removal of arsenic by live cell fabricated magnetic nanoparticles. *International Journal of Molecular Sciences*, *20*(14), 3566.
Kimber, R. L., Lewis, E. A., Parmeggiani, F., Smith, K., Bagshaw, H., Starborg, T., . . . Lloyd, J. R. (2018). Biosynthesis and characterization of copper nanoparticles using Shewanella oneidensis: Application for click chemistry. *Small*, *14*(10), 1703145.
Komarneni, S. (2003). Nanophase materials by hydrothermal, microwave-hydrothermal and microwave-solvothermal methods. *Current Science*, 1730–1734.
Kowshik, M., Ashtaputre, S., Kharrazi, S., Vogel, W., Urban, J., Kulkarni, S. K., & Paknikar, K. M. (2002). Extracellular synthesis of silver nanoparticles by a silver-tolerant yeast strain MKY3. *Nanotechnology*, *14*(1), 95.
Kumar, S. V., & Rajeshkumar, S. (2018). Plant-based synthesis of nanoparticles and their impact. In *Nanomaterials in Plants, Algae, and Microorganisms* (pp. 33–57). Academic Press.
Kumar, V. V., Anbarasan, S., Christena, L. R., SaiSubramanian, N., & Anthony, S. P. (2014). Bio-functionalized silver nanoparticles for selective colorimetric sensing of toxic metal ions and antimicrobial studies. *Spectrochimica Acta Part A: Molecular and Biomolecular Spectroscopy*, *129*, 35–42.
Lashkenari, M. S., Davodi, B., & Eisazadeh, H. (2011). Removal of arsenic from aqueous solution using polyaniline/rice husk nanocomposite. *Korean Journal of Chemical Engineering*, *28*, 1532–1538.
Latha, N., & Gowri, M. (2014). Biosynthesis and characterisation of Fe3O4 nanoparticles using Caricaya papaya leaves extract. *International Journal of Recent Scientific Research*, *3*(11), 1551–1556.
Latif, A., Sheng, D., Sun, K., Si, Y., Azeem, M., Abbas, A., & Bilal, M. (2020). Remediation of heavy metals polluted environment using Fe-based nanoparticles: Mechanisms, influencing factors, and environmental implications. *Environmental Pollution*, *264*, 114728.
Mahanty, S., Chatterjee, S., Ghosh, S., Tudu, P., Gaine, T., Bakshi, M., . . . Chaudhuri, P. (2020). Synergistic approach towards the sustainable management of heavy metals in wastewater using mycosynthesized iron oxide nanoparticles: Biofabrication, adsorptive dynamics and chemometric modeling study. *Journal of Water Process Engineering*, *37*, 101426.
Manquián-Cerda, K., Cruces, E., Rubio, M. A., Reyes, C., & Arancibia-Miranda, N. (2017). Preparation of nanoscale iron (oxide, oxyhydroxides and zero-valent) particles derived from blueberries: Reactivity, characterization and removal mechanism of arsenate. *Ecotoxicology and Environmental Safety*, *145*, 69–77.
Mohanty, A. S., & Jena, B. S. (2017). Innate catalytic and free radical scavenging activities of silver nanoparticles synthesized using Dillenia indica bark extract. *Journal of Colloid and Interface Science*, *496*, 513–521.
Momeni, S., & Nabipour, I. (2015). A simple green synthesis of palladium nanoparticles with Sargassum alga and their electrocatalytic activities towards hydrogen peroxide. *Applied Biochemistry and Biotechnology*, *176*, 1937–1949.

Mondal, P., & Purkait, M. K. (2019). Preparation and characterization of novel green synthesized iron-aluminum nanocomposite and studying its efficiency in fluoride removal. *Chemosphere*, *235*, 391–402.

Mukherjee, D., Ghosh, S., Majumdar, S., & Annapurna, K. (2016). Green synthesis of α-Fe2O3 nanoparticles for arsenic (V) remediation with a novel aspect for sludge management. *Journal of Environmental Chemical Engineering*, *4*(1), 639–650.

Nakbanpote, W., Goodman, B. A., & Thiravetyan, P. (2007). Copper adsorption on rice husk derived materials studied by EPR and FTIR. *Colloids and Surfaces A: Physicochemical and Engineering Aspects*, *304*(1–3), 7–13.

Narayanan, K. B., Park, H. H., & Sakthivel, N. (2013). Extracellular synthesis of mycogenic silver nanoparticles by Cylindrocladium floridanum and its homogeneous catalytic degradation of 4-nitrophenol. *Spectrochimica Acta Part A: Molecular and Biomolecular Spectroscopy*, *116*, 485–490.

Nath, D., & Banerjee, P. (2013). Green nanotechnology–a new hope for medical biology. *Environmental Toxicology and Pharmacology*, *36*(3), 997–1014.

Pan, G., Li, L., Zhao, D., & Chen, H. (2010). Immobilization of non-point phosphorus using stabilized magnetite nanoparticles with enhanced transportability and reactivity in soils. *Environmental Pollution*, *158*(1), 35–40.

Peiris, M. M. K., Gunasekara, T. D. C. P., Jayaweera, P. M., & Fernando, S. S. N. (2018). Ti_{O2} nanoparticles from Baker's yeast: A potent antimicrobial. *Journal of Microbiology and Biotechnology, 28*(10), 1664–1670.

Pete, A. J., Bharti, B., & Benton, M. G. (2021). Nano-enhanced bioremediation for oil spills: A review. *ACS ES&T Engineering*, *1*(6), 928–946.

Phenrat, T., Skácelová, P., Petala, E., Velosa, A., & Filip, J. (2020). Nanoscale zero-valent iron particles for water treatment: From basic principles to field-scale applications. *Advanced Nano-Bio Technologies for Water and Soil Treatment*, 19–52.

Priyadarshini, E., Priyadarshini, S. S., Cousins, B. G., & Pradhan, N. (2021). Metal-fungus interaction: Review on cellular processes underlying heavy metal detoxification and synthesis of metal nanoparticles. *Chemosphere*, *274*, 129976.

Qin, X., Lu, W., Asiri, A. M., Al-Youbi, A. O., & Sun, X. (2013). Microwave-assisted rapid green synthesis of photoluminescent carbon nanodots from flour and their applications for sensitive and selective detection of mercury (II) ions. *Sensors and Actuators B: Chemical*, *184*, 156–162.

Ragam, P. N., & Mathew, B. (2020). Unmodified silver nanoparticles for dual detection of dithiocarbamate fungicide and rapid degradation of water pollutants. *International Journal of Environmental Science and Technology*, *17*, 1739–1752.

Rathoure, A. K. (Ed.). (2018). *Biostimulation Remediation Technologies for Groundwater Contaminants*. IGI Global.

Remya, R. R., Rajasree, S. R., Suman, T. Y., Aranganathan, L., Gayathri, S., Gobalakrishnan, M., & Karthih, M. G. (2018). Laminarin based AgNPs using brown seaweed Turbinariaornata and its induction of apoptosis in human retinoblastoma Y79 cancer cell lines. *Materials Research Express*, *5*(3), 035403.

Salem, S. S., & Fouda, A. (2021). Green synthesis of metallic nanoparticles and their prospective biotechnological applications: An overview. *Biological Trace Element Research*, *199*, 344–370.

Samuel, M. S., Datta, S., Chandrasekar, N., Balaji, R., Selvarajan, E., & Vuppala, S. (2021). Biogenic synthesis of iron oxide nanoparticles using Enterococcus faecalis: Adsorption of hexavalent chromium from aqueous solution and in vitro cytotoxicity analysis. *Nanomaterials*, *11*(12), 3290.

San Keskin, N. O., Celebioglu, A., Sarioglu, O. F., Uyar, T., & Tekinay, T. (2018). Encapsulation of living bacteria in electrospun cyclodextrin ultrathin fibers for bioremediation of heavy metals and reactive dye from wastewater. *Colloids and Surfaces B: Biointerfaces*, *161*, 169–176.

Shaikh, W. A., & Sukalyan, C. (2018). UV-assisted photo-catalytic degradation of anionic dye (Congo red) using biosynthesized silver nanoparticles: A green catalysis. *Desalination and Water Treatment, 130*, 232–242.

Shin, W. K., Cho, J., Kannan, A. G., Lee, Y. S., & Kim, D. W. (2016). Cross-linked composite gel polymer electrolyte using mesoporous methacrylate-functionalized SiO2 nanoparticles for lithium-ion polymer batteries. *Scientific Reports, 6*(1), 26332.

Siddiqui, S. I., & Chaudhry, S. A. (2017). Iron oxide and its modified forms as an adsorbent for arsenic removal: A comprehensive recent advancement. *Process Safety and Environmental Protection, 111*, 592–626.

Siddiqui, S. I., & Chaudhry, S. A. (2018). Nigella sativa plant based nanocomposite-MnFe2O4/BC: An antibacterial material for water purification. *Journal of Cleaner Production, 200*, 996–1008.

Sigmund, W., Yuh, J., Park, H., Maneeratana, V., Pyrgiotakis, G., Daga, A., . . . Nino, J. C. (2006). Processing and structure relationships in electrospinning of ceramic fiber systems. *Journal of the American Ceramic Society, 89*(2), 395–407.

Singh, A. K., Talat, M., Singh, D. P., & Srivastava, O. N. (2010). Biosynthesis of gold and silver nanoparticles by natural precursor clove and their functionalization with amine group. *Journal of Nanoparticle Research, 12*, 1667–1675.

Singh, B. K. (2009). Organophosphorus-degrading bacteria: Ecology and industrial applications. *Nature Reviews Microbiology, 7*(2), 156–164.

Singh, B. K., & Walker, A. (2006). Microbial degradation of organophosphorus compounds. *FEMS Microbiology Reviews, 30*(3), 428–471.

Singh, J., Kumar, V., Kim, K. H., & Rawat, M. (2019). Biogenic synthesis of copper oxide nanoparticles using plant extract and its prodigious potential for photocatalytic degradation of dyes. *Environmental Research, 177*, 108569.

Singh, V., & Mishra, A. K. (2016). Green and cost-effective fluorescent carbon nanoparticles for the selective and sensitive detection of iron (III) ions in aqueous solution: mechanistic insights and cell line imaging studies. *Sensors and Actuators B: Chemical, 227*, 467–474.

Somu, P., & Paul, S. (2018). Casein based biogenic-synthesized zinc oxide nanoparticles simultaneously decontaminate heavy metals, dyes, and pathogenic microbes: A rational strategy for wastewater treatment. *Journal of Chemical Technology & Biotechnology, 93*(10), 2962–2976.

Stafiej, A., & Pyrzynska, K. (2007). Adsorption of heavy metal ions with carbon nanotubes. *Separation and Purification Technology, 58*(1), 49–52.

Subramaniyam, V., Subashchandrabose, S. R., Thavamani, P., Megharaj, M., Chen, Z., & Naidu, R. (2015). *Chlorococcum* sp. MM11 – a novel phyco-nanofactory for the synthesis of iron nanoparticles. *Journal of Applied Phycology, 27*, 1861–1869.

Sun, L., Singh, A. K., Vig, K., Pillai, S. R., & Singh, S. R. (2008). Silver nanoparticles inhibit replication of respiratory syncytial virus. *Journal of Biomedical Nanotechnology, 4*(2), 149–158.

Sunkar, S., & Nachiyar, C. V. (2012). Biogenesis of antibacterial silver nanoparticles using the endophytic bacterium Bacillus cereus isolated from Garcinia xanthochymus. *Asian Pacific Journal of Tropical Biomedicine, 2*(12), 953–959.

Tosco, T., Papini, M. P., Viggi, C. C., & Sethi, R. (2014). Nanoscale zerovalent iron particles for groundwater remediation: A review. *Journal of Cleaner Production, 77*, 10–21.

Tratnyek, P. G., & Johnson, R. L. (2006). Nanotechnologies for environmental cleanup. *Nano Today, 1*(2), 44–48.

Tripathi, R. M., Gupta, R. K., Singh, P., Bhadwal, A. S., Shrivastav, A., Kumar, N., & Shrivastav, B. R. (2014). Ultra-sensitive detection of mercury (II) ions in water sample using gold nanoparticles synthesized by Trichoderma harzianum and their mechanistic approach. *Sensors and Actuators B: Chemical, 204*, 637–646.

Tsekhmistrenko, S. I., Bityutskyy, V. S., Tsekhmistrenko, O. S., Horalskyi, L. P., Tymoshok, N. O., & Spivak, M. Y. (2020). Bacterial synthesis of nanoparticles: A green approach. *Biosystems Diversity*, *28*(1), 9–17.

Vahabi, K., Mansoori, G. A., & Karimi, S. (2011). Biosynthesis of silver nanoparticles by fungus Trichoderma reesei (a route for large-scale production of AgNPs). *Insciences Journal*, *1*(1), 65–79.

Varanasi, P., Fullana, A., & Sidhu, S. (2007). Remediation of PCB contaminated soils using iron nano-particles. *Chemosphere*, *66*(6), 1031–1038.

Veisi, H., Azizi, S., & Mohammadi, P. (2018). Green synthesis of the silver nanoparticles mediated by Thymbra spicata extract and its application as a heterogeneous and recyclable nanocatalyst for catalytic reduction of a variety of dyes in water. *Journal of Cleaner Production*, *170*, 1536–1543.

Venkateswarlu, S., Kumar, S. H., & Jyothi, N. V. V. (2015). Rapid removal of Ni (II) from aqueous solution using 3-mercaptopropionic acid functionalized bio magnetite nanoparticles. *Water Resources and Industry*, *12*, 1–7.

Verma, A., & Bharadvaja, N. (2022). Plant-mediated synthesis and characterization of silver and copper oxide nanoparticles: Antibacterial and heavy metal removal activity. *Journal of Cluster Science*, *33*(4), 1697–1712.

Vijayaraghavan, K., & Ashokkumar, T. (2017). Plant-mediated biosynthesis of metallic nanoparticles: A review of literature, factors affecting synthesis, characterization techniques and applications. *Journal of Environmental Chemical Engineering*, *5*(5), 4866–4883.

Wang, L., Zhang, C., Gao, F., Mailhot, G., & Pan, G. (2017). Algae decorated TiO2/Ag hybrid nanofiber membrane with enhanced photocatalytic activity for Cr (VI) removal under visible light. *Chemical Engineering Journal*, *314*, 622–630.

Wang, X., Zhang, D., Qian, H., Liang, Y., Pan, X., & Gadd, G. M. (2018). Interactions between biogenic selenium nanoparticles and goethite colloids and consequence for remediation of elemental mercury contaminated groundwater. *Science of the Total Environment*, *613*, 672–678.

Wu, Y., Pang, H., Liu, Y., Wang, X., Yu, S., Fu, D., . . . Wang, X. (2019). Environmental remediation of heavy metal ions by novel-nanomaterials: A review. *Environmental Pollution*, *246*, 608–620.

Yadav, K. K., Singh, J. K., Gupta, N., & Kumar, V. (2017). A review of nanobioremediation technologies for environmental cleanup: A novel biological approach. *Journal of Materials and Environmental Sciences*, *8*(2), 740–757.

Zhang, W. X. (2003). Nanoscale iron particles for environmental remediation: An overview. *Journal of Nanoparticle Research*, *5*, 323–332.

Zhou, C., Jiang, W., & Via, B. K. (2014). Facile synthesis of soluble graphene quantum dots and its improved property in detecting heavy metal ions. *Colloids and Surfaces B: Biointerfaces*, *118*, 72–76.

9 Energy Recovery from Municipal Solid Waste through Pyrolysis Techniques

Mukta Mayee Kumbhar, Sunanda Swain, Prajna Sarita Sethy, and Dilpreet Kaur

9.1 INTRODUCTION

The major problem faced by countries across the globe due to overpopulation is unchecked release of waste into the environment (Roy, 1988). Waste categorization is based on the sources of their generation, viz., industrial waste, municipal waste, hospital waste, agricultural waste, etc. As per the data provided by the CPCB, in India, 160,038 t of waste were produced in a day in the year 2021. Developed countries have advanced technologies, viz., landfilling, pyrolysis, incineration, and enough financial support to apply these methods, whereas the rest of the countries are facing the obstacle of discarding and recycling municipal solid wastes and, in the meantime, tackle with environmental issues. Additionally, the continuous assemblage of municipal solid waste also raises concentration of anthropogenic gases in the atmosphere. These factors also bring about an increase in global warming potential (GWP).

Moreover, the imbalance between energy demand and supply in many countries is also a rising concern, as these countries are unable to suffice the rapidly increasingly energy demand, and the population is also increasing day by day. Developed countries are finding solutions to maintain energy supply to cope up with increasing demand, while underdeveloped and developing countries are struggling in decreasing the gap between energy supply and energy demand. Even though India is the most populated and fastest-growing economy in the world, still, it could not reduce the energy production and demand gap over the last decade and still continues to encounter the same issue. Different types of discarding ways currently used for treating wastes are non-engineered, landfilling, biocomposting, biochemical conversion, and thermochemical conversion (Anuradda and Rangan, 2001; Babu, 2008). Amidst all the disposal methods, biochemical and thermochemical conversion methods contribute to value-added products, from which energy can be obtained. Therefore, development activities regarding biochemical and thermochemical conversion are

DOI: 10.1201/9781003517108-9

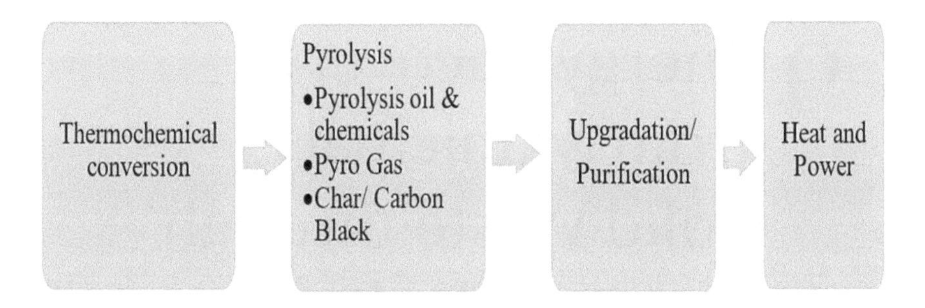

FIGURE 9.1 Route of thermochemical conversion and end product using pyrolysis.

most active in developed countries. The different ways of pyrolysis thermochemical conversion and product extraction are described in Figure 9.1.

One of the thermochemical conversion methods used is pyrolysis, which operates on the principle of breaking down complex hydrocarbons into value-added products in the presence of heat energy, with a slight amount or deficiency of oxygen (Younes and Mohammed, 2013). In this method, the organic feedstock is kept in an airtight container and heated for reaction purpose. After that, the volatile substance generated is collected and cooled to room temperature before obtaining the value-added products. For example, fuels are gained in different forms, such as liquids, gases, and solids. The advantage of using the pyrolysis technique is conversion of biological materials into value-added products. The value-added products are obtained in the form of either bio-oil, biogas, or biochar. Certain amount of energy is required, either in the form of combustion or electrical energy, to increase the percentage of volatile substance conversion into gaseous form from the waste. The current study briefly describes the different types of techniques and reactors used in pyrolysis and the value-added products extracted after process completion (Van de Velden et al., 2010).

9.2 TYPES OF PYROLYSIS TECHNIQUES

Pyrolysis technique is a promising thermochemical method for waste management. Pyrolysis techniques can be divided into many sorts according to the environment and waste type associated with the procedure. These three pyrolysis techniques are mentioned in Figure 9.2.

9.2.1 Slow Pyrolysis Techniques

Different heating rates can be employed with the pyrolysis process. The common method for producing biochar involves slow pyrolysis at temperatures about 300°C and heating rates that range from 0.1 to 0.8°C. Batch process is another name for this procedure. Solids and gas typically have longer residence durations, which can be from minutes to days (Lu et al., 2020). The majority of the literature was solely concerned with biochar, although some of it also looked at how slow pyrolysis could produce bio-oil (Stamatov et al., 2006). Another slow pyrolysis method is carbonization, which involves pyrolyzing a variety of feedstocks without condensing the

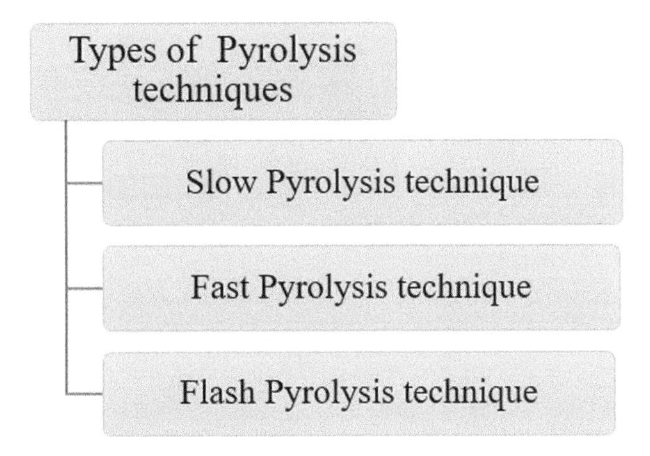

FIGURE 9.2 Types of pyrolysis techniques.

products. For many years, this method has been employed to create charcoal. The characteristics of the feedstock, such as its ash concentration, moisture content, fixed carbon content, process temperature, and volatile matter, have a major impact on the yields of bio-oil and biochar. There are also additional feedstocks available, including palm and cashew nut shells. Despite producing less wood, they are still used (Patwardhan et al., 2010).

9.2.2 Fast Pyrolysis Technique

Fast pyrolysis is regarded as a cutting-edge and environmentally beneficial technology. It has garnered a lot of interest because it produces a lot more bio-oil than previous technologies. At a specific temperature of between 400 and 700°C, it is a continuous procedure (Sipra et al., 2018). Fast pyrolysis often takes place in the reactor over a brief interval of time. Chemical kinetics, transition events, and heat transfer rates have a remarkable impact on the chemistry of pyrolysis end products. Up to 70% of the bio-oil can be produced by the fast pyrolysis method (Xiu and Shahbazi, 2012; Lehto et al., 2013). With rising temperatures, the amount of carbon in the char increases for both the slow and quick pyrolysis processes. However, due to longer maintenance times and a slower heating rate, slow pyrolysis results in char that contains less oxygen compared to that of quick pyrolysis (Velghe et al., 2011). The largest amount of bio-oil that can be formed, according to numerous research, is when the temperature of the reaction inside the reactor hits 500°C (Stevens et al., 2009; Demirbas and Balat, 2009). Additionally, this technique produces syngas and charcoal (Lehto et al., 2013).

9.2.3 Flash Pyrolysis Technique

The highly developed pyrolysis method known as "flash pyrolysis" is thought to be able to produce large quantities of high-quality syngas and bio-oil with very little

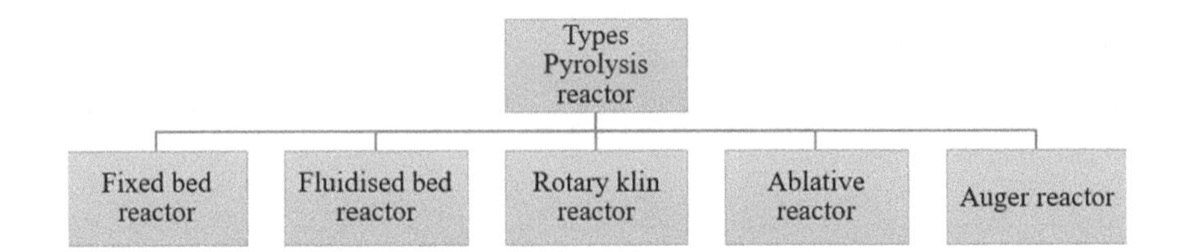

FIGURE 9.3 Pyrolysis reactor types.

water content (Yu et al., 2007). The high temperatures used for this pyrolysis process range from 700 to above 1,000°C. This pyrolysis technique uses high temperatures between 700 and above 1,000°C per second. Dwelling point is only somewhat longer as compared to that of the rapid pyrolysis technique, which requires less than 0.5 sec (Win et al., 2020). The feedstock should have the smallest feasible particle size because this recipe requires a very high heating rate (Kataki et al., 2018). With this technology, it is possible to convert about 80% of the feedstock's limited energy to various energy products, and the created products' energy density is higher than that of the raw feedstock. Moreover, it contains heavy metals and nitrogen pollutants, which require a massive amount of expensive hydrogen to be removed (Fang et al., 2018).

9.3 PYROLYSIS REACTOR TYPES

There are various pyrolysis reactor types used to produce bio-oil, with fluidized bed (bubbling and circulation), fixed-bed, rotary kiln, screw reactors, and ablative reactors being the most popular. Pyrolysis requires two key design considerations in order to be effective and capable: the reactor must enable significant heat transmission to the feedstock and a brief dwell period (Musale et al., 2013).

9.3.1 Fixed-Bed Reactor

This reactor is affordable and simple to construct when using uniform-sized solid waste as a feedstock. Heat is provided for the thermal collapse of solid waste in this reactor both externally and by allowing minimal incineration. The products are released out of the pyrolyzer, whereas the charcoal stays in the reactor because of the rise in the size of the gaseous substance. A sweep gas is widely employed in some of the most advanced fixed-bed reactors to eliminate syngas inside the reactor (Onay, 2007). This syngas is usually inert and deficient in oxygen. The primary output of this kind of reactor is biochar. When processing large amounts of sample, this type of reactor has several disadvantages, including a prolonged residence time, a slow heating rate, and an uneven temperature distribution inside the reactor. Figure 9.4 represents the process used in a fixed-bed reactor.

9.3.2 Fluidized Bed Reactor

It is a general-purpose reactor with a wide range of uses in both laboratories and large-scale industries. Typically, the chemical and oil industries use these reactors.

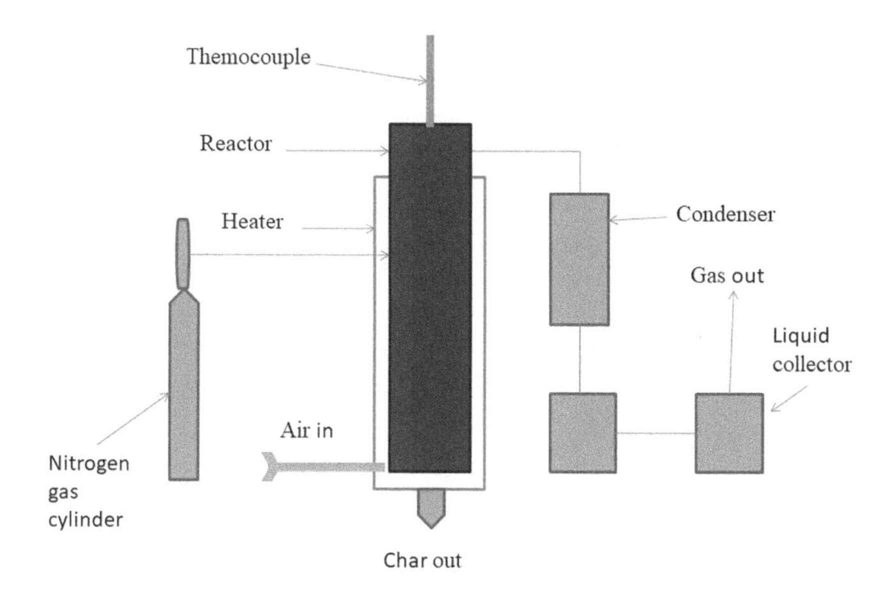

FIGURE 9.4 A fixed-bed pyrolysis reactor.

When bio-oil is fed and produced in a sustained manner, fluidized bed reactors are more effective. There are two variations of this type of reactor: bubbling and circulating.

A bubbling fluidized bed reactor receives shredded solid waste with a size range of 2 to 6 mm, which is then added to heated sand. The bed is usually fluidized using an inert gas. As long as it is ensured that the oxygen content is below stoichiometric levels and does not allow for total oxidation of the chemical, small amounts of air may also be employed. The inert bed solids must be mixed properly, and better, more consistent temperature regulation is required (Beheshti et al., 2015). The biochar in the bed needs to be separated in order to stop secondary cracking, because it serves as a catalyst for vapor cracking. Most frequently, biochar is separated from the produced gas using cyclones on one or more scales. Figure 9.5 illustrates the process used in a bubbling fluidized bed reactor.

Just like a bubbling fluidized bed reactor, a circulating fluidized bed reactor also operates on the same principle. The circulating type differs only in that solids are continually circulated via an internal loop comprised of a loop and a cyclone seal when the bed is firmly strained. The development of this reactor relies on a distinctive hydrodynamic technology known as fast bed (Grace and Bi, 1997). The fundamental advantage of this reactor design is the simplicity with which the embedded biochar can be separated. The bed solidifies after being recycled through the loop, seals to the reactor, and absorbs the heat produced by incineration. Figure 9.6 represents the process used in a circulating fluidized bed pyrolysis reactor.

9.3.3 Rotary Kiln Reactor

This kind of pyrolysis reactor heats the feedstock more efficiently than the fixed-bed reactors (Li et al., 2014). The dependable ceramic-coated cylinder and slow rotation

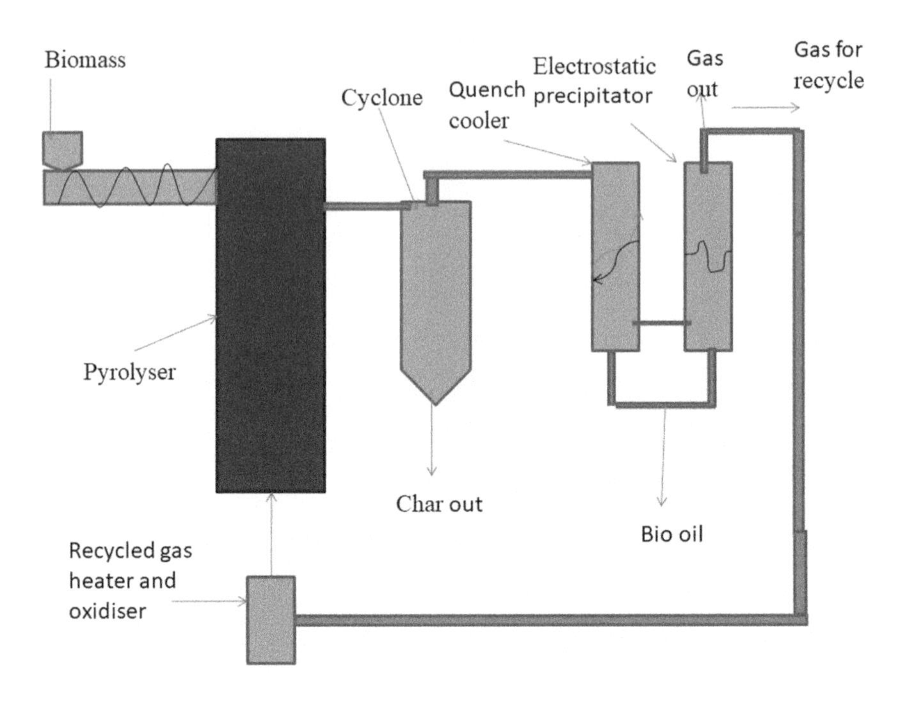

FIGURE 9.5 A bubbling fluidized bed reactor.

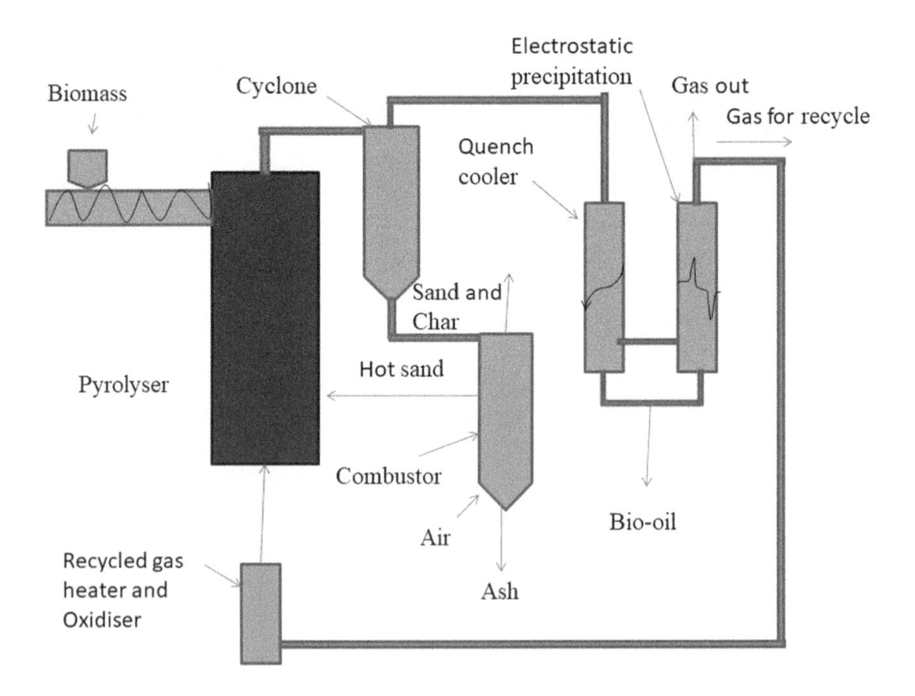

FIGURE 9.6 A circulating fluidized bed pyrolysis reactor.

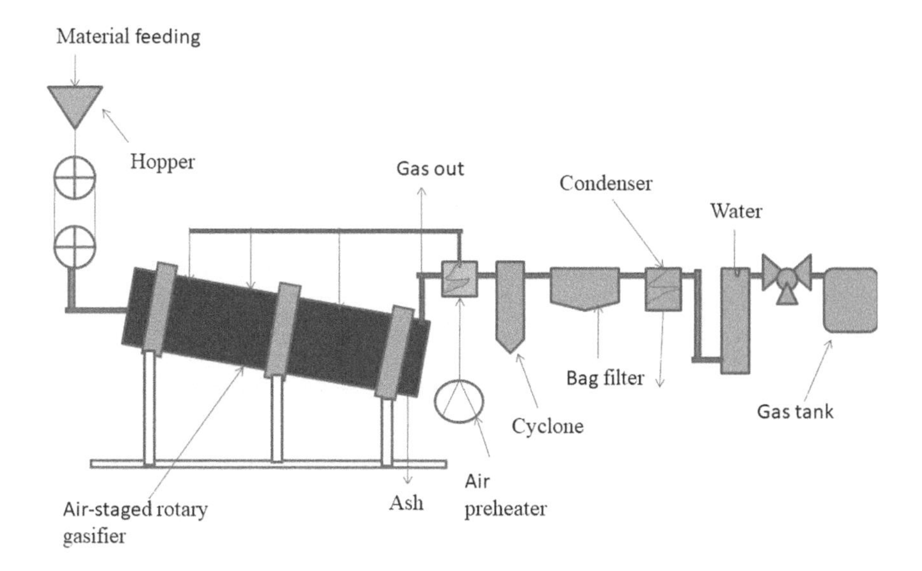

FIGURE 9.7 Rotating kiln pyrolysis reactor.

of the rotary kiln reactor effectively combine solid waste. It can be used for many different things, though this type of reactor employs a slow pyrolysis process, suggesting that the heating rate is also slow. The duration of stay is 1 h, and the maximum heating rate is 100°C/min (Fantozzi et al., 2007).

The rotation of the kiln facilitates the proper mixing of the solid waste. As pyrolytic gases are out as the mixture descends the cylinder and carbonizes, it is then gradually heated. Due to the presence of several advantages, the rotary kiln reactor is the most frequently and efficiently used reactor for the pyrolysis of solid (Chen et al., 2014). The ability to feed heterogeneous materials, the flexibility in adjusting the residence duration, the ability to mix solid waste properly, the lack of a need for pre-treatment of solid waste, and the ease of maintenance are some of the benefits of this reactor. Figure 9.7 illustrates the process used in a rotating kiln pyrolysis reactor.

9.3.4 Ablative Reactor

The pyrolysis process in this reactor is analogous to the melting of butter in a hot vessel. High pressure forms during the pyrolysis process among the feedstock particle and the heated wall of reactor in an ablative reactor. The liquid feedstock substance melts because the high pressure allows for unimpeded heat conduction from the feedstock particles. A thin film of liquid develops as the feedstock particles move up against the heated wall. The liquid sheet soon evaporates when heated, leaving the pyrolysis zone. This type of reactor typically has a high heat transfer rate and a brief residence duration (Peacocke et al., 1994). The ablative reactor may produce bio-oil with a yield of up to 80%. Figure 9.8 represents the process used in an ablative reactor.

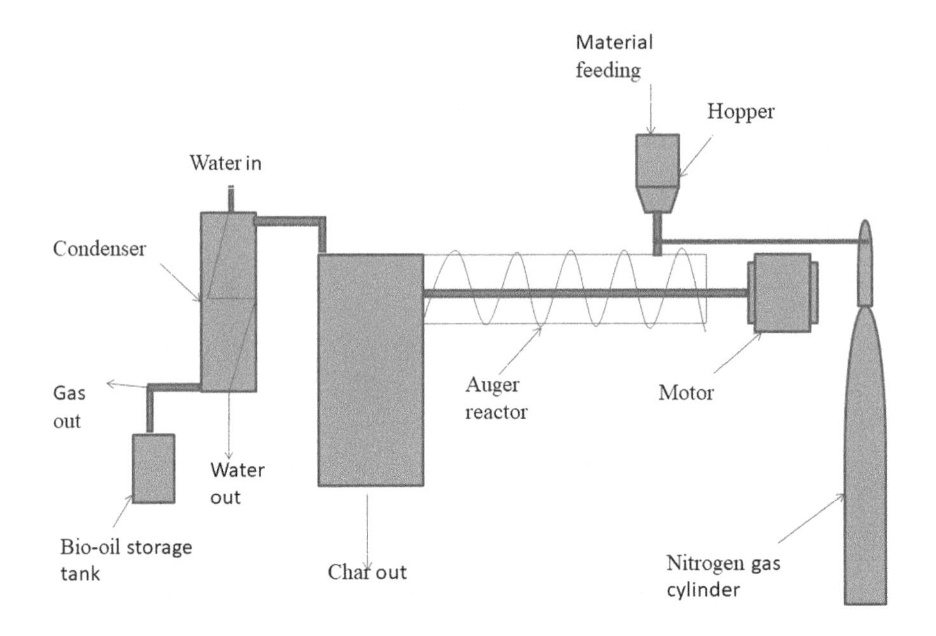

FIGURE 9.8 An ablative reactor.

9.3.5 Auger Reactor

This kind of reactor is frequently referred to as a screw reactor since it has a screw. The feedstock is lowered into the reactor with the help of the rotating screw, and the heat required for the process is given through the wall of the reactor. The screw frequently performs two tasks as a result: first, it mixes the feedstock, and second, it controls how long the feedstock stays in the reactor. An inert gas is frequently supplied into the feedstock hopper system to ensure that no oxygen enters the feedstock. Additionally, the reactor uses a minor amount of positive pressure produced by the inert gas to transmit the pyrolysis vapors. According to some reports, reactor tubes with a lower diameter have been observed to be sufficiently heated by external heat sources (Bortolamasi and Fottner, 2001).

Nevertheless, a heated solid carrier containing feedstock particles is required if large volumes of innovation are anticipated. Because of this, feedstock particles can interact more quickly as they move through the reactor tubes. Pellets of steel or ceramic can be used to create the solid carrier. The bio-oil is created by compressing the pyrolysis process–generated vapors in a condenser (Badger and Fransham, 2006). Figure 9.9 illustrates the process used in an auger reactor.

9.4 TYPES OF ENERGY RECOVERY FROM PYROLYSIS

9.4.1 Bio-Oil

India comes in 7th position in producing bio-oil or pyrolysis oil in the globe. Presently, India is producing roughly 450 to 500 MT (million tons) of biomass in a year.

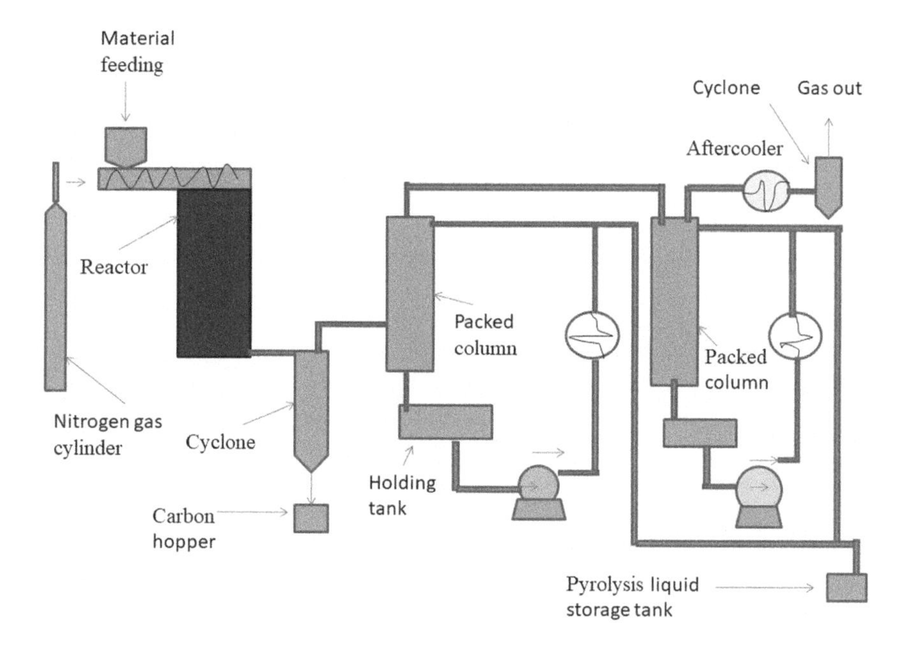

FIGURE 9.9 An auger reactor.

Chances of exploiting energy from organic matter is higher, compared to other renewable options, provided that thorough planning and execution is done (https://www.icac.org/projects/CommonFund/20/15chandra.pdf). Using organic matter and residues for heating purpose through ignition for cooking is still a common practice in most parts of rural areas in India (Jasvinder and Sai, 2010). Biochemical and thermochemical conversion could also be used for power generation, but due to financial hindrance, it is still not considered feasible to implement. Sugarcane, banana, rice, coconut, maize, and bajra are some of the prominent organic feedstock that can be used to harness renewable energy.

Research is going on in the area of pyrolysis process to check the effect of heating rate, heating time, etc. on bio-oil yield. Researchers in India are mainly focusing on the feedstocks. Some of the examples of organic materials are agricultural residue, wood, edible oil cakes, edible and non-edible oilseeds. Bagasse, husk, and straw are the residues of agriculture feedstocks which are available aplenty in India (Kumar et al., 2010). But these were not used for commercial purposes, as slow pyrolysis was performed in pyrolysis reactors having small capacity. The convenient sample size for pyrolysis treatment is either 25 g of powder or a maximum of 10 mm length of waste. The maximum temperature of reactor was set on 500°C. The bio-oil yield from wood and agriculture residue is extracted in the range of 20–50% (Patrick Horne and Williams, 1996). The extract of bio-oil can be raised by carrying the pyrolysis treatment in an airtight reactor. Using edible oils for biodiesel generation is not supported because India ships in edible oils from other countries. Very few works are reported regarding pyrolysis of oilseeds. This is because of the edible nature of oil cakes and their use as animal feed and manure (Sarkar and Chowdhury, 2014).

Some researchers are proposing digestion and gasification for extracting bio-oil from non-edible seed cakes, but others are interested in pyrolysis process under vacuum conditions for obtaining value-added products from waste.

Increase in vehicle utilization results in the accumulation of automobile tires in the open grounds unchecked, which causes severe environmental impacts. The dumping of tires in open areas is also expected to increase in the future. The scientific community is suggesting to exploit bio-oil from used cycle tubes and tires, which can be used as alternative propellant in internal combustion (IC) engines (Edwin et al., 2013).

Statistics shows that 34,69,780 t of plastic waste were generated in India in the year 2020 (Central Pollution Control Board, 2021). Plastics are used in agricultural, commercial, domestic, transport, and industrial sectors. In the era of fast-growing economy, cheaper plastics are easily accessible and affordable. But the drawback of using plastic is that plastics are non-decomposable and non-renewable. Examples of renewable plastic include polyethylene terephthalate (PET), high-density polyethylene (HDPE), low-density polyethylene (LDPE), polypropylene (PP), polyvinyl chloride (PVC), and polystyrene (PS), and non-renewable plastics are multilayer and laminated plastics, Bakelite, polycarbonate, melamine, and nylon (Valli et al., 2012). One of the methods of extracting energy and waste from recyclable fuels is pyrolysis technique. Researchers have made attempts to derive bio-oil and characterize and use it as a substitute propellant for compression ignition (CI) engines and spark ignition (SI) (Christine et al., 2013).

The prominent variables affecting yield rate and quality of the value-added products are the nature of feedstock, heating temperature, and vapor residency time, efficacy of liquefaction, reactor type, and kind of technique. Researchers claim that the bio-oil yield goes up to 60% through pyrolysis technique. Residues of ash produced from feedstock can have an impact on the pyrolysis products. Loss of heat due to radiation and conduction in the heater is unavoidable, but improvement in the condensation process can increase yield of bio-oil (Murugan and Gu, 2015).

Pyrolysis oils contains alkene/alkane, that is, hydrocarbons. Bio-oil recovered from organic feedstock does not contain aromatic compound. Bio-oil recovered from automobile tires and used plastics contains aliphatic and aromatic compounds (benzene). Hence, bio-oil mainly comprises organic compounds, such as -CHO, -OH, -COOH, and aromatics (Cordella et al., 2012; Lu Y et al., 2012). These oils are acidic in nature. Therefore, complications arise due to poor ignition attributes and immiscibility properties with hydrocarbons while using it as a substitute source of fuel for IC engine. Thus, further treatment is required to refrain combustion devices and storage tanks from corrosion (Beld et al., 2013). Density and viscosity of petrol and diesel are lower than bio-oil. Hence, the ignition value of bio-oil is lower than that of diesel, which is a major problem for transporting and storing oil. The durability of combustion devices is affected by the amount of ash content, acidity, and aromatic group present in the bio-oil. Bio-oil acquired from organic waste with denser texture, such as waste tires and rubber, forms steady solution when mixed with biodiesel, whereas bio-oil acquired from lignocellulose and cellulose cannot blend easily with diesel. The chemical properties of bio-oil can be altered to mix easily with petroleum-based fuels by implementing developed methods. Some of these advanced methods are hydro-treatment (Majhi et al., 2013), steam reforming (Trane et al., 2012), esterification (Cui et al., 2010), hydro-deoxygenation (Joshi and Lawal, 2012), and emulsification (Ikura et al., 2003). Phenols or amines present in them can prove to be helpful by improving the oxidation stability of biodiesel.

9.4.2 Biogas

Yield of the biogas is maximized when heat energy required is provided by the steam. This means that maximum generation of the biogas is affected at higher temperature (800°C) by whether the source of energy is heat or steam. While considering medium- and low-temperature treatments, it is seen that generation proportion is directly proportional to both retention time and temperature. If temperature is set at 600°C, for 180 min, then maximum biogas yield can be achieved. Biogas derived through this process can either be utilized directly as propellant or after separating it into methane and hydrogen gas.

1. *Biomethane.* The public transport buses uses 80 kg of methane approximately in a day (The Hindu, 2013). Roughly estimated, 346 buses can be supported from 1 mt of waste when treated for 120 min at 873 K. Likewise, auto rickshaws consume 6.5 kg CNG/day (200 km/day). If 1 mt of waste pyrolysis for 120 min at 873 K is converted into energy, that would support 4,597 auto rickshaws per day (Agarwal et al., 2013).
2. *Biohydrogen.* Biohydrogen is chosen as the alternative for fuel in the future because of zero emissions. Heat or electricity is released in the form of energy, along with pure water. It is mainly used in the chemical synthesis process or as a source of energy fuel. Conversion of biogas into energy is used in bi-fuel engines (Verhelst et al., 2009), fuel cell electrical vehicle (FCEV) (Kelly et al., 2011), residential applications (Kazempoor et al., 2009), bi-fuel generators (Sainz et al., 2011), whereas chemical synthesis process falls under hydrogenation of oils (Jianhua et al., 2012), fertilizer production, and ammonia synthesis by Haber's process.

9.4.3 Activated Carbons

To extract the biochar, raw waste materials are heated for 60 mins in a tube furnace to raise the derivative thermogravimetric (DTG) temperature. The biochar activation was done by increasing furnace temperature from 600°C to 900°C in the presence of nitrogen at 10°C/min. Again, reaction was continued while gas was changed to CO_2, and activation temperature was maintained for 30 min. After that, the temperature was brought down to room temperature in the presence of nitrogen to reduce the chances of oxidation. CO_2 acts as the activating agent of biochar (Rodríguez-Reinoso and Molina-Sabio, 1992).

9.4.4 Capability of Activated Carbons in Removing Organic Pollutants

To illustrate the capability of activated carbon in removing pollutants from the environment, methylene blue can be utilized as an example of organic substituent to quantify adsorption capacity of the sorbents. Brunauer-Emmett-Teller (BET) surface area analysis can be used to show the capacity of a sample to adsorb pollutants from water (Goldfarb et al., 2017). Hence, low-temperature treated municipal solid waste samples (activated at 600°C and pyrolyzed at 408°C) and high-temperature treated samples (pyrolyzed and activated at 900°C) can be allowed for adsorption testing. Functional groups like alcohols, carboxylic acids, carbonyl, and amino group increase the adsorption rate of pollutants like methylene blue on biochars; ClO, in particular, raises the adsorption rate of biochars toward heavy metals (Regmi et al., 2012). These functional groups, provided with higher surface area, were supposed to have greater adsorption capability for municipal solid waste activated and pyrolyzed at higher temperature. But observed result was different from the theoretical aspect. The sample having surface area 248.38 m^2/g, pyrolyzed at lower temperature of 408°C, and activated at higher temperature of 900°C holds the highest equilibrium adsorption capability for methylene blue, at 326.90 mgdye/gchar. The adsorption capability of the same waste sample pyrolyzed at 900°C was found to be 233.51 mgdye/gchar. The sample both activated and pyrolyzed at 900°C had the adsorption capacity of 243.92 mgdye/gchar. The reason behind it can most likely be the changes in volatile carbon content of biochars, which decreases upon raising the heat energy provided during pyrolysis (Istan et al., 2016). This data emphasizes on the urgency of balance surface areas of specific biochar to improve the generation process, as extreme carbonization process at higher temperature decreases both surface area and absorption capability. It can also consume more energy and proves to be an ineffective choice for the proposed integrated waste management activity.

9.5 IMPACT OF CATALYST ON THE YIELD OF PYROLYSIS PRODUCTS

Pyrolysis mainly produces liquids as its by-products. It is very important to convert liquid into gas. The reason is that the constituents of by-products are water and organic compounds, viz., alcohols, aldehydes, organic acids, ketones, phenols, and ethers that are rich in oxygen, and acidic radicals, henceforth, are advised not to be used as fuel. Research work is ongoing for the post-treatment of pyrolysis liquid in order to improve its quality. Catalysts can overcome the lower efficacy by reducing activation energy and improving decomposition of waste, resulting in an increase in the yield of value-added products. Synthetic catalysts are less beneficial in terms of high price. Therefore, reusing catalysts or synthesizing catalysts from natural minerals or using lesser amounts of catalyst is suggested (Lin et al., 2010). Fluid catalytic cracking (FCC) is now being used to reduce the cost on catalysts during pyrolysis treatment of solid waste. Biochar can be collected as the by-product after completion of the pyrolysis process, when all the volatile components are eliminated from the solids. The impact of catalysts on pyrolysis yield is described.

9.5.1 Dolomite

It is an inexpensive and easily available material. It is used in the form of calcined dolomite, in the presence of high temperature, to increase biogas generation, in the meantime reducing biochar yield. The reason is the formation of coke and removal of oxygen from water. Hence, in the presence of a catalyst, syngas synthesized is a value-added product which can be used as transportation fuel. Calcined dolomite can also be utilized to remove tar from the product gas. But it only functions above 500°C (Tursunov, 2014).

9.5.2 Zeolite

The structure of zeolite is microporous in nature, with higher BET surface area.

M. Rehan et al. (2017) added zeolite in powder form mixed with plastic in the reactor, generating different types of catalysts. It is found that nearly 99% of bio-oil generated comprises of aromatic hydrocarbons. Even though the catalytic process decreases oil production from 80.8 to 52%, the quality of the oil is enhanced. Refining is important in order to utilize the synthesized bio-oil. Natural zeolite is usually proposed by researchers to skip cost expenses on the synthetic catalysts. Activated zeolite is stable at higher temperature. The presence of activated zeolite through the pyrolysis process improves secondary pyrolysis of small hydrocarbon units. This indicates that zeolite shows splendid productivity of value-added products when added as a catalyst (Rehan et al., 2017).

9.5.3 Nickel (Ni) and Ruthenium (Ru)

These catalysts mainly focus on hydrogen gas generation. As plastic waste contains different kinds of chemical structures with different reactivity, henceforth, Namioka et al. (2011) supplied Ru as the catalyst to decrease fluctuation in between steam reforming reaction rates of polypropylene (PP) and polystyrene (PS). Earlier, nickel was used as a catalyst to extract hydrogen gas from plastic wastes. Presently, Ru has proven to be efficient in reducing the void between different chemical structures of plastic waste, and also reducing the requisite temperature below 200 K. This decreased temperature improves thermal efficacy and inhibits disintegration of catalysts in the presence of high temperature.

9.5.4 Other Catalysts

Usually, catalyst used in pyrolysis enhances the generation of product and decreases the energy used in the process (Usman et al., 2012). In addition to these catalysts, alternatively, economic improvement of the pyrolysis process through catalytic cracking is used to amalgamate waste with fluid catalytic cracking (FCC) system. These FCC catalyst systems boost market accessibility of transformation of solid waste into FCC unit. Nonetheless, equilibrium FCC catalysts are more effective in comparison to natural FCC used in pyrolysis.

9.6 ENVIRONMENTAL IMPACTS OF PYROLYSIS TECHNIQUE

Modern pyrolysis techniques make it somewhat simpler to lessen and get rid of corrosion and pollution, by stopping the development of big aromatic chlorinated compounds and maintaining heavy metals, chlorine, and sulfur in plant wastes. The pyrolysis method produces modest amounts of gas that allow lower-dimensional gas to be used for sinking costs, operating expenses, and cleaning equipment. The emission of ammonia (NH_3) started at 260°C and continued up to and above those temperatures, whereas the emission of hydrochloric acid (HCl) started at 230°C and terminated at 400°C, respectively. It was also observed that a significant amount of SO_2 was emitted when the temperature was held between 300°C and 600°C. As a result, there is a risk that NH_3 and SO_2 will contaminate gaseous and liquid materials (Zhan et al., 2019).

As feedstocks, mixtures of several forms of plastic trash were employed. The bio-oil and syngas generated by the pyrolysis contained a variety of metals, including P, Zn, K, Sb, Cr, Ca, S, Br, and Cl. When mixed solid waste was used as feedstock, more pollutants were created than when plastic waste was used. It is believed that PVC, a dangerous substance that is the primary catalyst for the production of HCl, is present in the plastic portion of solid waste. This HCl emission could contaminate syngas and bio-oil by inducing degradation (Yang et al., 2014). Additionally, during the pyrolysis of solid waste, HCl may be created at higher temperatures when NaCl combines with water to form NaOH and HCl. This is the rationale behind why different HCl emission patterns have been predicted by different studies. At around 850°C, two pollutants, CO and CO_2, began to form as the primary pollutants.

To produce energy recovery products that improve facilities, a single-element pyrolysis method should be carefully considered. For this reason, mutual pyrolysis, combustion, and gasification technologies were adopted in industrial pyrolysis plants. Moreover, syngas filtering emission control systems were created. There are numerous actions that can be taken to lessen the negative effects that pyrolysis has on the environment. Among such steps are the use of catalysts to improve the product quality, the interception of HCl, NH_3, and SO_2 in the gaseous phase, and the release of a few components from the feedstock. Miskolczi et al. (2013) conducted an experiment where two different types of catalysts, calcium- and iron-based, were employed. They asserted that iron-based catalysts generated the best results for absorbing bromine from deteriorating oil, whereas calcium-based catalysts demonstrated excellent efficacy in the elimination of chlorine. It was discovered the preceding two catalysts had no impact on nitrogen removal. Many scientists modified the pyrolysis process to eliminate a number of impurities from the waste products. Most often, this technique has been used in screw reactors. In the initial stage of this procedure, halogen hydrides are emitted in the low-temperature region of the pyrolysis reactor. In the second stage, polymer matrices degrade in the high-temperature region. The primary objective was to remove HCl more successfully by employing the lowest temperature feasible in order to stop HCl from entering pyrolysis (Yung et al., 2014).

It is seen that pollutants like Cl, N, and S can be transferred from solid waste to the pyrolysis products. It is also shown that food waste causes the rise in bio-oil and Cl concentrations in these products. The value of the biochar produced using the pyrolysis method from solid waste can be increased by controlling the pyrolysis at higher temperatures. It might help reduce the amount of volatile organic matter in biochar, increase its specific surface area, and increase the latent heat value of the material (Wang et al., 2012; Rajarao et al., 2014).

9.7 CONCLUSION

Solid waste management is currently a major concern for many developing nations. The quickest method to reduce trash volume and recover energy from waste is developed using thermochemical alteration technology. Thermochemical processes like gasification, incineration, and pyrolysis are examples. Pyrolysis is reported to be the most appropriate and effective technology. Pyrolysis of municipal solid waste is achieved with the help of various kinds of reactors. The most commonly practiced reactor types are fluidized bed reactors, fixed-bed reactors, rotary kiln reactors, screw reactors, and ablative reactors. Out of these, the rotary kiln reactor is the most promising reactor because of its capability to process various kinds of solid waste materials, mix these waste properly, and easily maintain the reactor. The pyrolysis oil is the most valuable and practical by-product of pyrolysis. The solid char is advertised as a potential fuel source; however, there is a danger in it, because it contains heavy metals and other organic contaminants. The parameters impacting the composition of the products must be developed by understanding the reaction kinetics in order to maximize the composition and yield of these desired products. While pyrolysis occurs in an oxygen-free environment, which lowers the carbon dioxide and nitrogen oxide emissions, it is more environmentally friendly than other treatment techniques. It is a simple and adaptable process, since the needed product dispersion may be achieved by varying parameters like temperature, residence time, and pressure. These are several interpretations of the chapter. Rotary pyrolysis is the most widely used method for managing solid waste, and it typically produces 25%, 27%, and 43% of syngas, biochar, and bio-oil, respectively.

9.8 FUTURE ASPECTS

Constituents of waste differ from area to area; hence, pre-treatment of waste through drying and shredding methods is important. An appropriate way should be implemented for pre-treating the municipal solid wastes in order to sort waste effectively (Zhou et al., 2010). At the same time, large-scale waste treatment plants should also be constructed to reduce the gap in between waste discard rate and pyrolysis of solid waste for energy recovery. Emphasis on the research work regarding different components of the waste and complementary effects of different waste components on pyrolysis product should be given. Especially how mixing of different components can affect quality and quantity of pyrolysis by-products. Co-pyrolytic products and the use of the pyrolysis product as an alternative for fuel should also be taken into consideration in the near future. In the meantime, it is also important to improve sorting methods for waste to remove moisture from the condensate. Additionally, these points require to be enhanced for more proficient future design of the plant:

1. Using thin bed reactors to reduce the energy required by the pyrolysis system.
2. Replacing the cistern turbine generator with a gas machine to restore the efficiency of power output.
3. Avoiding the heat from scaling up for straightforward and long-term activities.

REFERENCES

Agarwal M, Tardio J, Venkata MS. 2013. Critical analysis of pyrolysis process with cellulosic based municipal waste as renewable source in energy and technical perspective. *Bioresour Technol* 147:361–368. doi: 10.1016/j.biortech.2013.08.011.

Anuradda G, Rangan B. 2001. Biomass pyrolysis for power generation – a potential technology. *Renew Energy* 22:9–14.

Babu BV. 2008. Biomass pyrolysis: a state-of the-art review. *Biofuels Bioprod Biorefin* 2:393–414.

Badger PC, Fransham P. 2006. Use of mobile fast pyrolysis plants to densify biomass and reduce biomass handling costs – a preliminary assessment. *Biomass Bioenergy* 30:321–325.

Beheshti SM, Ghassemi H, Shahsavan-Markadeh R. 2015. Process simulation of biomass gasification in a bubbling fluidized bed reactor. *Energy Convers Manag* 94:345–352.

Beld BVD, Holle E, Florijn, J. 2013. The use of pyrolysis oil and pyrolysis oil derived fuels in diesel engines for CHP production. *Appl Energy* 102:190–197.

Bortolamasi M, Fottner J. 2001, September. *Design and Sizing of Screw Feeders: Annual Report 2019–20 on Implementation of Plastic Waste Management Rules, 2016.* New Delhi: Partec Central Pollution Control Board, pp. 1–91.

Central Pollution Control Board. 2021, *Annual Report 2020–21 on Implementation of Plastic Waste Management Rules*, 2016. 1–41. https://cpcb.nic.in/uploads/plasticwaste/Annual_Report_2020-21_PWM.pdf.

Chen D, Yin L, Wang H, He P. 2014. Pyrolysis technologies for municipal solid waste: a review. *Waste Manag* 34:2466–2486.

Christine C, Shijo T, Varghese S. 2013. Synthesis of petroleum-based fuel from waste plastics and performance analysis in a CI engine. *J Energy* 10, Article ID 608797.

Cordella M, Torri C, Adamiano A, Fabbri D, Barontini F, Cozzani V. 2012. Biooils from biomass slow pyrolysis: a chemical and toxicological screening. *J Hazard Mater* 231:26–35.

Cui HY, Wang JH, Wei SQ, Zhuo SP, Li ZH, Wang LH, Yi WM. 2010. Upgrading bio-oil by esterification under supercritical CO_2 conditions. *J Fuel Chem Technol* 38:673–678.

Demirbas MF, Balat M. 2009. Bio-oil from pyrolysis of black alder wood. *Energy Sources, Part A* 31(19):1719–1727.

Edwin RR, Robert KZ, Pillai BC. 2013. Optimization of process parameters in flash pyrolysis of waste tyres to liquid and gaseous fuel in a fluidized bed reactor. *Energy Convers Manag* 67:145–151.

Fang S, Gu W, Chen L, Yu Z, Dai M, Lin Y. 2018. Ultrasonic pretreatment effects on the co-pyrolysis of municipal solid waste and paper sludge through orthogonal test. *Bioresour Technol* 258:5–11.

Fantozzi F, Colantoni S, Bartocci P, Desideri U. 2007. Rotary kiln slow pyrolysis for syngas and char production from biomass and waste – part I: working envelope of the reactor. *J Eng Gas Turbines Power* 129:901–907.

www.globalmethane.org/documents/analysis_fs_en.pdf.

Goldfarb JL, Buessing L, Gunn E, Lever M, Billias A, Casoliba E, Schievano A, Adani F. 2017. Novel integrated biorefinery for olive mill waste management: utilization of secondary waste for water treatment. *ACS Sustain Chem Eng* 5(1):876–884. http://dx.doi.org/10.1021/acssuschemeng.6b02202.

Grace JR, Bi H. 1997. Introduction to circulating fluidized beds. In Grace JR, Avidan AA, Knowlton TM, editors. *Circulating Fluidized Beds*. Dordrecht, Netherlands: Springer, pp. 1–20.

The Hindu. 2013, March 4. APSRTC's CNG hit. *The Hindu*, Hyderabad.

Ikura M, Stancilescu M, Hogan E. 2003. Emulsification of pyrolysis derived bio-oil in diesel fuel. *Biomass Bioenergy* 24:221–232.

Istan S, Ceylan S, Topcu Y, Hintz C, Tefft J, Chellappa T, Guo J, Goldfarb JL. 2016. Product quality optimization in an integrated biorefinery: conversion of pistachio nutshell biomass to biofuels and activated biochars via pyrolysis. *Energy Convers Manag* 127:576–588. http://dx.doi.org/10.1016/j.enconman.2016.09.031.

Jasvinder S, Sai G. 2010. Biomass conversion to energy in India – a critique. *Renew Sustain Energy Rev* 14:1367–1378.

Jianhua G, Renxiang R, Ying Z. 2012. Hydrotreating of phenolic compounds separated from bio-oil to alcohols. *Ind Eng Chem Res* 51:6599–6604.

Joshi N, Lawal A. 2012. Hydrodeoxygenation of pyrolysis oil in a microreactor. *Chem Eng Sci* 74:1–8.

Kataki R, Bordoloi NJ, Saikia R, Sut D, Narzari R, Gogoi L. 2018. Waste valorization to fuel and chemicals through pyrolysis: technology, feedstock, products, and economic analysis. *Waste to Wealth* 477–514.

Kazempoor P, Dorer V, Ommi F. 2009. Evaluation of hydrogen and methane fuelled solid oxide fuel cell systems for residential applications: system design alternative and parameter study. *Int J Hydrogen Energy* 34:8630–8644.

Kelly NA, Gibson TL, Ouwerkerk DB. 2011. Generation of high pressure hydrogen for fuel cell electric vehicles using photovoltaic-powered water electrolysis. *Int J Hydrogen Energy* 36:15803–15825.

Kumar G, Panda AK, Singh RK. 2010. Optimization of process for the production of bio-oil from eucalyptus wood. *J Fuel Chem Technol* 38(2):162–167.

Lehto J, Oasmaa A, Solantausta Y, Kytö M, Chiaramonti D. 2013. *Fuel Oil Quality and Combustion of Fast Pyrolysis Bio-oils* (Vol. 87). Espoo, Finland: VTT.

Li SQ, Yao Q, Chi Y, Yan JH, Cen KF. 2014. Pilot-scale pyrolysis of scrap tires in a continuous rotary kiln reactor. *Ind Eng Chem Res* 43:5133–5145.

Lin H, Huang M, Luo J, Lin L, Lee C, Ou K. 2010. Hydrocarbon fuels produced by catalytic pyrolysis of hospital plastic wastes in a fluidizing cracking process. *Fuel Process Technol* 91:1355–1363.

Lu JS, Chang Y, Poon CS, Lee DJ. 2020. Slow pyrolysis of municipal solid waste (MSW): a review. *Bioresour Technol* 312:123615.

Lu Y, Wei X-Y, Cao J-P, Li P, Liu F-J, Zhao Y-P, Fan X, Zhao W, Rong, L-C, Wei Y-B, Wang S-Z, Zhou J, Zong Z-M. 2012. Characterization of bio-oil from pyrolysis of rice husk by detailed compositional analysis and structural investigation of lignin. *Bioresour Technol* 116:114–119.

Majhi A, Sharma YK, Bal R, Behera B, Kumar J. 2013. Upgrading of bio-oils over PdO/Al2O3 catalysts and fractionation. *Fuel* 107:131–137.

Miskolczi N, Ateş F, Borsodi N. 2013. Comparison of real waste (MSW and MPW) pyrolysis in batch reactor over different catalysts. Part II: contaminants, char and pyrolysis oil properties. *Bioresour Technol* 144:370–379.

Murugan S, Gu S. 2015. Research and development activities in pyrolysis – contributions from Indian scientific community – a review. *Renew Sustain Energy Rev* 46:282–295.

Musale H, Bhattacharyulu Y, Bhoyar R. 2013. Design consideration of pyrolysis reactor for production of bio-oil. *Int J Eng Trends Technol* 5.

Namioka T, Saito A, Inoue Y, Park Y, Min T-J, Roh S-A, Yoshikawa K. 2011. Hydrogen-rich gas production from waste plastics by pyrolysis and low-temperature steam reforming over a ruthenium catalyst. *Appl Energy* 88:2019–2026.

Onay O. 2007. Influence of pyrolysis temperature and heating rate on the production of bio-oil and char from safflower seed by pyrolysis, using a well-swept fixed-bed reactor. *Fuel Process Technol* 88:523–531.

Patrick Horne A, Williams Paul T. 1996. Influence of temperature on the products from the flash pyrolysis of biomass. *Fuel* 75(9):1051–1059.

Patwardhan PR, Satrio JA, Brown RC, Shanks BH. 2010. Influence of inorganic salts on the primary pyrolysis products of cellulose. *Bioresour Technol* 101:4646–4655.

Peacocke GVC, Madrali ES, Li CZ, Güell AJ, Wu F, Kandiyoti R, et al. 1994. Effect of reactor configuration on the yields and structures of pine-wood derived pyrolysis liquids: a comparison between ablative and wire-mesh pyrolysis. *Biomass Bioenergy* 7:155–167.

Rajarao R, Mansuri I, Dhunna R, Khanna R, Sahajwalla V. 2014. Characterisation of gas evolution and char structural change during pyrolysis of waste CDs. *J Anal Appl Pyrol* 105:14–22.

Regmi P, Garcia Moscoso JL, Kumar S, Cao X, Mao J, Schafran G. 2012. Removal of copper and cadmium from aqueous solution using switchgrass biochar produced via hydrothermal carbonization process. *J Environ Manage* 109:61–69. http://dx.doi.org/10.1016/j.jenvman.2012.04.047.

Rehan M, Miandad R, Barakat M, Ismail I, Almeelbi T, Gardy J, Hassanpour A, Khan M, Demirbas A, Nizami A. 2017. Effect of zeolite catalysts on pyrolysis liquid oil. *Int Biodeterior Biodegrad* 119:162–175.

Rodríguez-Reinoso F, Molina-Sabio M. 1992. Activated carbons from lignocellulosic materials by chemical and/or physical activation: an overview. *Carbon NY* 30(7):1111–1118. http://dx.doi.org/10.1016/0008-6223(92)90143-K.

Roy GK. 1988. Municipal solid waste recycle – an economic proposition for a developing nation. *Indian J Environ Prot* 8(1):50–54.

Sainz D, Dieguez PM, Urroz JC, Sopena C, Guelbenzu E, Perez-Ezcurdia A, Benito-Amurrio M, Marcelino-Sadaba S, Arzamendi G, Gandia LM. 2011. Conversion of a gasoline engine-generator set to a bi-fuel (hydrogen/gasoline) electronic fuel-injected power unit. *Int J Hydrogen Energy* 36:13781–13792.

Sarkar A, Chowdhury R. 2014. Studies on catalytic pyrolysis of mustard seed press cake with NaCl. *Int J Eng Sci Res Technol* 3(6):90–96, 2277–9655.

Sipra AT, Gao N, Sarwar H. 2018. Municipal solid waste (MSW) pyrolysis for bio-fuel production: a review of effects of MSW components and catalysts. *Fuel Process Technol* 175:131–147.

Stamatov V, Honnery D, Soria J. 2006. Combustion properties of slow pyrolysis bio-oil produced from indigenous Australian species. *Renew Energy* 31(13):2108–2121.

Stevens D, Kinchin C, Czernik S. 2009. *Production of Gasoline and Diesel from Biomass via Fast Pyrolysis, Hydrotreating and Hydrocracking: A Design Case*. Richland, WA: Pacific Northwest National Laboratory.

Trane R, Dahl S, Skjoth-Rasmussen MS, Jensen AD. 2012. Catalytic steam reforming of bio-oil. *Int J Hydrogen Energy* 37:6447–6472.

Tursunov O. 2014. A comparison of catalysts zeolite and calcined dolomite for gas production from pyrolysis of municipal solid waste (MSW). *Ecol Eng* 69:237–243.

Usman M, Alaje T, Ekwueme V, Adekoya T. 2012. Catalytic degradation of water sachet waste (LDPE) using mesoporous silica KIT-6 modified with 12-tung-stophosphoric acid. *Pet Coal* 54(2):85–90.

Valli MJ, Gnanavel G, Thirumarimurugan M, & Kannadasan T. 2012. Alternate fuel from synthetic plastics waste – review. *Int J Pharm Chem Sci* 1(3):720–724, 2277–5005.

Van de Velden M, Baeyens J, Brems A, Janssens B, Dewil R. 2010. Fundamentals, kinetics and endothermicity of the biomass pyrolysis reaction. *Renew Energy* 35(1):232–242.

Velghe I, Carleer R, Yperman J, Schreurs S. 2011. Study of the pyrolysis of municipal solid waste for the production of valuable products. *J Anal Appl Pyrol* 92(2):366–375.

Verhelst S, Maesschalck P, Rombaut N, Sierens R. 2009. Efficiency comparison between hydrogen and gasoline, on a bi-fuel hydrogen/gasoline engine. *Int J Hydrogen Energy* 34:2504–2510.

Wang Z, Chen D, Song X, Zhao L. 2012. Study on the combined sewage sludge pyrolysis and gasification process: mass and energy balance. *Environ Technol* 33:2481–2488.

Win MM, Asari M, Hayakawa R, Hosoda H, Yano J, Sakai S-I. 2020. Gas and tar generation behavior during flash pyrolysis of wood pellet and plastic. *J Mater Cycles Waste Manag* 22:547–555.

Xiu S, Shahbazi A. 2012. *Co-deoxy-liquefaction of Swine Manure and Waste Vegetable Oil for Bio-oil Production.* In 2012 Dallas, Texas, July 29–August 1, 2012 (p. 1). American Society of Agricultural and Biological Engineers.

Yang SI, Wu MS, Wu, CY. 2014. Application of biomass fast pyrolysis part I: Pyrolysis characteristics and products. *Energy* 66:162–171.

Younes C, Mohammed K. 2013. Thermal conversion of biomass, pyrolysis and gasification – a review. *Int J Eng Sci* 2(3):75–85.

Yu H, Liu Y, Dong W, Li W, Li R. 2007. *Experimental Study on the Biomass Flash Pyrolysis.* Berlin, Heidelberg: Springer, pp. 1152–1154.

Zhan H, Zhuang X, Song Y, Chang G, Wang Z, Yin X, et al. 2019. Formation and regulatory mechanisms of N-containing gaseous pollutants during stage-pyrolysis of agricultural biowastes. *J Clean Prod* 236:117706.

Zhou Y, Wu W, Qiu K. 2010. Recovery of materials from waste printed circuit boards by vacuum pyrolysis and vacuum centrifugal separation. *Waste Manag* 30:2299–2304.

10 Endophytic Microbes in Agarwood Oil Production from *Aquilaria malaccensis* Lam. Engendering Bio-Resources for Socioeconomic Development

Bipul Das Chowdhury, Abhijit Bhattacharjee, and Bimal Debnath

ABBREVIATIONS

AAO	*Aquilaria agallocha* oil
Agar-bit	Biologically agarwood-inducing technique
BCD	Burning chisel drilling
CA Kit	Cultivated agarwood kit
P450	Cytochrome P450–dependent mono-oxygenase
EAA	Ethyl acetate extract of *Aquilaria agallocha*
FPP	Farnesyl pyrophosphate
FPS	Farnesyl pyrophosphate synthase
FTPEC	Flindersia-type 2-(2-phenylethyl) chromone
FORDA	Forestry Research and Development Agency
MeJA	Methyl jasmonic acid
MEP	Methylerythritol phosphate pathway
MAP	Mevalonic acid pathway

DOI: 10.1201/9781003517108-10

MAPK	Mitogen-activated protein kinase
OMT	O-methyltransferase
PTP	Partial trunk pruning
PAMP	Pathogen-associated molecular pattern
PECs	2-phenylethyl chromones
PKs	Polyketide synthases
SCS	Selective cutting system
SesTP	Sesquiterpene synthase
TFs	Transcription factors
Agar-Wit	Whole-tree agarwood-inducing technique

10.1 INTRODUCTION

Agarwood is a highly valued, dark resin-infused heartwood. It is a rich source of incense fragrance. "Agarwood" comes from "aloes," first mentioned by "Kalidasa" during the 4th- and 5th-century common era (Lee and Mohamed, 2016). Agarwood is also known as "the wood of the gods" because of its vast array of uses. Agarwood essential oil is a vital ingredient in the perfume industry, having high demand. The chief source of agarwood is the *Aquilaria* spp. tree. In *Aquilaria* plants, formation of agarwood is generally associated with the wounding and fungal infection of plants. Changes in the chemical compounds of softwood lead to the formation of agarwood. Endophytic fungi are important biological agents involved in agarwood formation by establishing a symbiotic relationship with the host plant. These endophytes live inside plants either intercellularly or intracellularly without causing any harm to the plant (Bhore et al., 2013). All *Aquilaria* trees do not produce agarwood. Gibson (1977) estimated that approximately 10% of wild *Aquilaria* spp. produces resin. Biological agents like fungi and microbes are required to induce agarwood formation and expansion to other parts of the tree. *Acremonium* sp., *Diplodia* sp., *Fusarium* sp., *Libertella* sp., *Scytalidium* sp., *Thielaviopsis* sp., and *Trichoderma* sp. were reported from several fungal-infected trees of Indonesia. *Micrococcus endophyticus*, a new species, was discovered from the roots of surface-sterilized *Aquilaria sinensis* (Chen et al., 2009). The use of specific molecular methods divulged the succession patterns of fungi (*Cunninghamella bainieri*, *F. solani*, and *Lasiodiplodia theobromae*) associated with wound-induced agarwood in wild *Aquilaria malaccensis* (Mohamed et al., 2014). In Vietnam, fungi like *Aspergillus phoenicis*, *Penicilinum citrnum*, and *Penicillium* spp. were essential strains for forming agarwood in *A. crassna*. *Bacillus* spp. produce natural products like bacteriocins, antibiotics, different enzymes, sources of novel products. Hence, their role in the industry has become more pertinent than ever.

Agarwood is known to the world by different synonyms due to differential usage in different trade types, diverse socio-cultural backgrounds, and different languages and dialects. They are known as eaglewood and gaharu in Malaysia and Papua

New Guinea; oudh or attar in the Middle East; aloe wood in Indonesia; agilawood, garoowood, calambae, and chen xiang in China; chim-hyuang in Korea; jinkoh in Japan; and agar in India. Indonesia and Malaysia are the two major contributors of agarwood globally and are also known as the place of origin for agarwood; apart from these countries, *Aquilaria* plantations are found in some regions of Bangladesh, Bhutan, India, Laos, Myanmar, Papua New Guinea, Thailand, and Vietnam (Akter et al., 2013). The most valuable and highest quality of agarwood is produced from the species of *A. malaccensis*. Apart from this genus, agarwood is also produced from *Gyrinops*, *Aetoxylon*, *Claoxylon*, *Encleia*, *Gonystylus*, and *Wikstroemia*. India exports a significant portion of its agarwood products to Saudi Arabia and the United Arab Emirates. Although agarwood is a tree species, it is treated as a non-timber forest product (NTFP). Therefore, the harvesting system of the *Aquilaria* and *Gyrinops* species is not subjected to the regulations imposed in the selective cutting system (SCS), which is applied for a species of timber forest product.

What are the factors that trigger agarwood formation in some old *Aquilaria* trees? It has been a mystery, as the agarwood resin forms in the tree trunk due to the response of certain known and unknown factors. It is seen that when an entire tree is infected, high-quality dense resin and oil are formed in the tree. There are three hypotheses promulgated regarding the formation of agarwood. It is the result of pathological, wounding, or nonpathological processes. Gianno (1986) stated that trees with a diameter at breast height (DBH) >20 cm can produce at least one kg of agarwood per tree. Moreover, a combination of 2% sugar solution, *Acremonium* sp., and methyl jasmonic acid was reported to produce agarwood of kemedangan quality class IV.

Production of agarwood through wounding by using hammer nails and blades leads to the formation of an inferior quality, which does not comply with market quality and demand. Several mechanical approaches were probed by Pojanagaroon and Kaewrak (2003). It was found that agarwood formation was stimulated by making holes, inducing wounds, and removing bark. The formation of agarwood was accentuated by seasonal changes. Various researchers showed successful agarwood formation by artificial induction methods, such as different injection methods, using different fungal strains, and transferring the strain directly through liquid media. Many inoculation practices have been developed to date with variable degrees of success.

The resinous agarwood is a sesquiterpenoid that acts as a defense compound, and it is placed in the phytoalexin group. These secondary metabolites are synthesized and accumulated in the plant in response to microbial infection, physiological stimulation, and stress response. These resins get accumulated in the phloem; they are hardened there, turn blackish, and later produce fragrance.

Despite considerable advancements in organic synthesis, 75% of prescribed drugs and essence across the globe are still being derived directly from plants. Hence, plant species remain an essential source of new drugs for curing diseases like cancer. In India, several agarwood grades are distilled separately and are finally blended to produce "attar." Minyak attar, a water-based perfume containing agarwood oil, is traditionally used by the Muslim community. In several soaps and shampoo products, the essence of agarwood is used. Top luxury fashion houses use agarwood oil in their premium collections. Agarwood oil is yet to be synthesized chemically. Although a few chemical substitutes of agarwood oil are available in the market, they are nowhere close to the natural product in terms of quality. Luxury brands are always willing to pay a hefty amount for agarwood oil (Akter et al., 2013). The yield and qualities of resinous agarwood also vary considerably.

10.1.1 Distribution Status

The genus *Aquilaria* belongs to the dicotyledonous family Thymelaeaceae, and among the 21 *Aquilaria* species, 13 species have been recognized as agarwood-producing species (Lee and Mohamed, 2016). In India, three species of *Aquilaria* are found: mainly, *A. khasiana* Hall. is available in the northeastern states (NE) of India, especially in the Khasi Hills of Meghalaya (Kanjilal et al., 1982); *A. macrophyla* Miq. is found in the Nicobar Islands; and *A. malaccensis* is found in the natural forests and foothills of the northeastern states of India at an altitude of 1,000 m above in Nagaland, Mizoram, Manipur, Arunachal Pradesh, Tripura, and West Bengal, especially in Assam and Meghalaya (Barden et al., 2000). The International Union for Conservation of Nature (IUCN) red-listed *A. malaccensis* as critically endangered. In contrast, this tree was listed as a threatened species by the Convention on International Trade in Endangered Species of Wild Fauna and Flora (CITES) and endangered species in Appendix II. Since 1995, *A. malaccensis* has been the primary source of agarwood and agarwood oil. *A. malaccensis* is also reported in Bangladesh, Bhutan, Indonesia, Iran, Malaysia, Myanmar, Philippines, Singapore, and Thailand.

Aquilaria malaccensis (syn. *A. agallocha*) is considered an endemic tree to northeast India (Kanjilal et al., 1982). It is a noteworthy species for producing good-quality agarwood and commercially valuable species of northeast India, including Tripura. This species is known as the "sanchi plant" in North-East India. In India, *A. malaccensis* is considered "extinct in the wild," especially in Assam and Tripura. This depletion in the number of agar plants is based on immense value and high demand for quality agarwood.

10.1.2 Cultivation Status of Agarwood

Due to the conservation efforts of CITES, those countries that are at the top of the agarwood trade have decided to cultivate *Aquilaria* and to ensure a continuous supply of agarwood or agarwood products. Because *Aquilaria* is the major agarwood-producing genus, the plantation of *Aquilaria* is under regulation by CITES representatives in each top trade country. It is recorded that a total of 1.3×10^3 ha of land is covered with 1.2×10^6 *Aquilaria* trees in Malaysia (Hashim et al., 2016). The total planted agar plants are about 3.4×10^6, as reported by the Forestry Research and Development Agency, Indonesia (FORDA), representing one-third of the total provinces (Turjaman and Hidayat, 2017). Over 8.0×10^5 seedlings of *Aquilaria* were planted in Bangladesh in 2007 (Chowdhury et al., 2016). In Bhutan, the planted trees are nearly about 2.0×10^4, whereas India has planted over 1.0×10^7 trees, accounting for the home gardens of Assam. Like other Asian countries, Southeast Asian country Myanmar also planted about 1.0×10^6 trees. In the trade history of agarwood, Thailand has an important role and also active in agar plantation, but there is no report regarding the estimation of *Aquilaria* plantation size and the number of plants. The plantation of agar plants has been developed from home gardens or small land plantations toward large-scale plantations by enterprising or private companies or by government agencies with interests in trade.

10.1.3 Trade History of Agarwood and Wood-Derived Oil

Agarwood has been traded for more than 2,000 years across Europe and Asia, especially in Southwest Asian countries (Burkill, 1966). Among the agarwood-producing species of *Aquilaria* in the CITES trade database, *A. malaccensis* is considered the only true agarwood-producing tree and the most commonly utilized agarwood species (Lee and Mohamed, 2016). *Aquilaria malaccensis* was the first species subjected to trade regulations under CITES (Rasool and Mohamed, 2016). Agarwood is traded in the form of chips, oil, logs, and timber. The majority of the agarwood is processed to make oil used as raw material to develop perfumes and other cosmetic products, while agarwood chips are used to prepare a powder which is utilized as raw material for incense, and solid woods can be carved to make wooden sculptures, expensive jewelry, and other religious items. In Malaysia, indigenous people (primary collectors) sell agarwood products to local traders, wherefrom the product is transported to the national supermarkets for resale. In Indonesia, Sumatra and Kalimantan Islands are the leading growers of this species, contributing 85% of the country's wild agarwood (Turjaman and Hidayat, 2017).

Based on the occurrence of this species, the import and export of agarwood products are high in Bangladesh, India, Indonesia, Malaysia, Singapore, and Thailand. In 2003, Singapore and Malaysia reported the highest amount of imported and exported agarwood products, and the purpose of the import–export is mainly commercial-based, and the source of agarwood is wild. Chung and Purwaningsih (1999) and Barden et al. (2000) reported earlier that the world's significant agarwood product exporters are Malaysia and Indonesia. In Singapore markets, it is estimated that 70% of the agarwood is imported from Indonesia, Myanmar, Thailand, Laos, Vietnam, and Cambodia, whereas in the global trade of agarwood chips, higher import was reported in 2007, higher export was reported in 2003, and in the case of agarwood oil, higher import was reported in 2013 and export in 2015, respectively.

10.1.4 The Relationship between Host Plants and Endophytes

Every living organism needs a few basic requirements to live, namely, habitat, nutrition, and a relationship with their neighboring organisms. *Endophytes* are defined as microbes that colonize living plant tissues without harming the host. Endophytes depend on their whole or partial life cycle associated symbiotically (either intercellularly or intracellularly) and asymptomatically with the host. They have a vast range of diversity in the world, including mainly fungi and bacteria. According to Vanessa and Christopher (2004), there should be an endophytic microbe in the healthy tissues of all plants. Endophytes protect their host by increasing the additional intrinsic host defense mechanisms and enhancing competitive abilities against other pathogenic microbes and various abiotic stress, like drought, cold, and heat stress. Endophytes eventually compete with the pathogenic organisms of the host for their essential needs, that is, habitat, nutrients, etc., but extensive colonization can cause a "barrier effect"; as a result, other local endophytes and harmful pathogens are competed out. Environmental factors play an essential role in any organism's diversity, as endophytic diversity depends on the nature of the host plants, geographical location, and other environmental features. The comparison of endophytic diversity at geographic localities shows that the tropical regions are more diverse in endophytes than temperate areas, because of higher plant diversity in the tropics. Besides these factors, the frequency intensities and degrees of disturbance also influence the composition of endophytic mycota. According to Arnold et al. (2001), in tropical regions, the woody nature of the host plants is suited more for the invasion of fungal endophytes.

10.1.5 Agarwood: The History of Oleoresin Formation

Agarwood mainly forms in the woody parts of agar plants, like stem, branch, and roots, where the xylem structures have been modified. The plant's xylem structures are mostly responsible for synthesizing "a white-milky substance" called oleoresin through the compartmentalization process. These synthesizing substances may be stored and distributed to the wounded areas, then may fill up the compartments, and as a result, solid and impregnated wood is formed. The three essential players of the agarwood formation process are the host plant, an inducer, and the environment. The formation of agarwood is a non-specific host response to any type of wound and microbial infection. Inducement of agarwood formation means introducing an inducer for initiating and stimulating the oleoresin formation. Several environmental factors also influence agarwood formation, in which time is the significant factor that has influenced the formation of agarwood (Mohamed et al., 2014); others are the age of the host, season of infection, soil fertility, temperature, humidity, geographical location of the host, species of host, and pests. The timing of agarwood formation is crucial for disease transmission. Rahman and Khisa (1984) reported that artificial induction produced more resin in September than in May. Tropical insect invasion was also season-specific and generally longer than those in temperate regions (Mabberly, 1992). Most insects in forest undergrowth were active in low-light conditions, including heavy cloud cover during daytime.

10.2 AGARWOOD PRODUCTION AND SUSTAINABLE DEVELOPMENT

The quest for sustainable development, carbon sequestration, revenue generation, and growth of GDP has prompted us to explore the world around us and discover the available avenues. These avenues, if correctly tapped, can address the problem of unemployment across the globe. With the outburst of the population in recent years, unemployment and pollution have been burning issues requiring immediate attention. Production of newer alternatives needs much time and trial. Existing resources need to be tapped for a quick solution. The developing countries of the world are now focusing on plantation *Aquilaria* in recent years. This plant is primarily cultivated for its aroma and has found usage in the perfume industry. This chapter investigated the opportunities to exploit the true potential of *Aquilaria malaccensis*. The authors' curiosity led to numerous questions about which is the best technique for the induction of agarwood. Which are the most effective inducers for agarwood production? What is the multidimensional aspect of agarwood? What are the medicinal properties of agarwood? What are the possible pharmacological activities of agarwood oil? Are there any alternative ways to use agarwood oil in drug industries? What are the socioeconomic aspects of cultivating agarwood? What is the importance of agarwood in agroforestry? This chapter is the combined effort of the authors to highlight the utility of agarwood industrial applications and revenue generation. This chapter is the first attempt of its type, which categorizes the utilization of *Aquilaria malaccensis* through different methods of induction. An extensive literature study was taken up to acquire a broad idea of achieving economic development through the cultivation of this tree.

10.3 INDUCTION OF AGARWOOD PRODUCTION

The story of every induction method starts with physical injury and ends up with the formation of resin; the yield quality and quantity are different in different methods. However, agarwood trees are not able to produce resin in response to physical injury alone. The inducement practices are aimed at boosting the yield of agarwood production and the quality of agarwood in the shortest period. The resinous part of agarwood is formed in response to biotic and abiotic stress or foreign introducers. The impregnated resinous wood, agarwood, is a result of several sequential events. The intermediate steps between physical injury and the formation of resin vary in different methods. There are possibly two induction methods for agarwood formation in host plants: natural induction and artificial agarwood induction techniques.

10.3.1 Natural Induction

Naturally, agarwood formation in agar plants is a rare case and a very long process that can take up to several decades. Under natural conditions, the resin is found to be formed in 20-year-old trees, or more than 20 years; trees more than 50 years old are reported to have the highest concentration of resin and best yield quality. The causal agents for natural induction can be pests or insects, ants, and snails; physical wounds or damage on the host due to thunder strikes, animal grazing, broken branches, and infestations of disease by pests can also help initiate agarwood formation (Wu et al., 2017). A wound is a primary step for the natural induction method that can provide an entry point for various and diverse pathogenic microbes to enter the host and trigger its defense system (Mohamed et al., 2010). Although this practice has many advantages, this technique is free of cost; it is possible to obtain a high-quality grade of agarwood; no cultivation, plantation, or induction is required; and even the best quality of agarwood is produced by natural fungal infections. However, the resultant yield quantity is insufficient, unsustainable, and undetermined. Even in nature, the rate of agarwood production in *Aquilaria* plants is near about 10%. In India, especially in Upper Assam, the infection rate is much higher due to an attack by a stem borer (*Zeuzera conferta*). The larvae of stem borer make tunnels in the initial sites of infection, and then infection gradually spreads up. Through this process, oleoresins are accumulated in the infected areas.

10.3.2 Artificial Induction

The demand for agarwood has been increasing due to its high economic and medicinal value. As the production rate of agarwood in nature is slow, the global demand can be met through several artificial practices developed to lessen the time for high-quality agarwood formation (Pojanagaroon and Kaewrak, 2003). It is expected that artificial methods can provide higher agarwood yields compared to natural methods. The artificial induction of agarwood is categorized into conventional and non-conventional types (Table 10.1).

10.3.2.1 Conventional Practices

Conventional methods are those methods that are practiced by local cultivars traditionally for several generations. These methods are practiced physically or mechanically wound to the tree to trigger agarwood formation (Pojanagaroon and Kaewrak, 2003). Wounds are induced using nails, chisels, ax, or machete, but the black resinous agarwood is consumed only at the injured or wounded areas of the tree. Different physical and mechanical methods that are used to induce agarwood formation are burning chisel drilling (BCD), partial trunk pruning (PTP), nailing, cutting of branches, bark removal, and holing on trees by local cultivars. These conventional induction methods are time-consuming and highly risky due to yielding agarwood's unpredictable and uncertain quality. As local growers mainly apply the nailing method, the scent of harvested agarwood oil is inferior because of contamination from the smell of rusty nails. Conventional induction methods do not fulfill the desired quality demands.

10.3.2.2 Non-Conventional Practices

Non-conventional methods are advanced practices where an inducer or a catalyst can apply to the host to stimulate agarwood formation. At least a minor physical wound is required as an entry point for the applying inducer or catalyst. Inducers are either chemical or biological in characteristics.

10.3.2.2.1 Chemical Inducers

This induction method is applied to stimulate agarwood formation in the whole tree. Chemical inducers act rapidly, are fast, and are easy to apply. Chemical compounds, strong acids, or bases are applied in this method. Such inducers are sodium chloride, methyl jasmonic acid, hydrogen peroxide, formic acid, acetic acid, salicylic acid, ferric chloride, plant hormones, and biologically derived chemicals (Thanh et al., 2015). The different concentrations of such chemicals are applied at the wound site of the tree. With the transpiration of water, inducers are spread to all parts of the tree and initiate agarwood formation. The optimum strength of every applied chemical must

TABLE 10.1
Agarwood-Inducing Methods with Their Advantages and Disadvantages

Category	Type	Inducer	Applied methods	Time taken	Yield quality	Advantages	Disadvantages	References
Natural induction	Natural wounding	Pest attack/ wounding of the tree by natural calamities	Infestation of pests, insects, ants; broken branches; accidental wound	Unpredictable	High	Free of cost; no cultivation, plantation, and induction required	Extremely low yield quantity; unsustainable and undetermined	Mohamed et al., 2010
Artificial induction	Conventional methods	Physical/ mechanical wounds	Burning-chisel drilling; partial trunk pruning; bark removal; holing; nailing; branch cutting; aeration	Longer time	Uncertain	Economically cost-effective	Laborious; specificity in agarwood formation	Rasool and Mohamed, 2016; Wu et al., 2017
	Non-conventional methods	Chemical	CA kit method; agar-wit method; agar-bit method; AINM method	Fast	Consistent and better	Easy to apply; no specificity on the formation of agarwood; high yield quantity	Optimum strength of inducers is needed to know; toxicity of chemicals and their doubtful impact on human and environment	Chong et al., 2015; Tang and Liu, 2016
		Biological/ organic	Fungal infection method	Less than conventional methods	Average	Low cost; easily available from natural sources; safe to handle; eco-friendly	Laborious and time-consuming for making holes; long incubation rate	Chong et al., 2015; Sangareswari Nagajothi et al., 2016
		Chemical and biological	Pinhole method	Fast	High	High yield quantity	Laborious for making holes	Tian et al., 2013

be known by the keeper who has developed the inducer; otherwise, an excess volume of chemicals may kill the tree. This is how a chemical inducer is commercialized after several field trials.

Some chemicals are very toxic to the environment and especially to humans. Hence, the choice of chemicals is one of the most critical parts of this method, mainly when the harvested agarwood is used as a raw material in the fragrance industry, although the elimination of such toxic compounds or the selection of non-toxic chemicals can make this method valuable. Several induction methods are developed on the basic principle of the chemical induction concept. Cultivated agarwood kit (CA kit) is a proper combination of physical and chemical inductions that can improve the yield quantity and quality of agarwood. Whole-tree agarwood-inducing technique (agar-wit) is a transpiration-dependent chemical induction method where chemical inducers are transported via host plant transpiration. Similar to agar-wit, in biologically agarwood-inducing technique (agar-bit) method, inducers are distributed in the whole plant by plant transpiration (Wu et al., 2017).

10.3.2.2.2 Biological Inducers

Biological inducers like fungi, bacteria, and yeasts are called organic inducers. Tunstall reported the inoculation of fungi from *A. agallocha* Roxb. (Gibson, 1977). Biological inducers are introduced by isolating endophytic microbes either from healthy or diseased agarwood, then pure strains of those specific microbes are introduced into a healthy tree to trigger agarwood formation. Unlike the chemical induction method, an entry point for the microbe is essential to infect the host and initiate the process in the biological method; therefore, the host plant must be wounded. Compared to mechanical induction methods, fungal inoculated agarwood contains a higher concentration of sesquiterpenes (Chhipa et al., 2017). All fungal species cannot induce agarwood formation. Some dominant endophytic fungal genera that can induce agarwood formation are *Fusarium*, *Trichoderma*, *Penicillium*, *Alternaria*, *Aspergillus*, *Epicoccum*, and *Curvularia* (Table 10.2).

The role of fungi in the production of resinous wood was suggested by the isolation of *Epicoccum granulatum* from the infected *A. agallocha* wood (Bhattacharyya et al., 1952). Some researchers have used glucose as an inducer in wounded tree holes before applying fungal inoculum for agarwood formation, but in maximum cases, glucose was used by the host tree rather than the fungal strain to cure the wound (Nobuchi and Siripatanadilok, 1991). A positive result was shown by *Chaetomium globosum* for agarwood production in *A. agallocha* (Tamuli et al., 2005). Different species of *Acremonium* were isolated from *A. malaccensis* (Rahayu, 2008). In Indonesia, a team of FORDA identified *F. solani*, one of the most effective biological agents in agarwood formation. Tian et al. (2013) developed a pinhole infusion method to improve the biological induction, and chemical and biological inducers indicated improvement in resin formation.

10.4 MULTIDIMENSIONAL ASPECTS OF AGARWOOD

There are several characteristics of agarwood that are of economic importance and have a very high agroeconomic value.

TABLE 10.2
Artificially Inoculated Microbes with Their Infectious Zone and Duration of Infection

No.	Name of Microbe	Source	Infection (cm) Syringe	Stick	Bottle Drip	Infection Duration (Month)	References
1.	*Alternaria lini*	–	*	*	–	03	
2.	*Alternaria* sp.	Soil	*	*	–		Chhipa and Kaushik, 2017
3.	*Aspergillus flavipes*	NA	2–8.5	*	–		
4.	*Aspergillus flavus*	Wood	8.5*	2.83*	–		
5.	*Aspergillus oryzae*	–	2–8.5*	*	–		
6.	*Aspergillus sydowii*	–	2–8.5*	*	–		
7.	*Bacillus anthricus*	Stem	2.91*	–	–		
8.	*Bacillus cereus*	Stem	3.45*	–	–		
9.	*Bacillus megaterium*	Stem	2.35*	–	–		
10.	*Botryodiplodia theobromae*	Stem	–	–	–	–	Rahman and Basak, 1980
11.	*Chaetomium globosum*	Stem	–	–	–	01	Tamuli et al., 2005
12.	*Fusarium oxysporum*	Stem	–	–	–		
13.	*Fusarium solani*	Wood	–	–	–	03	Putri et al., 2017
14.	*Fusarium proliferatum*	Wood	*	*	–		Chhipa and Kaushik, 2017, 2020
15.	*Pantoea dispersa*	Stem	1.26*	–	–		
16.	*Penicillium citrinum*	Wood	2–10*	*	–		
17.	*Penicillium polonicum*	Stem	2.60*	–	–		
18.	*Penicillium polonicum*	Wood	2.6*	10.00*	–		
19.	*Penicillium aethiopicum*	Stem	3.00*	–	–		
20.	*Penicillium aethiopicum*	Wood	2–10*	*	–		
21.	*Pichia kudriavzevii*	–	*	*	–		
22.	*Rhizopus oryzae*	Stem	2.00*	–	–		
23.	*Syncephalastrum racemosum*	Soil	2.60*	–	–		
24.	*Talaromyces aculeatus*	Soil	*	*	–		
25.	*Trichoderma harzianum*	Wood	*	*	–		
26.	*Trichoderma koningii*	Wood	*	*	–		
27.	*Trichoderma asperellum*	Soil	3.00*	–	–		
28.	*Phomopsis aquilariae*	Stem	–	–	–	–	Rahman and Basak, 1980
29.	*Trichoderma spirale* and *Aspergillus* sp.	Wood	–	–	6.2–10.9	03	Justin et al., 2020
30.	*Fusarium* and *Penicillium* sp.	Wood	–	–	5.8–9.5		
31.	*Trichoderma harzianum, Lasiodiplodia* sp. and *Curvularia* sp.	Wood	–	–	6.3–12.1		

10.4.1 Economic Aspect

Wood chips, wood dust or powder, and agar oil are the three primary forms of tradable agarwood in the world. In Upper Assam (India) home gardens, the agar plant was preferred as a "cash crop" because agar-based home gardens were financially benefited and earned a significant amount of money. The quality grade of agarwood oil mostly depends on the grade of wood used in making oil, and the length of distillation is also dependable for good-quality oil (Chetpattananondh, 2012). The uses of agarwood oil may be multidimensional, such as body fragrance, incense of soaps and shampoos, and medical purposes. The "most expensive wood," agarwood has a price ranging from a few to thousands of dollars per kilogram for the top quality of wood (Barden et al., 2000). The price of agarwood starts from US$ 20/kg to US$ 6×10^3/kg globally, depending on the resin quality of wood chips; the value of agarwood oil is much more than that of wood, that is, US$ 3×10^4/kg. The highest and best quality-grade agarwood oil is produced in China, and second in Vietnam. Singapore trades agarwood for US$ 1.2×10^8/year (Hansen, 2000). The value of agarwood oil was US$ 3×10^4 to 4×10^4 in 2014, and chips ranged between US$ 30 and 1×10^4/kg in 2015 (Chowdhury et al., 2017). The current global market for agarwood chips and oil and other agarwood products is estimated to range from US$ 6×10^9 to US$ 8×10^9 (Akter et al., 2013). The world's most expensive perfume, "Shumukh," which currently holds two Guinness World Records, comprises only natural ingredients, that is, sandalwood, rare pure Indian agarwood oil, and Turkish rose.

10.4.2 Medicinal Properties

Agarwood has been used in Ayurveda and traditional East Asian medicine, aromatherapy, and pharmaceutical tinctures. In Thailand, *A. malaccensis* is used as a blood and heart tonic. The resinous part of this plant is precious and used in the curation of stimulants, cardiac tonics, and carminatives. In Chinese traditional medicines, this active ingredient is used to formulate different cough drugs and those for acroparalysis, anti-histamine, and asthma. Traditionally, the leaves, bark, and root of the heartwood of *Aquilaria agallocha* are used as a cardiotonic, carminative, anodyne, aphrodisiac, aromatic, astringent and are bitter, stimulant, and fragrant. It is also used as a mouth freshener and appetizer and relieves itching in pruritus. Chakrabarty et al. (1994) reported that the uninfected wood is used as *Kayu gaharu lemppong* by Malayans (an Indian tribal community) and Chinese to treat jaundice and body pain.

10.4.3 Socioeconomic Aspect

The socioeconomic value of any material mainly depends on the social and economic improvement of a particular community directly associated with that material. An individual agar plant gives a one-time income to a cultivar. Many researchers reported that agarwood has an economically high potential compared to other home garden products because of the low input management, lack of site-specificity, and intercropping adaptation of agar plants. These reasons made agarwood a "cash crop" in the home gardens of NE states of India. International traders play an essential role in importing and exporting agarwood and agarwood products; they have received the most outstanding share of the distribution chain. In Bangladesh, about 350–400 nationally registered agar-based factories or enterprises located in Barlekha Upazila of Moulvibazar are significant contributors to regional employment and gross domestic product. Home gardens of Upper Assam can be considered a tool for farm conservation of agarwood, and in other parts of Northeastern India, agarwood is introduced as a potential economic crop for developing the rural economy. In Thailand, agarwood products were necessary as a high-income source, and therefore, many villagers (38%) were attracted to this for livelihood activity (Jha, 2014). In a survey conducted by Shahidullah and Haque (2010), 53% of agar growers agreed that their livelihood improved by cultivating agar plants, and that the profit margin of mediators increased by 59–139%, and 22–90% for wholesalers. Overall, agarwood products positively impacted the livelihood of local agarwood producers.

Traditional knowledge-holders act as local "experts" who provide basic information on induction methods. Their information is considered unscientific as these methods are not standardized scientifically. The pieces of information gathered are used to generate the hypotheses necessary for further future research. Traditional knowledge serves as an information resource for starting points to develop a new paradigm and evolve concepts about the conservation and development of tropical forests with side-by-side wild species of *Aquilaria*.

10.4.4 Agarwood and Agroforestry System

In the agroforestry farms or home gardens of Vietnam, agar had been widely grown and intercropped with a combination of crops and fruit trees, such as upland rice, cassava, beans, sweet potato, yam, banana, pineapple, jackfruit, and many more. Thus, oil processing technologies were also promoted as a livelihood program for farmers and rural communities. In the northeastern region of India, patchouli plants are highly suitable for planting under the shade of agar plants, and this is referred to as the patchouli-based agroforestry model. Blanchette et al. (2015) reported that *Aquilaria* had been grown with rubber, teak, banana, and even oil palm in Southeast Asia; agar plants were facilitated as a shade tree in tea plantations and increased agarwood production. In Malaysia, *A. malaccensis* is often cultivated in both monoculture and crop rotation systems. Agar plants can be planted on hilly slopes and down areas, which can help reduce soil erosion and land sliding during the rainy season. These plants can also be planted in agricultural areas without affecting the field crops and on borders of gardens, school or college campuses, office compounds, or other public residential sites.

10.4.5 Pharmacological Activity

Agarwood oil of *A. agallocha* was reported as in vivo and in vitro anti-inflammatory agent. Dash et al. (2008) have shown human pathogenic bacteria are inhibited by the different extracts of *A. agallocha* Roxb. The anti-inflammatory activity of *A. agallocha* oil (AAO) was described by carrageenan-induced edema in the animal model (paw of rats) and human RBC membrane stabilization method. The percentage reduction in paw volume was 58.59% at 50 mg/kg and 62.11% at 100 mg/kg. Membrane stabilization at 0.1, 0.25, and 0.5 mg/ml concentration showed 39.66%, 62.94%, and 78.50%, respectively, as compared with standard diclofenac. The ethyl acetate extract of *A. agallocha* wood has shown analgesic activity in model organism mice (*Mus musculus*). Rahman et al. (2013) investigated the antibacterial activity of *A. agallocha* oil by agar well diffusion method against *Escherichia coli*, *Enterococcus faecalis*, *Staphylococcus aureus*, and *Pseudomonas aeruginosa*. Spontancous motility, prolonged hexobarbiturate-induced sleeping time, and decreased rectal temperature are induced by applying benzene extract of *A. malaccensis* agarwood to the central nervous system of mice. The methanol and water extracts at 0.01 mg/ml concentration enhanced glucose uptake activity in rat adipocytes. Ethyl acetate extract of *A. agallocha* (EAA) has been investigated as an antioxidant agent at different concentrations on human blood hemolysate; the resultant strong antioxidant effect of EAA has been shown to be in a concentration range of 0.5–3.5 mg/ml. Free radical scavenging activity of *A. agallocha* oil was studied and found to have IC_{50} values equal to 2,190 mg/ml and 47,180 mg/ml, respectively.

10.5 ENDOPHYTES AND AGARWOOD PRODUCTION

Endophytes are such organisms that play an essential role in plant physiology and help the plant to grow, to tolerate biotic and abiotic stress; they are not detrimental to the host plant. The role of endophytic fungi has been documented as the most valuable player of agarwood formation in *Aquilaria* sp. Some endophytes showed pathogenicity in the host plant, but the pathogenicity property of endophytes could be an economical source to produce resinous oleoresin after induction of such endophytes (Barden et al., 2000). *Fusarium oxysporum* and *Chaetomium globosum* can induce agarwood formation successfully in *A. agallocha* (Tamuli et al., 2005). *Fusarium* is recorded as the most dominant fungal taxa for agarwood production, followed by *Lasiodidiplodia* (35%) and *Trichoderma* (13%) (Mohamed et al., 2010). Comparatively, resinous wood contained higher diversity of endophytes than healthy and non-resinous wood of *Aquilaria*. Premalatha and Kalra (2013) have isolated *Preussia* sp. from *A. malaccensis* wood for the first time; *Trichoderma*, *Alternaria*, *Cladosporium*, *Davidiella*, *Massarina*, *Curvularia*, *Hypocrea*, *Fusarium*, *Pichia*, and *Phaeoacremomium* were found in infected resinous wood. *Epicoccum granulatum* was also isolated from *A. agallocha* (Bhattacharyya et al., 1952). *Cytosphaera mangiferae* was isolated from the fungal association in *A. malaccensis* from Bangladesh. Nobuchi and Siripatanadilok (1991) reported *F. bulbigenum* and *F. Lateritium* from *Aquilaria* sp. Hypocreaceae was most dominant (23.8%), followed by Nectriaceae (13.5%) and Botryospaeriaceae (13.2%), in wood samples of *Aquilaria*. However, in the case of bacteria, Bacillaceae was found to be the most dominant family, with 62.5% richness (Chhipa and Kaushik, 2017). They have also compared source-depending fungal and bacterial dominance between soil and stem samples, and the highest result is measured in the soil in both fungal and bacterial dominance cases. Agarospirol, the most responsible compound for imparting agarwood's aroma is found in the woods infected with *Pantoea dispersa*, *Penicillium polonicum*, *Syncephalastrum racemosum*, *Penicillium aethiopicum*, and *T. asperellum*, respectively. Bacterial isolates (31%) were more able to produce agarospirol than fungal isolates (23%) in agarwood by artificial infection (Chhipa and Kaushik, 2017). Mochahari et al. (2020) have reported that the association of fungi in agarwood was more dormant and symbiotic than pathogenic; they have also isolated four endophytic fungi from juvenile *A. malaccensis* (*Alternaria* sp., *Curvularia* sp., *Rhizopus* sp., and *Sterilia* sp.).

The present study highlights 114 fungal and 26 bacterial strains reported from the wood samples of *A. malaccensis* (Table 10.3, **10.4;** Figure 10.1–10.4). Isolated strains are reported from India, Malaysia, and Indonesia. Most of the fungal endophytes are

TABLE 10.3
Endophytic Fungi Isolated from Wood Samples of *Aquilaria malaccensis* Plant

SL No.	Endophytic Fungi	Source	Country Name	Reference(s)
1.	*Alternaria tenuissima*	Wood*	Assam, India	Chhipa and Kaushik (2017)
2.	*Alternaria* sp.	Wood*	Assam, India	Premalatha and Kalra (2013)
3.	*Alternaria* sp.	Soil*	Assam, India	Chhipa and Kaushik (2017)
4.	*Arthrinium* sp.	Wood*	Assam, India	
5.	*Aschersonia* sp.	Wood*	Assam, India	
6.	*Ascomycota* sp.	Wood#	Assam, India	Premalatha and Kalra (2013)
7.	*Aspergillus aculeatus*	Wood*	Assam, India	Chhipa and Kaushik (2017)
8.	*Aspergillus flavus*	Wood*	Assam, India	
9.	*Aspergillus nidulans*	Wood*	Assam, India	
10.	*Aspergillus niger*	Wood#	Assam, India	Tamuli et al. (2005)
11.	*Aspergillus* sp.	Wood#	Bangka Belitung Islands, Indonesia	Hartono et al. (2019)
12.	*Aspergillus* sp.	Wood/soil*	Assam, India	Chhipa and Kaushik (2017)
13.	*Aspergillus* sp. (yellow)	Wood#	TN, Assam and Kerala, India	Sangareswari Nagajothi et al. (2016)
14.	*Botryodiplodia* sp.	Wood-	Selangor, Malaysia	Mohamed et al. (2014)
15.	*Chaetomium* sp.	Wood#	TN, Assam and Kerala, India	Sangareswari Nagajothi et al. (2016)
16.	*Chaetomium globosum*	Wood#	Assam, India	Tamuli et al. (2000)
17.	*Cladosporium cladosporoides*	Wood#	Assam, India	Premalatha and Kalra (2013)
18.	*Cladosporium* sp.	Wood#	Assam, India	
19.	*Cochliobolus lunatus*	Wood#	Selangor, Malaysia	Mohamed et al. (2010)
20.	*Cochliobolus lunatus*	Wood*	Assam, India	Chhipa and Kaushik (2017)
21.	*Cochliobolus lunatus*	Wood*#	Selangor, Malaysia	Mohamed et al. (2010)
22.	*Cunninghamella bainieri*	Wood*	Selangor, Malaysia	
23.	*Cunninghamella* sp.	Wood*	Assam, India	Chhipa and Kaushik (2017)
24.	*Curvularia verruculosa*	Wood*	Assam, India	

TABLE 10.3 *(Continued)*
Endophytic Fungi Isolated from Wood Samples of *Aquilaria malaccensis* Plant

SL No.	Endophytic Fungi	Source	Country Name	Reference(s)
25.	*Curvularia lunata*	Wood*	Assam, India	
26.	*Curvularia* sp.	Wood#	Bangka Belitung Islands, Indonesia	Hartono et al. (2019)
27.	*Curvularia* sp.	Wood*	Assam, India	Chhipa and Kaushik (2017)
28.	*Davidiella tassiana*	Wood*	Assam, India	Premalatha and Kalra (2013)
29.	*Diplodia* sp.	Wood-	Selangor, Malaysia	Mohamed et al. (2014)
30.	*Diaporthe phaseolorum*	Wood*	Assam, India	Chhipa and Kaushik (2017)
31.	*Diaporthe* sp.	Wood*	Assam, India	
32.	*Epicoccum granulatum*	Wood#	India	Bhattacharyya et al. (1952)
33.	*Epicoccum nigrum*	Wood*	Assam, India	Chhipa and Kaushik (2017)
34.	*Epicoccum sorghinum*	Wood*	Assam, India	
35.	*Epicoccum* sp.	Wood*	Assam, India	
36.	*Fusarium solani*	Wood#	Assam, India	Premalatha and Kalra (2013)
37.	*Fusarium solani*	Wood*	Assam, India	Chhipa and Kaushik (2017)
38.	*Fusarium solani*	Wood#	Selangor, Malaysia	Mohamed et al. (2010)
39.	*Fusarium fujikuroi*	Wood*	Assam, India	Chhipa and Kaushik (2017)
40.	*Fusarium proliferatun*	Wood*	Assam, India	
41.	*Fusarium equiseti*	Wood*	Assam, India	
42.	*Fusarium bulbiferum*	Wood-	Selangor, Malaysia	Mohamed et al. (2014)
43.	*Fusarium laterium*	Wood-	Selangor, Malaysia	
44.	*Fusarium oxysporum*	Wood*	Selangor, Malaysia	
45.	*Fusarium oxysporum*	Wood*	Assam, India	Chhipa and Kaushik (2017)
46.	*Fusarium oxysporum*	Wood#	Assam, India	Tamuli et al. (2000)
47.	*Fusarium* sp.	Wood*#	Selangor, Malaysia	Mohamed et al. (2010)
48.	*Fusarium* sp.	Wood#	Bangka Belitung Islands, Indonesia	Hartono et al. (2019)
49.	*Fusarium* sp.	Wood/Soil*	Assam, India	Chhipa and Kaushik (2017)
50.	*Fusarium* sp.	Wood*	Assam, India	Premalatha and Kalra (2013)
51.	*Fusarium* sp. (ash-colored)	Wood#	TN, Assam and Kerala, India	Sangareswari Nagajothi et al. (2016)
52.	*Fusarium* sp. (colorless)	Wood#	TN, Assam and Kerala, India	
53.	*Fusarium* sp.	Wood#	Assam, India	Tamuli et al. (2005)
54.	*Gibberella intermedia*	Wood*	Assam, India	Chhipa and Kaushik (2017)
55.	*Gibberella* sp.	Wood*	Assam, India	

(Continued)

TABLE 10.3 *(Continued)*
Endophytic Fungi Isolated from Wood Samples of *Aquilaria malaccensis* Plant

SL No.	Endophytic Fungi	Source	Country Name	Reference(s)
56.	*Hypocrea caerulescens*	Wood*	Assam, India	
57.	*Hypocrea fairnosa*	Wood#	Assam, India	Premalatha and Kalra (2013)
58.	*Hypocrea lixii*	Wood*#	Selangor, Malaysia	Mohamed et al. (2010)
59.	*Hypocrea rufa*	Wood*	Assam, India	Chhipa and Kaushik (2017)
60.	*Hypocrea* sp.	Wood*	Assam, India	
61.	*Hypocreales* sp.	Wood*	Assam, India	
62.	*Hypocreales* sp.	Wood*	Assam, India	Premalatha and Kalra (2013)
63.	*Lasiodiplodia theobromae*	Wood#	Selangor, Malaysia	Mohamed et al. (2010)
64.	*Lasiodiplodia pseudotheobromae*	Wood*	Assam, India	Chhipa and Kaushik (2017)
65.	*Lasiodiplodia theobromae*	Wood*	Assam, India	
66.	*Lasiodiplodia* sp.	Wood*	Assam, India	
67.	*Lasiodiplodia* sp.	Wood#	TN, Assam and Kerala, India	Sangareswari Nagajothi et al. (2016)
68.	*Lasiodiplodia* sp.	Wood#	Selangor, Malaysia	Mohamed et al. (2010)
69.	*Lodderomycetes* sp.	Wood*	Assam, India	Premalatha and Kalra (2013)
70.	*Massarina albocarnis*	Wood#	Assam, India	
71.	*Microdiplodia* sp.	Wood*	Assam, India	Chhipa and Kaushik (2017)
72.	*Meyerozyma guillermondil*	Wood*	Assam, India	
73.	*Mucor circinelloides*	Wood*	Assam, India	
74.	*Mucor fragilis*	Wood#	Selangor, Malaysia	Mohamed et al. (2010)
75.	*Mucor* sp.	Wood*	Assam, India	Chhipa and Kaushik (2017)
76.	*Nigrospora* sp.	Wood*	Assam, India	
77.	*Paecilomyces* sp.	Wood*	Assam, India	
78.	*Penicillium citrinum*	Wood*	Assam, India	
79.	*Penicillium chrysogenum*	Wood*	Assam, India	
80.	*Penicillium granulatum*	Wood*	Assam, India	
81.	*Penicillium polonicum*	Wood*	Assam, India	
82.	*Penicillium aethiopicum*	Wood*	Assam, India	
83.	*Penicillium* sp.	Wood*	Assam, India	
84.	*Penicillium* sp.	Wood#	Bangka Belitung Islands, Indonesia	Hartono et al. (2019)
85.	*Penicillium* sp. (orange-red)	Wood#	TN, Assam and Kerala, India	Sangareswari Nagajothi et al. (2016)
86.	*Penicillium* sp.	Wood#	Assam, India	Tamuli et al. (2005)
87.	*Peniophora* sp.	Wood*#	Bangka Belitung Islands, Indonesia	Hartono et al. (2019)

TABLE 10.3 *(Continued)*
Endophytic Fungi Isolated from Wood Samples of *Aquilaria malaccensis* Plant

SL No.	Endophytic Fungi	Source	Country Name	Reference(s)
88.	*Phaeoacremonium* sp.	Wood*	Assam, India	Premalatha and Kalra (2013)
89.	*Pichia* sp.	Wood*	Assam, India	
90.	*Pichia* sp.	Wood*	Assam, India	Chhipa and Kaushik (2017)
91.	*Pleosporales* sp.	Wood#	India	Premalatha and Kalra (2013)
92.	*Preussia africana*	Wood#	Selangor, Malaysia	Mohamed et al. (2010)
93.	*Preussia africana*	Wood#	Assam, India	Premalatha and Kalra (2013)
94.	*Preussia* sp.	Wood#	Selangor, Malaysia	Mohamed et al. (2010)
95.	*Preussia* sp.	Wood#	Assam, India	Premalatha and Kalra (2013)
96.	*Rhizopus oryzae*	Wood*	Assam, India	Chhipa and Kaushik (2017)
97.	*Saccharomycetes* sp.	Wood*	Assam, India	Premalatha and Kalra (2013)
98.	*Sordariomycetes* sp.	Wood#	Assam, India	
99.	*Sterilia* sp.	Chips*#	Assam, India	Mochahari et al. (2020)
100.	*Syncephalastrum racemosum*	Wood*	Assam, India	Chhipa and Kaushik (2017)
101.	*Syncephalastrum* sp.	Soil*	Assam, India	
102.	*Talaromyces* sp.	Soil*	Assam, India	
103.	*Trichoderma asperellum*	Wood*	Assam, India	
104.	*Trichoderma harzianum*	Wood*	Assam, India	
105.	*Trichoderma koningii*	Wood*	Assam, India	
106.	*Trichoderma longibrachiatum*	Wood*	Assam, India	
107.	*Trichoderma reesei*	Wood*	Assam, India	
108.	*Trichoderma virens*	Wood*	Assam, India	
109.	*Trichoderma* sp.	Soil*	Assam, India	
110.	*Trichoderma* sp.	Wood*	Assam, India	Premalatha and Kalra (2013)
111.	*Trichoderma* sp.	Wood*#	Bangka Belitung Islands, Indonesia	Hartono et al. (2019)
112.	*Trichoderma* sp.	Wood*#	Selangor, Malaysia	Mohamed et al. (2010)
113.	*Trichoderma* sp.	Wood#	Assam, India	Tamuli et al. (2005)

NB: *natural healthy plants; #naturally induced plants; - isolate source did not mentioned.

reported from India individually, followed by Malaysia and Indonesia. Fifty-seven species are reported by Chhipa and Kaushik (2017), which is about 50% of all isolated endophytic fungi, whereas in the case of bacterial endophytes, reported countries are India (73%) and Malaysia (27%). In fungal isolates, most of the fungi are derived from Ascomycota and Zygomycota. *Aspergillus* sp., *Cochliobolus lunatus*, *Curvularia* sp., *F. solani*, *F. oxysporum*, *Fusarium* sp., *Lasiodiplodia theobromae*, *Penicillium* sp., *Pichia* sp., and *Trichoderma* sp. are reported from more than one

TABLE 10.4
List of Endophytic Bacterial Strains Isolated from Wood and Soil Samples of *A. malaccensis*

SL No.	Bacterial strains	Source	Country Name	Reference
1.	*Achromobacter* sp.	Stem*	Assam, India	Chhipa and Kaushik (2017)
2.	*Bacillus anthricus*	Stem*	Assam, India	
3.	*Bacillus arbutinivorans*	Stem/leaf#	Selangor, Malaysia	Bhore et al. (2013)
4.	*Bacillus cereus*	Stem/leaf#	Selangor, Malaysia	
5.	*Bacillus cereus*	Stem/soil*	Assam, India	Chhipa and Kaushik (2017)
6.	*Bacillus licheniformis*	Stem/leaf#	Selangor, Malaysia	Bhore et al. (2013)
7.	*Bacillus megaterium*	Stem/soil*	Assam, India	Chhipa and Kaushik (2017)
8.	*Bacillus mycoides*	Stem*	Assam, India	
9.	*Bacillus pumilus*	Stem/leaf#	Selangor, Malaysia	Bhore et al. (2013)
10.	*Bacillus pumilus*	Soil*	Assam, India	Chhipa and Kaushik (2017)
11.	*Bacillus subtilis*	Stem/leaf#	Selangor, Malaysia	Bhore et al. (2013)
12.	*Bacillus subtilis*	Stem*	Assam, India	Chhipa and Kaushik (2017)
13.	*Bacillus stratosphericus*	Stem/leaf#	Selangor, Malaysia	Bhore et al. (2013)
14.	*Bacillus* sp.	Stem/soil*	Assam, India	Chhipa and Kaushik (2017)
15.	*Lysinibacillus sphaericus*	Soil*	Assam, India	
16.	*Lysinibacillus* sp.	Soil*	Assam, India	
17.	*Paenibacillus alvei*	Soil*	Assam, India	
18.	*Pantoea dispersa*	Stem*	Assam, India	
19.	*Pantoea* sp.	Stem*	Assam, India	
20.	*Pseudomonas aeruginosa*	Stem/soil*	Assam, India	
21.	*Pseudomonas* sp.	Stem*	Assam, India	
22.	*Roseomonas aerophila*	Stem*	Assam, India	
23.	*Roseomonas* sp.	Stem*	Assam, India	
24.	*Stenotrophomonas* sp.	Stem*	Assam, India	
25.	*Streptomyces* sp.	Stem*	Assam, India	
26.	*Vibrio cholera*	Stem/leaf#	Selangor, Malaysia	Bhore et al. (2013)

NB: *natural healthy plants; #naturally induced plants

country. The fungal endophytes are divided into 35 genera, 22 families, 12 orders, eight classes, and five phyla, where the dominant genera are *Fusarium*, followed by *Trichoderma*, *Penicillium*, *Aspergillus*, and *Lasiodiplodia*. Nectriaceae is found to be the most dominant fungal family, followed by Trichocomaceae, Hypoceaceae, and Pleosporaceae (Figure 10.2).

Other infectious organisms also existed and were isolated from *A. malaccensis*. Those infectious organisms are found in the natural forest of agarwood and isolated from naturally induced agarwood samples. As discussed, *Zeuzera conferta*, in the natural induction method, a borer insect, is reported for initiating agarwood formation.

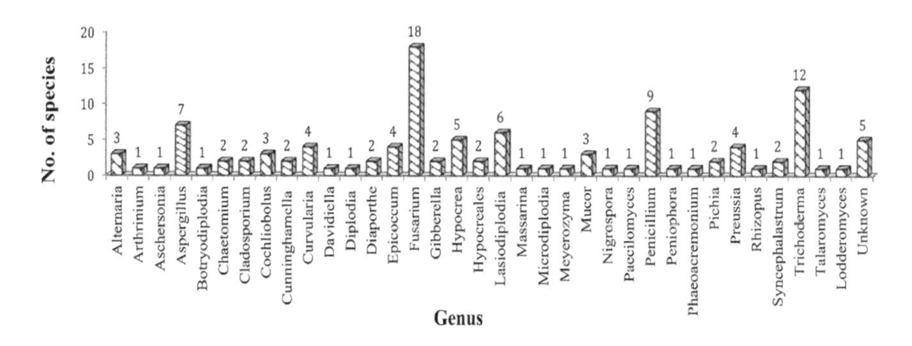

FIGURE 10.1 Distribution of endophytic fungi in different genera isolated from *A. malaccensis.*

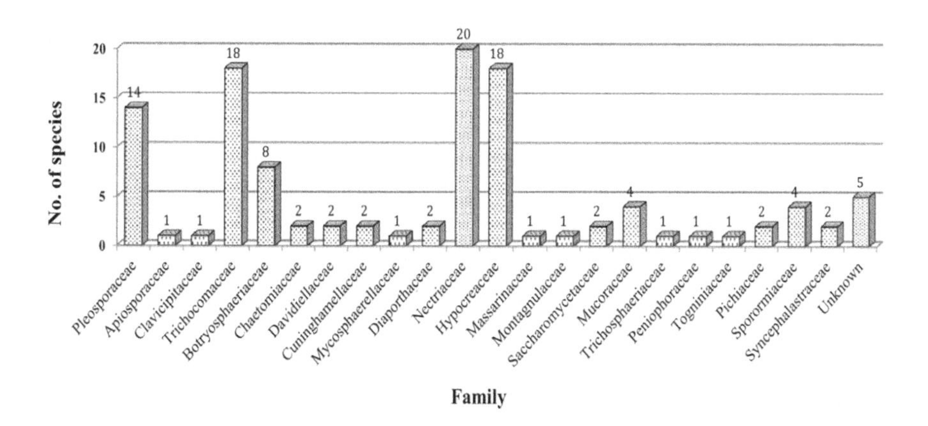

FIGURE 10.2 Diversity in the family of fungal isolates of *A. malaccensis.*

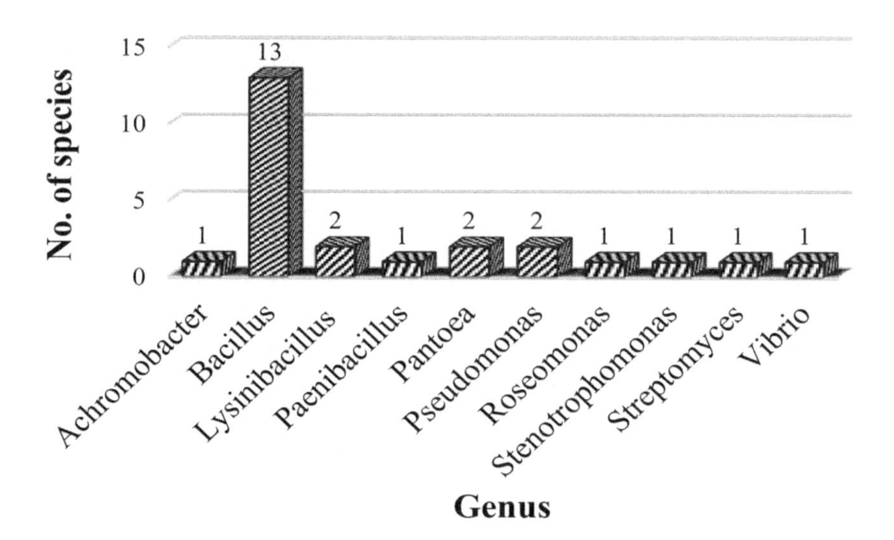

FIGURE 10.3 Diversity of different endophytic bacteria isolated from *A. malaccensis.*

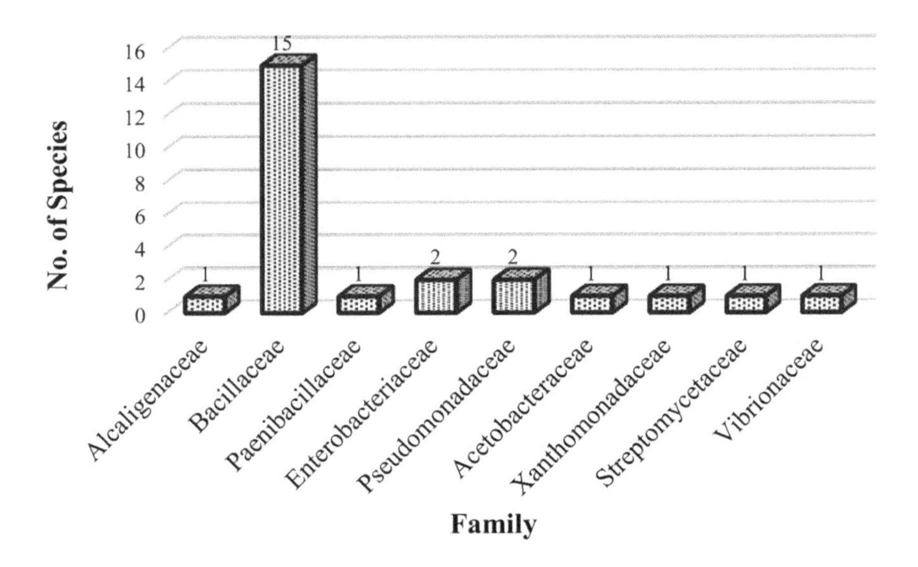

FIGURE 10.4 Distribution of endophytic bacteria in different families isolated from *A. malaccensis*.

10.5.1 Phytoconstituents of Agarwood and Wood-Derived Oil

The value and price of agarwood or agarwood oil are determined by quality grade, and the yield quality is dependent on the metabolic constituents of agarwood. Therefore, the quality of agarwood is correlated to its yield of resin and metabolic constituents (Pasaribu et al., 2015). Sesquiterpenes and phenylethyl chromone (PEC) derivatives are the main active compounds found in resinous agarwood (Naef, 2011), primarily sesquiterpenoids (Hashim et al., 2014; Jayachandran et al., 2014). These major compounds are responsible for the unique, fragrant, and sweet-smelling agarwood. The distribution of these major constituents is sometimes species-specific, such as some sesquiterpene derivatives like jinkoh-eremol and epi-γ-eudesmol found only in *A. malaccensis*, whereas in *A. sinensis* and *A. crassna,* baimuxinal does exist (Naef, 2011; Hashim et al., 2016). Yoneda et al. (1984) reported that jinkohol I and jinkohol II were characteristic features of *A. malaccensis.* (-)-Guaia-1(10), 11-dien-15-al is the most important component for the gorgeous and elegant characteristic of Kanankoh (*A. agallocha* Roxb.). Aromadendrene (Pasaribu et al., 2015) and valencene (Jayachandran et al., 2014) are suggested as effective and additional chemical markers which can indicate the higher grade of agarwood oil. Sometimes, the number of metabolite constituents and differences in agarwood oil composition can vary depending on the method of extraction and source of agarwood (Pasaribu et al., 2015).

10.5.2 Biosynthesis Pathways of Major Agarwood Constituents

The agar plant's biotic and abiotic stresses triggered the self-defense mechanism by initiating the production of many secondary metabolite biosyntheses, and as a result, agarwood resin accumulated. Biosyntheses of sesquiterpene and chromone derivatives in agar plants effectively induce agarwood formation. Generally, through the mevalonic acid pathway (MAP) and methylerythritol phosphate (MEP) pathway, the terpene derivatives are produced from the precursors (isoprenoid) in the plant cell cytosol and plastids. The precursor molecules are 5-carbon units (C-5). These molecules were converted into 15-carbon units, then named farnesyl pyrophosphate (FPP) with the help of the farnesyl pyrophosphate synthase (FPS) enzyme. FPS is the key limiting enzyme responsible for the biosynthesis and formation of sesquiterpene derivatives. The enzymes associated with sesquiterpene biosynthesis and formation are grouped into two classes, that is, cytochrome P450-dependent mono-oxygenases (P450) and sesquiterpene synthases. P450 oxygenase enzymes are essential for novel chiralities and modification of sesquiterpene derivatives. Kumeta and Ito (2010) stated that the initialization of agarwood formation was found due to sesquiterpene synthase involvement. Sesquiterpene synthase genes (SesTP), namely, the guaiene (AmGuaiS1) and sesquiterpene synthase (AmSesTPS1), were described as the temporal and spatial expression in *A. malaccensis* (Azzarina et al., 2016). AmSesTPS1 and AmGuaiS1 were more expressed in wounded wood at a magnitude of 18 times and 5.5 times higher than in unwounded wood. Many sesquiterpenoids have been reported earlier from agarwood. The significance of SesTP genes in agarwood production has yet to be ascertained.

Chromones are the second largest group of active secondary metabolites found in agarwood and wood-derived oil. Chromones have a wide range of potential therapeutic properties, like anti-cancerous, anti-immunomodulatory, anti-inflammatory, anti-diabetic, in neurological conditions, antibacterial, and antiviral activities (Khadem and Marles, 2011; Tawfik et al., 2014). Chromones are polycyclic organic compounds formed by the benzopyran ring and have a keto group that substitutes its oxime ring. Chromone derivatives are synthesized by several biosynthetic pathways involving the pentapeptide pathway, shikimic acid pathway, and the addition of nitrogenous moiety from amino acids or other sources (Khadem and Marles, 2011). Chromones with bicyclic rings have been used as privileged scaffolds in the formulation of new drugs (Reis et al., 2017). The 2-phenylethyl chromone (PEC) is a small class of chromones, which holds a phenylethyl group at the C_2 of the benzopyran ring of the chromone (Ibrahim and Mohamed, 2015). The formation of flindersia-type 2-(2-phenylethyl) chromones (FTPECs) is further described to be possibly catalyzed by type-III polyketide synthase (PKs) through condensation of the dihydrocinnamoyl-CoA analogs and malonyl-CoA with 2-hydroxy-benzoyl-CoA to produce a PEC scaffold. It is further catalyzed by hydroxylases or O-methyltransferases (OMTs) to form structurally diverse FTPECs, respectively (Liao et al., 2018).

The probability of getting agarwood-containing *Aquilaria* trees is extremely low in natural conditions (1–2%). It can only be found on either wounded or pathogenically infected trees (Chhipa and Kaushik, 2017). The mitogen-activated protein kinase (MAPK) signaling pathway is proposed as a wound-induced signaling mechanism for the agarwood formation in *A. sinensis*, which phosphorylates the downstream transcription factors (TFs), which ultimately leads to the expression of sesquiterpene synthase genes (Xu et al., 2013). The MAPK signaling cascade is a highly conserved signaling mechanism seen in eukaryotes that mediates extracellular signals to downstream responsive genes (Sinha et al., 2011). A total of 41 unigenes was reported from the transcriptome analysis of wounded *A. sinensis* and annotated as being related to the MAPK signaling pathway, and 25 to calcium signaling pathways, which may play roles in wound-induced agarwood formation (Xu et al., 2013). In plants, calcium ions (Ca^{2+}) play an essential role in intracellular secondary messenger molecules by regulating many signal transduction pathways reacting to external stimuli (Tuteja and Mahajan, 2007).

To increase the sesquiterpene content, using elicitor methyl jasmonic acid (MeJA) was found to be effective. Application of MeJA has been shown by Xu and co. earlier in 2013 in *A. sinensis.* They observed that the expression of several genes related to wound signal and regulation of signaling molecules (like C_2H_4, and H_2O_2) were increased by MeJA. However, MeJA treatment significantly upregulated the H_2O_2-producing NADPH oxidase noxB (Xu et al., 2013; Gong et al., 2017). MeJA was found to trigger H_2O_2 production in plants (Orozco-Cardenas et al., 2001; Hung et al., 2006). Programmed cell death (PCD) induction and sesquiterpene synthesis due to the elevation of expression of sesquiterpene synthase genes due to the endogenous accumulation of salicylic acid (SA) were evidenced in the suspension cultures of *A. sinensis.* Expression of these genes influenced by the jasmonic acid (JA) biosynthesis along with allene oxide cyclase (AOC), allene oxide synthase (AOS), lipoxygenase (LOX), and 12-oxophytodienoate reductase 3 (OPR3) genes can be increased by using heat shock, which can accumulate sesquiterpene in the *A. sinensis* through the production of JA. The differential stress responses of miRNAs from conserved miRNAs of *A. sinensis* affected the duration of wounding periods and also reflected the post-transcriptional regulation of wound response. The miR398 is down-regulatory stress-responsive miRNA. This miRNA can regulate the pathogen-associated molecular pattern (PAMP) and negatively triggered callose deposition and play an important role in the innate immune system of plants against bacteria (Li et al., 2010). Unlike miR398, miR160 and miR398 are also essential regulators in *A. sinensis* that have a role in agarwood formation.

For the protection and growth of this plant, federal governments and state governments must come forward and make agarwood policies for agarwood plantation zones. In India, the Assam Agarwood Policy 2020 came to effect on January 1, 2021, and the Tripura Agarwood Policy 2021 was also notified on July 19, 2021. These are two noble initiatives where various aspects of agarwood sectors for the profit of stakeholders are taken care of to make Assam and Tripura the "international hub" for the agarwood trade.

10.6 CONCLUSION

The extensive literature survey shows that the *Fusarium* genus is dominant in inducing fungal induction of *A. malaccensis*, followed by *Trichoderma* and *Penicillium.* Sesquiterpene (47.45%) and chromone derivatives (16.86%) are the main constituents of agarwood oil of *A. malaccensis.* Sesquiterpenes are accountable for the fragrance of the oil. The biological induction method is more profitable than physical or chemical induction methods of agarwood formation because the composition of oleoresin induced by biological inducers showed to be quite similar to a naturally produced oleoresin. There is a vast scope to explore biological inducers responsible for the production of quality-grade agarwood, as few biological inducers are reported for their role in agarwood development. In the future, agarwood oil can become a source of steady income generation without causing any intentional harm to nature. It can quickly become the startup material for small-scale industries, which, in turn, can help uplift the economy of developing nations.

10.7 CHALLENGES AND FUTURE PERSPECTIVES

Most traditional and local agarwood growers are not aware of non-conventional and modern techniques. They are practiced conventional or physical induction methods, resulting in the yield quality and quantity being inferior and low. If they adopt modern techniques, they must get more profit and better yield quality of agarwood. Improvement of induction efficiency and selection of high-yield lines of *Aquilaria* for resin production are the two crucial aspects for the future improvement of agarwood induction technology. The concept of improvement of induction efficiency can be summarized as either providing external stimuli to activate the production of signaling molecules or bypassing the external stimuli for the direct introduction of signaling molecules to the plants. The degree of plant response to stimuli is mainly dependent on the genetic makeup of the selected *Aquilaria* line. Moreover, a combined approach of genetic engineering with tissue culture can provide a high-yield *Aquilaria* line. In the future, the yield stability of agarwood quality and its market value will be controlled by the essential compounds of agarwood production and the biosynthesis mechanism of resin (Tan et al., 2019).

Due to the wide adaptability of *Aquilaria*, large-scale plantation of agarwood plants will help in carbon sequestration, protection of wild agar plants from extinction, and better utilization through agarwood oil extraction units and processing units would elevate the economy of developing nations. Agarwood policies would go a long way in enhancing the ease of doing business. More research and development for artificial induction of agarwood formation, sustainable harvesting, and improvement in the quality would lead to the formation of "green currency"; enhancement of quantity would lead to another "green revolution" in any country that adapts a meticulous thought and a well-planned policy. More agar-producing countries need to come forward and make a sustainable framework and adopt agarwood policies in their respective countries within the provisions of CITES.

ACKNOWLEDGMENTS

This work is done under the project support of DBT, New Delhi, Govt. of India (project no. BT/PR16867/NER/95/327/2015), dated: January 13, 2017. The authors are thankful to the DBT, New Delhi, Govt. of India, for research support.

REFERENCES

Akter, S., Islam, M.T., Zulkefeli, M., and Khan, S.I. 2013. Agarwood production-a multidisciplinary field to be explored in Bangladesh. *Int. J. Pharma Life Sci.* 2(1):22–32.

Arnold, A.E., Maynard, Z., and Gilbert, G.S. 2001. Fungal endophytes in dicotyledonous neotropical trees: patterns of abundance and diversity. *Mycol. Res.* 105:1502–1507.

Azzarina, A.B., Mohamed, R., Lee, S.Y., and Nazre, M. 2016. Temporal and spatial expression of terpene synthase genes associated with agarwood formation in *Aquilaria malaccensis* Lam. *NZ J. For. Sci.* 46:12.

Barden, A., Anak, N.A., Mulliken, T., and Song, M. 2000. *Heart of the Matter: Agarwood Use and Trade in CITES Implementation for Aquilaria Malaccensis*. Traffic International, Cambridge, England.

Bhattacharyya, B., Datta, A., and Baruah, H.K. 1952. On the formation and development of agaru in *Aquilaria agallocha*. *Sci. Cult.* 18:240–243.

Bhore, S.J., Preveena, J., and Kandasamy, K.I. 2013. Isolation and identification of bacterial endophytes from pharmaceutical agarwood-producing *Aquilaria* species. *Phcog. Res.* 5:134–137.

Blanchette, R.A., Jurgens, J.A., and Beek, H.H.V. 2015. Growing *Aquilaria* and production of agarwood in hill agroecosystems. In *Integrated Land Use Management in the Eastern Himalayas*, ed. K. Eckman and L. Ralte. Akhansha Publishing House, New Delhi, 66–82.

Burkill, I.H. 1966. *A Dictionary of the Economic Products of the Malay Peninsulam*, 2nd edn. Ministry of Agriculture and Cooperatives, Kaula Lumpur, Malaysia.

Chakrabarty, K., Kumar, A., and Menon, V. 1994. *Trade in Agarwood*. TRAFFIC India and WWF-India, New Delhi, India.

Chen, H.H., Zhao, G.Z., Park, D.J., Zhang, Y.Q., Xu, L.H., Lee, J.C., Kim, C.J., and Li, W.J. 2009. *Micrococcus endophyticus* sp. nov., isolated from surface-sterilized *Aquilaria sinensis* roots. *Int. J. Syst. Evol.* 59:1070–1075.

Chetpattananondh, P. 2012. *Overview of the Agarwood Oil Industry*. International Conference Proceedings of IFEAT on Essential Asia held at Singapore, pp. 131–138.

Chhipa, H., Chowdhary, K., and Kaushik, N. 2017. Artificial production of agarwood oil in *Aquilaria* sp. by fungi: a review. *Phytochem. Rev.* 16:835–860.

Chhipa, H., and Kaushik, N. 2017. Fungal and bacterial diversity isolated from *Aquilaria malaccensis* tree and soil, induces Agarospirol formation within 3 months after artificial infection. *Front. Microbiol.* 8:1286.

Chhipa, H., and Kaushik, N. 2020. Combined effect of biological and physical stress on artificial production of agarwood oleoresin in *Aquilaria malaccensis*. *BioRxiv*: 1–15.

Chong, S.P., Osman, M.F., Bahari, N., Nuri, E.A., Zakaria, R., and Abdul-Rahim, K. 2015. Agarwood inducement technology: A method for producing oil grade agarwood in cultivated Aquilaria malaccensis Lamk. *J. Agrobiotech.* 6:1–16.

Chowdhury, M., Hussain, M.D., Chung, S., Kabir, E., and Rahman, A. 2016. Agarwood manufacturing: a multidisciplinary opportunity for economy of Bangladesh-a review. *Int. Agric. Eng. J. CIGR J.* 18:171–178.

Chowdhury, M., Rahman, A., Hussain, M.D., and Kabir, E. 2017. The economic benefit of agarwood production through aeration method into the *Aquilaria malaccensis* tree in Bangladesh. *Bangladesh J. Agric. Res.* 42:191–196; ISSN: 0258–7122 (Print), 2408–8293 (Online).

Chung, R.C.K., and Purwaningsih. 1999. *Aquilaria malaccensis* Lamk. In *Plant Resources of South-East Asia No. 19: Essential-Oil Plants*, ed. L.P.A. Oyen and N.X. Dung, Backhuys Publishers, Leiden.

Dash, M., Patra, J.K., and Panda, P.P. 2008. Phytochemical and antimicrobial screening of extracts of *Aquilaria agallocha* Roxb. *Afr. J. Biotechnol.* 7(20):3531–3534; ISSN: 1684–5315.

Gianno, R. 1986. The exploitation of Resinous products in a low land Malayan forest. *Wallaceana* 43:3–6.

Gibson, I.A. 1977. The role of fungi in the origin of oleoresin deposits (agaru) in the wood of *Aquilaria agallocha* Roxb. *Bano Biggyan Patrika* 6:16–26; Corpus ID: 82304030.

Gong, B., Yan, Y., Wen, D., and Shi, Q. 2017. Hydrogen peroxide produced by NADPH oxidase: a novel downstream signaling pathway in melatonin-induced stress tolerance in *Solanum lycopersicum*. *Physiol. Plant.* 160:396–409.

Hansen, E. 2000. The hidden history of a scented wood. *Saudi Aramco World* 51:1–13.

Hartono, H., Wibowo, A., and Priyatmojo, A. 2019. Isolation, identification and the abilities of fungi associated with agarwood from Bangka Belitung Island to induce agarwood compounds. *J. Perlindungan Tanaman Indonesia* 23(1):94–108.

Hashim, Y.Z., Ismail, N., and Abbas, P. 2014. Analysis of chemical compounds of agarwood oil from different species by gas chromatography-mass spectrometry (GCMS). *IIUM Eng. J.* 15:55–60.

Hashim, Y.Z., Kerr, P.G., Abbas, P., and Mohd Salleh, H. 2016. *Aquilaria* spp. (agarwood) as source of health beneficial compounds: a review of traditional use, phytochemistry and pharmacology. *J. Ethnopharmacol.* 189:331–360.

Hung, K.T., Hsu, Y.T., and Kao, C.H. 2006. Hydrogen peroxide is involved in methyl jasmonate-induced senescence of rice leaves. *Physiol. Plant.* 127:293–303.

Ibrahim, S.R., and Mohamed, G.A. 2015. Natural occurring 2-(2-phenylethyl) chromones, structure elucidation and biological activities. *Nat. Prod. Res.* 29:1489–1520.

Jayachandran, K., Sekar, I., Parthiban, K.T., Amirtham, D., and Suresh, K.K. 2014. Analysis of different grades of agarwood (*Aquilaria malaccensis* Lamk.) oil through GC-MS. *Indian J. Nat. Prod. Resour.* 5:44–47; IPC code, Int. Cl. (2013.01)-A61K 36/00.

Jha, K.K. 2014. Agarwood plantation and products as livelihood strategy: a case study from Ban Khlong Sai Village, Northeast Thailand. *Int. J. Rural Dev. Manag. Stud.* 8:45–59.

Justin, S., Lihan, S., Elvis-Sulang, M.R., and Chiew, T.S. 2020. Formulated microbial consortium as inoculant for agarwood induction. *J. Trop. For. Sci.* 32(2):161–169.

Kanjilal, U.N., Kanjilal, P.C., Dey, R.M., and Das, A. 1982. *Flora of Assam*, Vol. IV (reprinted). Government of Assam, Assam, India.

Khadem, S., and Marles, R.J. 2011. Chromone and flavonoid alkaloids: occurrence and bioactivity. *Molecules* 17:191–206.

Kumeta, Y., and Ito, M. 2010. Characterization of δ-guaiene synthases from cultured cells of *Aquilaria*, responsible for the formation of the sesquiterpenes in agarwood. *Plant Physiol.* 154:1998–2007.

Lee, S.Y., and Mohamed, R. 2016. The origin and domestication of *Aquilaria*, an important agarwood-producing genus. In *Agarwood: Science Behind the Fragrance*, ed. R. Mohamed, Springer, Berlin, 1–20.

Li, Y., Zhang, Q., Zhang, J., Wu, L., Qi, Y., and Zhou, J.M. 2010. Identification of microRNAs involved in pathogen-associated molecular pattern-triggered plant innate immunity. *Plant Physiol.* 152(4):2222–2231.

Liao, G., Dong, W.H., Yang, J.L., Li, W., Wang, J., Mei, W.L., et al. 2018. Monitoring the chemical profile in agarwood formation within one year and speculating on the biosynthesis of 2-(2-phenylethyl)chromones. *Molecules* 23:1261.

Mabberly, D.J. 1992. *Tropical Rain Forest Ecology*, 2nd edn. Blackie and Son Ltd., London, UK; ISBN: 0216931479.

Mochahari, D., Kharnaior, S., Sen, S., and Thomas, S.C. 2020. Isolation of endophytic fungi from juvenile *Aquilaria malaccensis* and their antimicrobial properties. *J. Trop. For. Sci.* 32(1):97–103.

Mohamed, R., Jong, P.L., and Kamziah, A.K. 2014. Fungal inoculation induces agarwood in young *Aquilaria malaccensis* trees in the nursery. *J. For. Res.* 25(1):201–204.

Mohamed, R., Jong, P.L., and Zali, M.S. 2010. Fungal diversity in wounded stems of *Aquilaria malaccensis*. *Fungal Divers.* 43:67–74.

Naef, R. 2011. The volatile and semi-volatile constituents of agarwood, the infected heartwood of *Aquilaria* species: a review. *Flavour Fragr. J.* 26(2):73–87.

Nobuchi, T., and Siripatanadilok, S. 1991. Preliminary observation of *Aquilaria crassna* wood associated with the formation of aloeswood. *Bull. Kyoto Univ. For.* 63:226–235.

Orozco-Cardenas, M.L., Narvaez-Vasquez, J., and Ryan, C.A. 2001. Hydrogen peroxide acts as a second messenger for the induction of defense genes in tomato plants in response to wounding, systemin, and methyl jasmonate. *Plant Cell* 13:179–191.

Pasaribu, G.T., Waluyo, T.K., and Pari, G. 2015. Analysis of chemical compounds distinguisher for agarwood qualities. *Indonesian J. For. Res.* 2:7.

Pojanagaroon, S., and Kaewrak, C. 2003. Mechanical methods to stimulate aloes wood formation in Aquilaria crassna Pierre ex H. Lec. (Kritsana) trees. *Conser. Culti. Sust. Use MAPs, WOCMAP* 3(2):161–166.

Premalatha, K., and Kalra, A. 2013. Molecular phylogenetic identification of endophytic fungi isolated from resinous and healthy wood of *Aquilaria malaccensis*, a red-listed and highly exploited medicinal tree. *Fungal Ecol.* 6(3):205–211.

Putri, N., Karlinasari, L., Turjaman, M., Wahyudi, I., and Nandika, D. 2017. Evaluation of incense-resinous wood formation in Agarwood (*Aquilaria malaccensis* Lam.) using sonic tomography. *Agric. Nat. Resour.* 51(2):84–90.

Rahayu, G. 2008. *Increasing fragrance and terpenoid production in Aquilaria crassna by multi-application of methyl jasmonate comparing to single induction of Acremonium sp.* Paper Was Presented in International Conference on Microbiology and Biotechnology, Jakarta, 11–12 November.

Rahman, M.A., and Basak, A.C. 1980. Agar production in agar tree by artificial inoculation and wounding. *Bano Biggyan Patrtka* 9(1–2):87–93.

Rahman, M.A., and Khisa, K.S. 1984. Agar production in agar tree by artificial inoculation and wounding. Part II. Further evidence in favour of agar formation. *Bano Biggyan Patrika* 9:88–93; ISSN-0254–4539.

Rahman, H., Vakati, K., Eswaraiah, M.C., and Dutt, A.M. 2013. Evaluation of hepatoprotective activity of ethanolic extract of *Aquilaria agallocha* leaves (EEAA) against CCl4 induced hepatic damage in rat. *Sch. J. Appl. Med. Sci.* 1(1):9–12.

Rasool, S., and Mohamed, R. 2016. Understanding agarwood formation and its challenges. In *Agarwood: Science Behind the Fragrance*, ed. R. Mohamed, Springer, Berlin.

Reis, J., Gaspar, A., Milhazes, N., and Borges, F. 2017. Chromone as a privileged scaffold in drug discovery: recent advances. *J. Med. Chem.* 60:7941–7957.

Sangareswari Nagajothi, M., Parthiban, K.T., Kanna, S.U., Karthiba, L., and Saravanakumar, D. 2016. Fungal microbes associated with agarwood formation. *Am. J. Plant Sci.* 7:1445–1452.

Shahidullah, A.K.M., and Haque, C.E. 2010. Linking medicinal plant production with livelihood enhancement in Bangladesh: implications of a vertically integrated value chain. *J. Transdiscipl. Environ. Stud.* 9(2); ISSN: 1602–2297.

Sinha, A.K., Jaggi, M., Raghuram, B., and Tuteja, N. 2011. Mitogen activated protein kinase signaling in plants under abiotic stress. *Plant Signal. Behav.* 6(2):196–203; PMID: 21512321; PMCID: PMC3121978.

Tamuli, P., Boruah, P., Nath, S.C., and Leclercq, P. 2005. Essential oil of eaglewood tree: a product of pathogenesis. *J. Essent. Oil Res.* 17:601–604.

Tamuli, P., Boruah, P., Nath, S.C., and Samanta, R. 2000. Fungi from disease agarwood tree (*Aquilaria agallocha* Roxb): two new records. *Adv. For. Res. India* 22:182–187.

Tan, C.S., Isa, N.M., Ismail, I., and Zainal, Z. 2019. Agarwood induction: current developments and future perspectives. *Front. Plant Sci.* 10:122.

Tang, J., and Liu, Y. 2016. *A Natural Incense Inducer and Methods of Producing Agarwood*, CN103858689A. State Intellectual Property Office of the P.R.C., Beijing, China, 1–5.

Tawfik, H.A., Ewies, E.F., and El-Hamouly, W.S. 2014. Synthesis of chromones and their applications during the last ten years. *Int. J. Res. Pharm. Chem.* 4:1046–1085.

Thanh, L., Van Do, T., Son, N.H., Sato, T., and Kozan, O. 2015. Impacts of biological, chemical and mechanical treatments on sesquiterpene content in stems of planted *Aquilaria crassna* trees. *Agroforest. Syst.* 89:973–981.

Tian, J.J., Gao, X.X., Zhang, W.M., Wang, L., and Qu, LH. 2013. Molecular identification of endophytic fungi from *Aquilaria sinensis* and artificial agarwood induced by pinholes-infusion technique. *Afr. J. Biotechnol.* 12(21):3115–3131.

Turjaman, M., and Hidayat, A. 2017. Agarwood-planted tree inventory in Indonesia. *IOP Conf. Ser. Earth Environ. Sci.* 54:012–062.

Tuteja, N., and Mahajan, S. 2007. Calcium signaling network in plants: an overview. *Plant Signal. Behav.* 2(2):79–85.

Vanessa, M.C., and Christopher, M.M.F. 2004. Analysis of the endophytic actinobacterial population in the roots of wheat (*Triticum aestivum* L) by terminal restriction fragment length polymorphism and sequencing of 16S rRNA clones. *Appl. Environ. Microbiol.* 70:31787–31794.

Wu, Z.Q., Liu, S., Li, J.F., Li, M.C., Du, H.F., Qi, L.K., et al. 2017. Analysis of gene expression and quality of agarwood using Agar-bit in *Aquilaria sinensis. J. Trop. For. Sci.* 29:380–388.

Xu, Y., Zhang, Z., Wang, M., Wei, J., Chen, H., Gao, Z., et al. 2013. Identification of genes related to agarwood formation: transcriptome analysis of healthy and wounded tissues of *Aquilaria sinensis*. *BMC Genom.* 14(1):227.

Yoneda, K., Yamagata, E., Nakanishi, T., Nagashima, T., Kawasaki, I., Yoshida, T., Mori, H., and Miura, I. 1984. Sesquiterpenoids in two different kinds of agarwood. *Phytochemistry* 23(9):2068–2069.

11 Recent Trends in Conversion of Agro-Waste to Value-Added Green Products

Deepika Devadarshini, Swati Samal, Pradip Kumar Jena, and Deviprasad Samantaray

11.1 INTRODUCTION

Nature is the ultimate source of food, clothing, shelter, energy, and an unremitting resource for bio-economy. The extraction of the resource for different product formations and economic development has continuously increased to satisfy the needs of a booming economy. As a result, a broad range of renewable (sunlight, water, wind, and vegetation) and non-renewable sources (oil, natural gas, coal, and nuclear energy) is used for the production of various products (Zaman, 2014). The conversion and processing of these resources to consumables, such as food, clothing, furniture, paper, fuel, petroleum, ornaments, and other articles, generates a large amount of waste to nature. The unwanted by-product generated during domestic, commercial, agricultural, and industrial operations is termed *waste*. Amongst all, the agricultural sector contributes approximately 350–990 (Mt) (metric tons) agro-waste (AW) per year (Koul et al., 2022). *AW*s refer to the residue obtained after agricultural operations, such as processing and production of agro-products, including vegetables, crops, meat, poultry, fruits, and dairy products. Moreover, inappropriate waste management strategies pose a serious threat to the environment as well as living beings (Akande and Olorunnisola, 2018; Ali and Kumar, 2019).

The current practices of waste management strategies, such as incineration, composting, soil mulching, and uncontrolled landfilling, lead to environmental pollution. Further, manure, animal carcasses, and food processing wastes, which contain excess nitrates or carbon, lead to water pollution. Lignocellulosic biomass (LCB) is an abundantly available agro-industrial waste; however, its improper management causes the accumulation of non-biodegradable lignin in soil, leading to pollution. Other agro-industrial wastes, viz., bagasse, husks, fruit peels, and seedpods, upon burning, release greenhouse gases (GHGs), such as carbon dioxide, nitrous oxide, and methane, thereby leading to air pollution and climate change (Amran et al., 2021). As per the recent data available in the public domain, the air quality index of New Delhi was 250 μgm^{-3} during the period of October 2021 to January 2022, due

 DOI: 10.1201/9781003517108-11

to stubble burning in neighboring, states such as Punjab, Haryana, and Uttar Pradesh (Khan et al., 2022). Hence, the strategic management of AWs like valorization for the generation of bioproducts, like volatile fatty acids (VFAs) through acidogenic fermentation, and bioethanol, biopolymer, and SCP via syngas fermentation, is an economic and eco-friendly alternative to develop the bio-economy as well as to reduce carbon footprint.

11.2 WASTE IN GENERAL AND ITS CLASSIFICATION

Waste refers to any unwanted material which no longer suits its intended use and originates from households, institutes, electronic gadgets, medical, industries, mining, and the agricultural sector. It is solid, liquid, or gaseous in nature and contain a wide range of substances, from organic matter to hazardous chemicals. Some of the waste materials are biodegradable, while others are non-biodegradable. Biodegradable waste includes organic matter, viz., food waste, domestic waste, ash, and manure, which can be decomposed into carbon dioxide, water, methane, and humus. Non-biodegradable waste may be recyclable and non-recyclable waste, which consists of inorganic matter, viz., plastic, glass materials, and metals, which cannot be decomposed by biological processes or microbes (Dey et al., 2021). Further, waste can be classified into several categories based on their source, composition, and impact on the environment and public health.

11.2.1 Municipal Solid Waste (MSW)

It contains household, office, hotel, and institutional waste, such as food wastes, plastics, glass, papers, metals, and other materials (demolition and construction debris), generated in urban areas.

11.2.2 Industrial Waste

This includes a wide range of materials, like scrap metals, sludge, chemicals, solvents, resins, paints, ceramics, stones, rubber, leather, and abrasives, of varying environmental toxicity, generated during manufacturing, construction, mining, and other industrial operations.

11.2.3 Electronic Waste

It comprises appliances/gadgets like computers, televisions, cell phones, and other electronic devices that are intended for refurbishment, reuse, resale, recycling through material recovery, or disposal.

11.2.4 Medical Waste

Waste generated from healthcare facilities, such as hospitals, dental practices, blood banks, veterinary hospitals, medical research, and diagnostic laboratories, including needles, syringes, samples, blood, chemicals, pharmaceuticals, medical devices.

11.2.5 Radioactive Waste

Waste generated from nuclear power plants, mining, defense, and certain types of research facilities that contain radioactive materials, which are detrimental to our health and environment.

11.2.6 Agricultural Wastes (AWs)

AWs are residues such as crops, vegetables, fruits, and livestock operations predominantly obtained from the agricultural sector (Pattanaik et al., 2019). Broadly, AWs can be categorized into field residues (stems, stalks, leaves, roots, peels, husk, bark, wood fractions), agro-industry wastes (leaves, stalks, seed, seedpods, stems, roots, molasses, husks, bagasse, straw, stalk, shell, pulp, stubble, peel), livestock wastes (feathers, waste feed, manure, carcasses, residual milk, water), and hazardous wastes (pesticides, insecticides, herbicides). It is one of the leading sectors generating massive amounts of AWs, which accumulate indiscriminately and pose a global threat to health and food security (Dey et al., 2021; Adejumo and Adebiyi, 2020). Such large quantities of wastes are generated during preparation, production, storage, processing, and consumption of agricultural crops.

11.3 AGRO-WASTES

Agricultural productivities coupled with food security for fast-growing population have generated huge amount of wastes. Globally, 998 MT (million tons) of AWs are produced, among which India contributes approximately 350 MT/year (Obi et al., 2016). These are generated from various sources, like field residues, agro-industries, livestock, and hazardous wastes (Koul et al., 2022). The livestock industry occupies the first position in generating AWs of about 59.7%, while 35.6% of crop residues and 4.7% of agro-industries residues occupy the second and third positions, respectively (Bilala and Iqbalb, 2019; Ezekiel et al., 2021). Field residues include stems, stalks, leaves, roots, peels, husk, bark, and wood fractions generated after harvesting of crops (Adhikari et al., 2018). *Agro-industrial wastes* refer to agricultural and industrial residues, such as leaves, stalks, seed, seedpods, stems, roots, molasses, husks, bagasse, straw, stalk, shell, pulp, stubble, peel (Sadh et al., 2018). Livestock wastes consist of excreta, bedding material, residual milk, water, hair, feathers, or other debris which are left after animal waste handling procedure (Parihar et al., 2019). Hazardous wastes like pesticides, insecticides, herbicides, etc. are produced from routine agricultural practices. These are generally used for animal feed, fertilizers, and production of various value-added products.

Agricultural activities, including production and processing, generate a large quantity of Aws, depending upon the product formation and usage causing major threat to global health and food security. Almost every sector generates wastes, which are the vital factor for environmental pollution. On the other hand, the management of these wastes results in the massive downfall of bio-economy. In many developing countries, AWs are buried or burnt, thereby resulting in air pollution, harmful gas, dust, smoke, and emission, soil contamination, and the remains are concentrated into water, causing aquatic pollution (Adejumo and Adebiyi, 2020). Pesticides and fertilizers containing mercury, arsenic, lead, and cadmium are sprayed into the field and accumulate in the soil (Gontard et al., 2018; Jain et al., 2022; Sindhu et al., 2019). Further, these are washed into waterways, leach into the ground, and then get absorbed by plants, which are eventually consumed by animals and humans, causing serious health problems and premature deaths. Toxic metals can also result in crop failure and livestock poisoning from the contaminated soil, water, or food. Moreover, livestock are grown in confined conditions with improper diets, and the resulting residual animal feed causes environmental pollution. AW management is a serious concern; however, improper management strategies and lack of knowledge about their risks and benefits have drawn attention for the development of economic, eco-friendly, and sustainable management strategies. Further, it can broaden livelihood options and significantly contribute to green growth of economy (Dutta et al., 2022).

11.4 CONVENTIONAL STRATEGIES OF AGRO-WASTE MANAGEMENT

The increase in crop and livestock production generates large quantities of AWs, which cause environment and human health hazard. Herein, the different traditional physical, biological, thermal, and chemical strategies that have been used for AW management are summarized.

11.4.1 Physical Methods

Several physical methods, including briquetting, landfilling, and preparation of animal feed, have been used for AW management.

11.4.1.1 Briquetting

Briquetting is a technique of densifying waste materials, followed by compaction and binding. Due to their tremendous thermal efficiency, agro-waste and coal agro-waste combinations/blends have gained attention among different forms of briquettes. The primary technique involves the combustion of coal and biomass, and its lignin content has significant calorific value which helps in the binding of particles, resulting in the formation of briquettes or pellets. The ignition temperature of agro-waste is lower than that of coal; as a result, a considerable amount of heat is generated instead of smoke, which partially minimizes CO_2 emissions along with other greenhouse gases (Ganesh et al., 2021).

11.4.1.2 Landfilling

Dumping/landfills are considered as one of the most common methods for solid waste/residue management. These nutrient-rich wastes are biodegradable and form leachates that have adverse effects on the soil quality and pose significant environmental concern (Lou et al., 2015). In this method, biogas like methane and carbon dioxide are generated from landfills or dumped as organic matter via anaerobic digestion, causing environmental pollution (Esparza et al., 2020).

11.4.1.3 Animal Feed

The preparation of animal feed from AWs is the most common and sustainable technology of waste management. This process converts a substantial portion of AWs to animal feed. In general, crop residues, viz., straw, husks, and spent grain, are processed and used as animal feed. The residues/wastes are chopped and mixed with other feed ingredients, such as grains and supplements, to provide a balanced diet for livestock (Esparza et al., 2020). However, the residual animal feed is also a potential source for environmental pollution.

11.4.2 Biological Methods

Biological degradation and the transformation of waste under natural conditions are key strategies used for the AW management. In this method, microbes are used to degrade, transform, and stabilize AWs. Thus, various biologicals, such as composting, anaerobic digestion, mulching, and phytoremediation, are used for AW management.

11.4.2.1 Anaerobic Digestion

Anaerobic digestion of AWs is a booming technology which expedites the natural decomposition of organic materials without oxygen and energy. The methane-rich biogas generated from anaerobic digestion of AWs is used for cooking, heating, and lighting, while the slurry is used as liquid fertilizer/soil conditioner (Esparza et al., 2020). Earlier, an anaerobic digestion method was basically used for biogas generation from AWs, but for other value-added product formation.

11.4.2.2 Composting

Composting is the controlled microbe-mediated decomposition of organic (animal and plant) materials, such as crop residues, animal manure, and yard waste, into smaller nutrient-rich biodegradable components that can be used as soil conditioner to improve crop yield (Acevedo et al., 2022). In this method, organic materials are processed for composting with a bulking agent, such as straw or sawdust, in the presence of aeration and moisture. The rate of organic matter degradation is directly regulated by several factors, including particle size, temperature, moisture content, aeration, pH, and C–N ratio. Numerous aerobic microbes like *Trichoderma harzianum*, *Streptomyces aurefaciens*, *Bacillus subtilis*, *Trichoderma viridie*, and *Bacillus licheniformis* are present naturally or added artificially as inoculant, where they synergistically degrade organic matter into different products (Koul et al., 2021). During composting, various gases are released to the atmosphere, leading to environmental pollution.

11.4.2.3 Mulching

Mulching is a process of spreading AWs like crop residues or leaves on the soil surface. The residues gradually decompose and release nutrients into the soil, reducing soil erosion. It also helps to maintain moisture levels in the soil, reduces weed growth, and promotes the growth of soil microflora, increasing soil organic carbon content, thereby improving soil health and fertility. Mulching favors the growth of potato, chili, sugarcane, wheat, soybean, and sunflower by maintaining wet conditions and low temperature in the soil (Koul et al., 2021). Nevertheless, it increases pest infestation and microplastic contamination in the soil.

11.4.3 Thermal Methods

Thermal methods involve the usage of heat for treatment of AWs in an eco-friendly way. Incineration and pyrolysis are the common thermal methods used for thermal conversion of AWs to produce biofuel, biochar, and electrical energy etc.

11.4.3.1 Incineration

Incineration is a conventional method that uses high temperatures to convert waste materials into carbon dioxide, water, and energy. This process involves the combustion of AWs at temperatures ranging from 800 to 1,200°C in the presence of oxygen. Waste materials are burned in a controlled environment, and the generated heat is used for the production of electricity. Further, the remaining solid waste is processed for the removal of toxic metals or dioxins (Usmani et al., 2021). Overall, this process emits toxins and various pollutants that reduce air quality.

11.4.3.2 Pyrolysis

In this method, AWs are treated at 400–600°C in the absence of oxygen using a reactor. A wide array of reactors, including fluidized bed reactors and fixed-bed reactors, are commonly used for pyrolysis. It is completely regulated by different factors, like temperature, heating intensity, reactor zone residence time, and material size. Further, value-added products like biochar and bio-oil are produced through pyrolysis. However, high operational costs hold back its successful utilization in AW management.

11.4.4 Chemical Methods

AWs are basically composed of lignocellulosic materials, which pose a great challenge for their efficient conversion. Thus, to get rid of the issue, several chemical methods have been developed for AW management. In this approach, complex AWs are subjected to various chemical treatments to produce a simpler compound. Chemical substances such as acid, alkali, oxidative chemical substances, and organic solvents are employed for its treatment. Acid hydrolysis involves the treatment of AWs with different acids, like H_2SO_4 or HCl, to break down lignocellulosic materials, thereby releasing cellulose and hemicellulose, which are further converted to fermentable sugars by enzymatic hydrolysis. Similarly, alkaline pre-treatment of AWs by NaOH or NH_3 results in cellulose and hemicellulose compounds. Subsequently, on enzymatic hydrolysis, these compounds produce fermentable sugars. Moreover, oxidative (H_2O_2 or O_3) and organic solvent (ethanol or methanol) pre-treatment of AWs also produces fermentable sugars from lignocellulosic compounds. Further, these simple sugars can be used for the synthesis of different industrial (ethanol, methanol, butanol, acetic acid, lactic acid, butyric acid, polyhydroxy-alkanoic acids, etc.), pharmaceutical (antibiotics, amino acids, vitamins, enzyme, hormones, single-cell protein, single-cell oil, etc.), and agricultural (biofertilizer, biopesticide, bioherbicide, etc.)

products through the fermentation process. Despite several advantages, conversion efficiency and economic feasibility hold back its successful commercial utilization.

11.5 RECENT TRENDS IN AGRO-WASTE MANAGEMENT

Global population explosions, coupled with rapid industrial activities, have generated adequate quantities of AWs. Traditional methods of AW management are not effective to knob the interminably mounting waste load. Relatively, it is eternally concerned with eco-pollution and human health hazard (Singh and Dubey, 2022; Surendra et al., 2021). Therefore, a paradigm shift in the traditional AW management strategies is highly essential to pave the way for green environment and the development of bio-economy. Hence, emphasis has been given for strategic management of AWs through syngas fermentation and acidogenic fermentation to produce different value-added products (Figure 11.1).

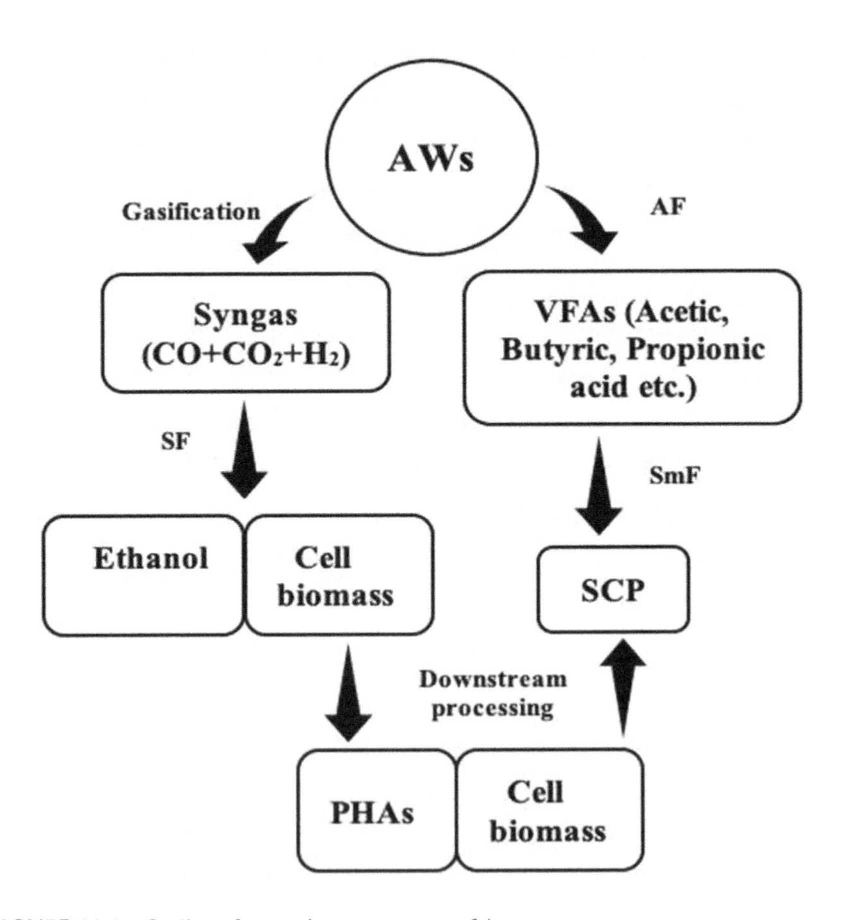

FIGURE 11.1 Outline of strategic management of Aws.

11.5.1 Syngas Fermentation

In the present context, AWs are the suitable substrates for the production of syngas/synthesis gas. In this process, carbonaceous materials like AWs thermochemically (gasification) converted to unicarbon compounds like CO, H_2, CO_2, and CH_4 in a gasifier under controlled conditions. However, syngas production and its composition are entirely regulated by operational conditions, type of gasifier, and various feedstocks used in the process. The syngas can be utilized for the production of bioethanol or other bioproducts through the fermentation process (Afolalu et al., 2021). Syngas fermentation is the recent advancement in bioprocess technology used for the bioconversion of unicarbon compounds for the production of commercially viable products. The bioconversion process is catalyzed by a specific group of anaerobic bacteria known as acetogens. They are widely distributed in diverse habitats, such as rumen liquor, compost, chicken manure, and biogas plant waste. Acetogens like *Clostridium autoethanogenum*, *Clostridium ljungdahlii*, *Clostridium ragsdalei*, *Clostridium coskatii*, and *Clostridium carboxidivorans* have the ability to reduce unicarbon compounds of syngas via the Wood–Ljungdahl (WL) pathway to produce different bioproducts, including biofuel (ethanol, butanol, and hexanol), 2,3-butanediol, and organic acids (acetic acid, butyric acid, and lactic acid), depending on the fermentation conditions and syngas composition (Ahuja et al., 2023; Afolalu et al., 2021). Moreover, its kinetics and process dynamics depends upon solubility of the syngas in the liquid phase. Among various products, bioethanol is the major product which can be used as fuel additives, in beverages, and in healthcare. The high-energy-density bioethanol replaces petroleum additives like gasoline and can be used as an alternate transportation fuel for future endeavor.

On the other hand, Fluchter et al. (2019) conducted an experiment to produce biopolymer like polyhydroxybutyrate (PHB) through syngas fermentation. It was observed that the recombinant strain *Clostridium coskatii* (p83-tcb) produced 0.98 ± 0.12 mM and 21.7 ± 0.27 mM of 3-HB autotrophically and heterotrophically, respectively. These biomaterials depict similar structural, thermal, and mechanical properties of synthetic plastic. Additionally, it is biodegradable under natural conditions and non-cytotoxic in nature (Devadarshini et al., 2022). Hence, it can be used for manufacturing household plastics, photographic films, biomedical devices, drug delivery carriers, chemicals, and nutritional supplements. Moreover, it is a suitable alternative to synthetic plastic and can address issues pertaining to white pollution (Maity et al., 2020). However, very few research reports are available in the public domain regarding biopolymer production via syngas fermentation. Despite several advantages, higher bioethanol and biopolymer yield through syngas fermentation is still facing a setback due to the lack of suitable high-yielding strains and techno-economic improvements (Vaishnavi et al., 2023). Therefore, in the present context, attention has been given for the integrated production of bioethanol, biopolymers, and SCP production through syngas fermentation to make the bio-process economic and eco-friendly.

11.5.2 Acidogenic Fermentation

Besides syngas fermentation, AWs can be used as a potential substrate for acidogenic fermentation (AF) to produce several value-added products. This process occurs during anaerobic digestion (AD) of AWs or organic matter in anoxic condition. Moreover, the microbe-mediated AD is carried out through four different steps, viz., hydrolysis, acidogenesis, acetogenesis, and methanogenesis. Herein, three different groups of microbes, like acidogens, acetogens, and methanogens, are syntrophically involved to produce different products. In AF, the complex organic matter of AWs is hydrolyzed by acidogens to simple sugar, short-chain polypeptides, and fatty acids and then fermented to produce acetate, lactate, and various VFAs. Then, the products of AF are oxidized to alcohols, H_2, and CO_2 by acetogens (Park et al., 2015; Portela-Grandio et al., 2021). Subsequently, the generated unicarbon compound CO_2 is reduced to CH_4 and CO_2 by methanogens. Among all the products of AF, VFAs (acetic acid, butyric acid, propionic acid, valeric acid, and caproic acid) are gaining utmost importance as a substrate for different industrial product formation via submerged or solid-state fermentation (Zewde et al., 2023; Zabaleta et al., 2021; Portela-Grandio et al., 2021). Thus, prior to hydrolysis, AWs are pre-treated with different alkali, like NaOH and $Ca(OH)_2$, to increase the rate of conversion of complex to simple matter. Further, the addition of 2-bromoethane sulfonic acid (BES) to the hydrolyzed matter (before fermentation) inhibits methanogens, thereby enhancing VFA production (Jayakrishnan et al., 2020). The VFAs can be used as a precursor of polyhydroxyalkanoates (PHAs), biogas, bio-hydrogen, biodiesel, and bio-electricity production. However, the cost-competitiveness of these products is the major factor for its successful commercialization. Thus, academia and industry are shifting their attention towards SCP production using VFAs based on the rising protein demand (Pesante et al., 2022). *SCP* refers to dead and dried cells of microbes, including bacteria, algae, fungi, and yeasts, rich in proteins, lipids, and vitamins, which can be used as a protein supplement for human food and animal feed (Tripathi et al., 2019). Herein, the probabilities of cost-competitiveness are higher, as microbial cells themselves act as a product. Therefore, SCP production using VFAs as a substrate can match the rising protein demand, develop bio-economy, and reduce environmental pollution through strategic management of AWs.

11.6 CONCLUSION

In conclusion, this chapter represents an overview of the generation of AWs from different sources and traditional and recent strategies used for their management for sustainable environment. Strategic management such as syngas fermentation and acidogenic fermentation of AWs produces various value-added products, like green chemicals (biofertilizers, nanomaterials, bioactive compounds, biomaterials, pharmaceutical and medicinal agents, and cultivation media) and energy (biohydrogen, bioethanol, biogas, and biofuel) for decarbonization of society. Recent trends in research also established that AW management has enormous potential to reduce greenhouse gas emission. Hence, proper AW management can help mitigate eco-pollution and climate change and produce innovative biomaterials for a circular bio-economy. However, before pilot-scale operation, its techno-economic feasibility should be critically evaluated for sustainable development.

REFERENCES

Acevedo MD, Lancellotti I, Andreola F, Luisa Barbieri L, Urena LJB and Ferre FC. 2022. Management of agricultural waste biomass as raw material for the construction sector: An analysis of sustainable and circular alternatives, *Environmental Sciences Europe*, 34(70). https://doi.org/10.1186/s12302-022-00655-7.

Adejumo IO and Adebiyi OA. 2020. Agricultural solid wastes: causes, effects, and effective management, in: Saleh HM (ed) *Strategies of Sustainable Solidwaste Management*. IntechOpen, pp. 1–20.

Adhikari S, Nam H and Chakraborty JP. 2018. Conversion of solid wastes to fuels and chemicals through pyrolysis, in: Bhaskar T, Pandey A, Mohan SV, Lee DJ and Khanal SK (eds) *Waste Biorefinery*. Elsevier, pp. 239–263.

Afolalu SA, Salawu EY, Ogedengbe TS, Joseph OO, Okwilagwe O, Emetere ME, Yusuf OO, Noiki AA and Akinlabi SA. 2021. Bio-agro waste valorization and its sustainability in the industry: a review, *IOP Conference Series: Materials Science and Engineering*, 1107. https://doi.org/10.1088/1757-899X/1107/1/012140.

Ahuja V, Bhatt AK, Ravindran B, Yang YH and Bhatia SK. 2023. A mini-review on syngas fermentation to bio-alcohols: current status and challenges, *Sustainability*, 15(4). https://doi.org/10.3390/su15043765.

Akande OM and Olorunnisola AO. 2018. Potential of briquetting as a waste-management option for handling market-generated vegetable waste in Port Harcourt, Nigeria, *Recycling*, 3(2). https://doi.org/10.3390/recycling3020011.

Ali SR and Kumar R. 2019. Strategic framework and phenomenon of zero waste for sustainable future, *Contaminants in Agriculture and Environment: Health Risks and Remediation*, 1. https://doi.org/10.26832/AESA-2019-CAE-0167-015.

Amran MA, Palaniveloo K, Fauzi R, Satar NM, Mohidin TBM, Mohan G, Razak SA, Arunasalam M, Nagappan T and Seelan JSS. 2021. Value-added metabolites from agricultural waste and application of green extraction techniques, *Sustainability*, 13. https://doi.org/10.3390/su132011432.

Bilala M and Iqbalb HMN. 2019. Sustainable bioconversion of food waste into high-value products by immobilized enzymes to meet bio-economy challenges and opportunities – a review, *Food Research International*, 123:226–240. https://doi.org/10.1016/j.foodres.2019.04.

Devadarshini D, Mohapatra S, Pati S, Maity S, Rath CC, Jena PK and Samantaray DP. 2022. Evaluation of PHAs production by mixed bacterial culture under submerged fermentation, *Biologia*, 78. https://doi.org/10.1007/s11756-022-01302-5.

Dey T, Bhattacharjee T, Nag P, Ritika, Ghati A and Kuila A. 2021. Valorization of agro-waste into value added products for sustainable development, *Bioresource Technology Reports*, 16. https://doi.org/10.1016/j.biteb.2021.100834.

Dutta J, Sen T, Sen P, Tiwari PK, Panda AK and Dutta J. 2022. Creating wealth from agrowaste, in: Singh S, Singh P, Sharma A and Choudhury M (eds) *Agriculture Waste Management and Bioresource*. Wiley. https://doi.org/10.1002/9781119808428.ch14.

Esparza I, Moreno NJ, Bimbela F, Azpilicueta CA and Gandia LM. 2020. Fruit and vegetable waste management: conventional and emerging approaches, *Journal of Environmental Management*, 265. https://doi.org/10.1016/j.jenvman.2020.110510.

Ezekiel AA, Ayinde OA, Oladeebo JO and Adeyanju JA. 2021. Poultry waste management techniques, the implication on environment and agricultural productivity in Afijo local government area, Oyo State, Nigeria, *International Journal of Waste Resources*, 11(2):1–8.

Fluchter S, Follonier S, Schiel-Bengelsdorf B, Bengelsdorf FR, Zinn M and Durre P. 2019. Anaerobic production of poly(3-hydroxybutyrate) and its precursor 3-hydroxybutyrate from synthetic gas by autotrophic Clostridia, *Biomacromolecules*, 20. https://doi.org/10.1021/acs.biomac.9b00342.

Ganesh KS, Sridhar A and Vishali S. 2021. Utilization of fruit and vegetable waste to produce value-added products: conventional 2 utilization and emerging opportunities-a review, *Chemosphere*, 287. https://doi.org/10.1016/j.chemosphere.2021.132221.

Gontard N, Sonesson U, Birkved M, Majone M, Bolzonella D, Celli A, Coussy HA, Jang GW, Verniquet A, Broeze J, Schaer B, Batista AP and Sebok A. 2018. A research challenge vision regarding management of agricultural waste in a circular bio-based economy, *Critical Reviews in Environmental Science and Technology*, 48(6):614–654. https://doi.org/10.1080/10643389.2018.1471957.

Jain A, Sarsaiya S, Awasthi MK, Singh R, Rajput R, Mishra UC, Chen J and Sh J. 2022. Bioenergy and bio-products from bio-waste and its associated modern circular economy: Current research trends, challenges, and future outlooks, *Fuel*, 307. https://doi.org/10.1016/j.fuel.2021.121859.

Jayakrishnan U, Deka D and Das G. 2020. Regulation of volatile fatty acid accumulation from waste: effect of inoculum pretreatment, *Water Environment Research*, 93. http://doi.org/10.1002/wer.1490.

Khan S, Anjum R, Raza ST, Bazai NA and Ihtisham. 2022. Technologies for municipal solid waste management: current status, challenges, and future perspectives, *Chemosphere*, 288. https://doi.org/10.1016/j.chemosphere.2021.132403.

Koul B, Yakoob M and Shah MP. 2021. Agricultural waste management strategies for environmental sustainability, *Environmental Research*, 206. https://doi.org/10.1016/j.envres.2021.112285.

Koul B, Yakoob M and Shah MP. 2022. Agricultural waste management strategies for environmental sustainability, *Environmental Research*, 206. https://doi.org/10.1016/j.envres.2021.112285.

Lou Z, Wang M, Zhao Y and Huang R. 2015. The contribution of biowaste disposal to odor emission from landfills, *Journal of Air and Waste Management Association*, 65(4). https://doi.org/10.1080/10962247.2014.1002870.

Maity S, Das S, Mohapatra S, Tripathi AD, Akhtar J, Pati S, Pattnaik S and Samataray DP. 2020. Growth associated polyhydroxybutyrate production by the novel Zobella tiwanensis strain DD5 from banana peels under submerged fermentation, *International Journal of Biological Macromolecules*, 153:461–469. http://doi.org/10.1016/j.ijbiomac.2020.03.004.

Obi FO, Ugwuishiwu BO and Nwakaire JN. 2016. Agricultural waste concept, generation, utilization and management, *Nigerian Journal of Technology*, 35(4):957–964. http://dx.doi.org/10.4314/njt.v35i4.34.

Parihar SS, Saini KPS, Lakhani GP, Jain A, Roy B, Ghosh S and Aharwal B. 2019. Livestock waste management: a review, *Journal of Entomology and Zoology Studies*, 7(3):384–393.

Park GW, Seo C, Jung K, Chang HN, Kim W and Kim YC. 2015. A comprehensive study on volatile fatty acids production from rice straw coupled with microbial community analysis, *Bioprocess and Biosystems Engineering*, 38. http://doi.org/10.1007/s00449-015-1357-z.

Pattanaik L, Pattnaik F, Saxena DV and Naik SN. 2019. Biofuels from agricultural wastes, in: Angelo Basile A and Dalena F (eds) *Second and Third Generation of Feedstock*. Elsevier, pp. 103–142.

Pesante G, Zuliani A, Bolzonella D and Frison N. 2022. Biological conversion of agricultural wastes into microbial proteins for aquaculture feed, *Chemical Engineering Transactions*, 92:391–396. http://doi.org/10.3303/CET2292066.

Portela-Grandio A, Lagoa-Costa B, Kennes C and Veiga MC. 2021. Polyhydroxyalkanoates production from syngas fermentation effluents: effect of nitrogen availability, *Journal of Environmental Chemical Engineering*, 9(6). https://doi.org/10.1016/j.jece.2021.106662.

Sadh PK, Duhan S and Duhan JS. 2018. Agro-industrial wastes and their utilization using solid state fermentation: a review, *Bioresource and Bioprocessing*, 5(1). https://doi.org/10.1186/s40643-017-0187-z.

Sindhu R, Gnansounou E, Rebello S, Binod P, Varjani S, Thakur IS, Nair RB and Pandey A. 2019. Conversion of food and kitchen waste to value-added products, *Journal of Environmental Management*, 241:619–630. https://doi.org/10.1016/j.jenvman.2019.02.053.

Singh A and Dubey GM. 2022. Agricultural waste management: problems and treatments, *International Journal for Innovative Research in Multidisciplinary Field*, 8(4):148–152.

Surendra KC, Angelidaki I and Khanal SK. 2022. Bioconversion of waste-to-resources (BWR-2021): Valorization of industrial and agro-wastes to fuel, feed, fertilizer, and biobased products, *Bioresource Technology*, 347. https://doi.org/10.1016/j.biortech.2022.126739.

Tripathi N, Hills CD, Singh RS and Atkinson CJ. 2019. Biomass waste utilisation in low-carbon products: harnessing a major potential resource, *Climate and Atmospheric Science*, 2(35):1–36.

Usmani Z, Sharma M, Awasthi AK, Sivakumar N, Lukk T, Pecoraro L, Thakur VK, Roberts D, Newbold J and Gupta VK. 2021. Bioprocessing of waste biomass for sustainable product development and minimizing environmental impact, *Bioresource Technology*, 322. https://doi.org/10.1016/j.biortech.2020.124548.

Vaishnavi J, Osborne WJ and Samuel J. 2023. Microorganism in waste valorization and its impact on the environment and economy, in: Samuel J, Kumar A and Singh J (eds) *Relationship Between Microbes and Environment for Sustainable Ecosystem Services: Microbial Tools for Sustainable Ecosystem Services*, 1st edn. Elsevier. https://doi.org/10.1016/C2020-0-02876-8.

Zabaleta MP, Atasoy M, Khatami K, Eriksson E and Cetecioglu Z. 2021. Bio-based conversion of volatile fatty acids from waste streams to polyhydroxyalkanoates using mixed microbial cultures, *Bioresource Technology*, 323. http://doi.org/10.1016/j.biortech.2020.124604.

Zaman AU. 2015. A comprehensive review of the development of zero waste management: lessons learned and guidelines, *Journal of Cleaner Production*, 91:12–25. http://dx.doi.org/10.1016/j.jclepro.2014.12.013.

Zewde A, Li Z and Zhou X. 2023. A review on the use and applications of volatile fatty acids on fecal sludge sanitization, *Journal of Water, Sanitation & Hygiene for Development*, 13(3). http://doi.org/10.2166/washdev.2023.252.

12 Green Inventory and Carbon Emissions

A Review

Bhabani Shankar Mohanty

12.1 INTRODUCTION

This review chapter is aimed at sustainable inventory control, which indicates deciding on inventory size, replenishment time, inventory cost, and environmental, social, and economic impact. It integrates ecological parameters by focusing on optimum inventory cost, size, and replenishment time. Carbon emissions and global warming are the primary concern of developing and developed countries. Several regulating bodies are putting efforts into reducing carbon footprints and supporting sustainable developments globally. This work reviews investment in green technology to formulate an inventory model with various environmental considerations. A carbon tax is a policy tool that governments use to address the issue of carbon emissions and climate change. This parameter influences the total cost in the green inventory model and encourages the control of carbon emissions. Green inventory and carbon emissions are also directly linked with supply chain management. The researchers (Ahi and Searcy, 2013) have broadly examined sustainable supply chain management. Carbon gas is produced during different phases of the supply chain model, like transportation, holding, and processing of commodities. There are a lot of sources of generation of greenhouse gas emissions in the supply chain as well as the inventory model. Generally, emissions are from the transportation of goods, employee commuting, business travel, organization operations, industrial processes, and indirect emissions from the generation of purchased electricity. In addition to carbon emissions, a green inventory may track other environmental parameters, such as water usage, waste generation, or air pollutants, depending on the organization's sustainability goals and priorities. By maintaining a green inventory, organizations can identify opportunities for reducing emissions, prioritize sustainability initiatives, comply with reporting requirements, demonstrate their commitment to environmental responsibility, and contribute to the efforts to mitigate climate change and promote sustainable practices. Carbon emissions are a crucial component of many inventory models, particularly in supply chain management and sustainability. Incorporating carbon emissions into inventory models allows organizations to assess and optimize their environmental impact and make informed decisions to reduce their carbon footprint. Organizations can calculate the carbon footprint of their inventory items by considering the emissions associated with their production, transportation, and use. This

DOI: 10.1201/9781003517108-12

involves gathering data on energy consumption, fuel usage, and emissions factors for different activities in the supply chain. By quantifying the carbon emissions of individual products or inventory categories, organizations can prioritize low-carbon alternatives and optimize their inventory strategies accordingly. Incorporating carbon emissions into inventory models allows organizations to evaluate their suppliers' environmental performance. By considering the emissions associated with the production and transportation of goods, organizations can make informed decisions about supplier selection and work collaboratively with suppliers to improve sustainability performance. Carbon emissions can be factored into transportation decisions within the inventory model. By considering emission factors for different transportation modes, distances, and volumes, organizations can optimize routes, consolidate shipments, and select more sustainable transportation options to minimize carbon emissions. Carbon pricing/taxes/trade assigns a cost to carbon emissions, and organizations can incorporate this cost into their inventory models. By considering the financial implications of carbon emissions, organizations can make more informed decisions about inventory levels, transportation modes, and sourcing strategies. Green inventory and carbon taxation are powerful tools for promoting sustainability, reducing emissions, and mitigating climate change. Combining the insights gained from green inventory with the economic incentives of carbon taxation enables organizations to make informed decisions, drive innovation, and contribute to a greener and more sustainable future. By integrating carbon emissions into inventory models, organizations can gain insights into the environmental impact of their operations and supply chains. This enables them to identify opportunities for improvement, prioritize sustainable practices, and make informed decisions that contribute to reducing carbon emissions and addressing climate change.

12.2 CONTRIBUTIONS

Kuo (2011) proposed an inventory model to reduce electronic waste using a manufacturing/remanufacturing simulation model. Hua et al. (2011) proposed a single-product optimal economic-order quantity model with a carbon emissions trading mechanism.

Bouchery et al. (2012) feel the importance of a multi-objective inventory model to control carbon emissions. Ji et al. (2012) proposed an integer linear programming model that considers transportation, inventory, and reproduction control tactics when computer manufacturers implement a green supply chain strategy.

Palak et al. (2013) aim to minimize total costs and carbon emissions by developing a bi-objective inventory replenishment decisions model for age-dependent deteriorating items. Chen and Hsieh (2013) developed a new vendor-managed inventory model for the apparel industry that considers different planning horizons in an uncertain environment. Urvashi et al. (2013) studied an integrated buyer–supplier green inventory model with remanufacturing for deteriorating products.

Bozorgi et al. (2014) extended a new inventory model with exact algorithms to minimize total cost and emissions. Kuo et al. (2014) considered a vehicle-routing model for suppliers so that carbon emissions reach a certain accuracy level. Sazvar et al. (2014) discussed a stochastic inventory model for deteriorating items with transportation costs, uncertain demand, and various greenhouse gas levels.

Fichtinger et al. (2015) studied the environmental impact on warehouse emissions from heating, cooling, air-conditioning, and lighting due to inventory holding. Bazan et al. (2015) proposed an optimal inventory model for the manufacturing and remanufacturing industry, considering greenhouse gas emissions and penalty tax. Glock and Kim (2015) developed an inventory model using heterogeneous vehicles in a supply chain inventory model. This model minimizes both operating costs and carbon emissions from transportation. Konur and Schaefer (2015) studied multi-item inventory control systems with economic and environmental objectives. Roozbeh Nia et al. (2015) examined a multi-item optimal inventory model with shortages for a single-buyer single-supplier with green vendor management. The total cost of the supply chain is minimized using nonlinear integer programming. Soysal et al. (2015) suggested a multi-period perishable model for the inventory-routing problem that includes truckload-dependent distribution costs to comprehensively evaluate CO_2 emissions with uncertain demand.

Ugarte et al. (2016) proposed a green supply chain simulation model to investigate the empirical impact of three lean logistics practices. The study proved that just-in-time inventory management significantly increases greenhouse gas emissions. Chibeles-Martins et al. (2016) presented a multi-objective meta-heuristics algorithm to overcome the need for a multi-objective approach for designing and planning supply chains. Franco et al. (2016) investigated green bi-objective inventory routing problems with carbon emissions due to transportation and holding inventory. Khan et al. (2016) developed a two-level sustainable supply chain economic order quantity model to prove that information sharing is the prime way to reduce total costs in a sustainable supply chain. Konur et al. (2016) gave a bi-objective continuous review stochastic inventory control model to show that the delivery policy and supplier selection significantly impact the economic and ecological balance. Tang et al. (2016) investigated a multi-objective mixed-integer model for the trade-off between costs and carbon emissions with consumer environmental behaviors.

Bazan et al. (2017) suggested a two-level closed-loop supply chain model with a facility to remanufacture used items for manufacturers and retailers. The model minimizes greenhouse emissions from production and transportation activities. Karimi et al. (2017) presented a two-stage newsboy problem in a green supply chain for a manufacturing vendor if stocks are provided at the start of the selling period. Marklund and Berling (2017) developed a sustainable inventory model to reduce emissions substantially with relatively small increases in total costs. Rahimi et al. (2017) extended the classical inventory routing problem to a dynamic model that integrates service level and green considerations.

Considering the interior point solution method, Niknamfar et al. (2018) developed a green supply chain model for a bi-objective series-parallel inventory–redundancy allocation problem. Kang et al. (2018) investigated a sustainable inventory-allocation planning model with carbon emissions and defective product disposal under an uncertain environment. Lee et al. (2018) formulated a green supply chain to minimize total cost by considering carbon emissions, transporting cost, demand rate, vehicle travel period, and backlogging constraints. Rau et al. (2018) developed a green cyclic inventory routing problem with means to solve it by particle swarm optimization. Sustainable inventory management aims to reduce the environmental impact without changing the profit. Tiwari et al. (2018) presented a single-vendor and single-buyer inventory for perishable items that considered carbon emissions. Taleizadeh et al. (2018) suggested a green inventory model with joint pricing, carbon emissions, and planned discounts.

Paam et al. (2019) developed a supply chain model that challenges the apple industry's economic and environmental sustainability. The model reduces fruit loss and total storage and processing costs. Halat and Hafezalkotob (2019) formulated a multi-stage green supply chain with carbon emissions to maximize social welfare and minimize inventory costs. Chen et al. (2019) suggested a sustainable inventory model for online pharmacies. The model deals with visual-attention-dependent rates to characterize customer demand. Al-Aomar and Alshraideh (2019) considered a service-oriented green optimization model for effective material management in the service area. Castellano et al. (2019) developed a single-vendor/multiple-buyer supply chain with the problem of coordinating production, inventory replenishment, routing decisions, and greenhouse gas emissions. George and Pillai (2019) produced a method to measure the performance of the supply chain in different situations. Shen et al. (2019) suggested a production-deteriorating inventory model with preservation technology and carbon tax. Kovacs and Illes (2019) introduced a sustainable green supply chain to optimize total cost and lead time with various constraints. Li et al. (2019) developed cold chain inventory routing models with the constraint of carbon emission regulations to reduce carbon emissions.

Jemai et al. (2020) and Jauhari (2020) suggested a green supply chain for deteriorating items with shortages and a green inventory model to optimize total cost and carbon emissions, respectively. The green supply chain model (Ajay et al. 2020a, 2020b, 2020c) is developed by considering first-in-first-out and last-in-first-out queuing patterns for stocks using a different algorithm. Zic and Zic (2020) constructed a multi-criteria simulation-based inventory optimization model to raise awareness about operational decisions, stochastic demand, target fill rates, and environmental impact. Mohammadnazari and Ghannadpour (2020) presented a sustainable green supply chain model that focuses on material provision and resource planning under uncertainty and the green transportation of material.

Dey et al. (2021) proposed a smart supply chain management under a controllable production rate and advertised dependent demand with variable lead time. Yadav et al. (2021), Shah et al. (2021), and Singh et al. proposed a green supply chain with preservation technology, emission-dependent demand, and a two-level credit policy, respectively.

Paul et al. (2022) formulated a green inventory deteriorating model to optimize profit for price- and green-sensitive demand. Gautam et al. (2022) proposed a production inventory model for defective items with greening degree-dependent demand and salvage cost. Shah et al. (2022b) formulated an optimal economic-order quantity model for deteriorating items with price and stock-dependent demand rates and greening efforts. The model maximizes profit concerning cycle time, selling price, and greening efforts. Shah et al. (2022a) suggested a green deteriorating inventory model with the carbon tax and trade.

12.3 DISCUSSION WITH TABLES AND FIGURES

A summarized table with some vital objective functions is provided in Table 12.1. Inventory models have single or multiple objectives. It is observed that most of the researchers (70%) are interested in minimizing the model's total cost. Researchers (53%) are also concerned about the environment and wish to reduce carbon emissions.

The distribution of publications per year is presented in Figure 12.1. In 2019, the highest percentages (16%) of research papers were published.

TABLE 12.1
Decision/Objective Functions Considered in Different Papers

Reference	Minimize Cost	Maximize Profit	Minimize Lead Time	Minimize Carbon Emissions	Minimize Cycle Time
Ajay et al. (2020a)	✓				
Ajay et al. (2020b)	✓				
Ajay et al. (2020c)	✓				
Al-Aomar and Alshraideh (2019)	✓				
Bazan et al. (2015)	✓				
Bazan et al. (2017)	✓			✓	
Bouchery et al. (2012)	✓			✓	
Bozorgi et al. (2014)	✓			✓	
Castellano et al. (2019)	✓				
Chen and Hsieh (2013)	✓				
Chen et al. (2019)		✓			
Chibeles-Martins et al. (2016)		✓		✓	
Dey et al. (2021)	✓				
Fichtinger et al. (2015)	✓			✓	
Franco et al. (2016)	✓			✓	
Gautam et al. (2022)		✓		✓	
Glock and Kim (2015)	✓				✓
Halat and Hafezalkotob (2019)	✓			✓	
Hua et al. (2011)	✓			✓	
Jauhari (2020)	✓			✓	
Jemai et al. (2020)	✓			✓	
Ji et al. (2012)		✓			
Kang et al. (2018)	✓				
Karimi et al. (2017)	✓			✓	
Khan et al. (2016)		✓			
Konur et al. (2016)	✓				
Konur and Schaefer (2015)	✓			✓	
Kovacs and Illes (2019)	✓		✓		
Kuo (2011)	✓				
Kuo et al. (2014)	✓			✓	
Lee et al. (2018)	✓				

TABLE 12.1 *(Continued)*
Decision/Objective Functions Considered in Different Papers

Reference	Minimize Cost	Maximize Profit	Minimize Lead Time	Minimize Carbon Emissions	Minimize Cycle Time
Li et al. (2019)	✓			✓	
Marklund and Berling (2017)	✓			✓	
Mohammadnazari and Ghannadpour (2020)	✓				
Niknamfar et al. (2018)		✓		✓	
Paam et al. (2019)	✓				
Palak et al. (2013)	✓			✓	
Paul et al. (2022)		✓			✓
Rahimi et al. (2017)		✓		✓	
Rau et al. (2018)	✓			✓	
Roozbeh Nia et al. (2015)	✓			✓	
Sazvar et al. (2014)	✓			✓	
Shah et al. (2022a)		✓		✓	✓
Shah et al. (2022b)					
Shah et al. (2021)	✓				✓
Shen et al. (2019)		✓		✓	
Singh et al. (2021)	✓				
Soysal et al. (2015)	✓			✓	
Taleizadeh et al. (2018)		✓			✓
Tang et al. (2016)	✓			✓	
Tiwari et al. (2018)	✓			✓	
Ugarte et al. (2016)				✓	
Urvashi et al. (2013)		✓			
Yadav et al. (2021)		✓		✓	
Zic and Zic (2020)	✓			✓	

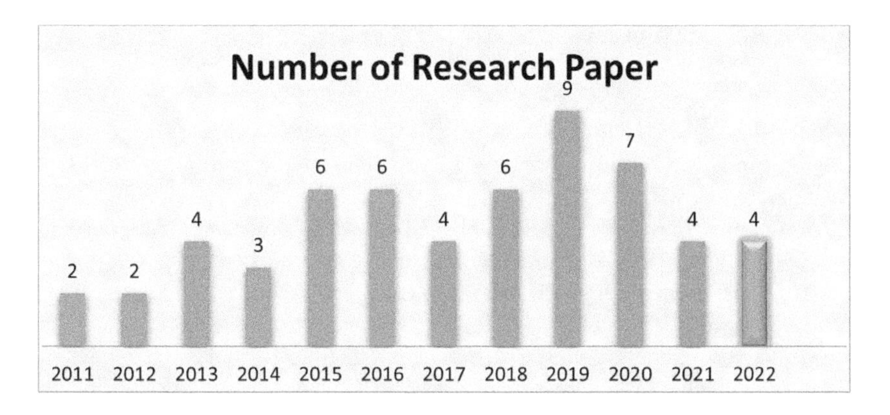

FIGURE 12.1 Frequency distribution of publications per year.

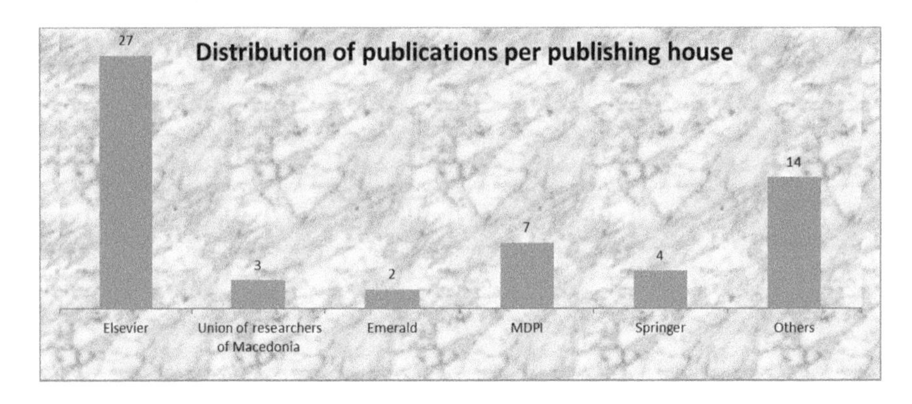

FIGURE 12.2 Distribution of publications per publishing house.

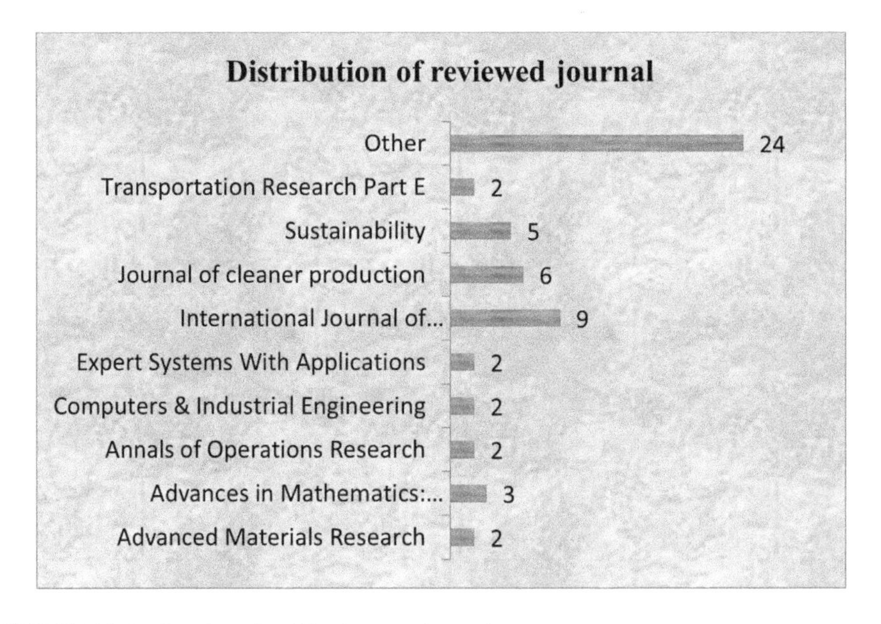

FIGURE 12.3 Number of publications per journal.

Figure 12.2 shows the frequency distribution of reviewed publications per publishing house. From the reviewed articles, 47% are published in journals of the Elsevier publishing company.

Reviewed research articles are published in various international journals of repute. Out of these, 16%, 11%, and 9% of articles are published in the *International Journal of Production Economics*, *Journal of Cleaner Production*, and *Sustainability*, respectively.

Generally, researchers consider different parameters to make the model more realistic and valuable for application in industry. Table 12.2 gives information about

TABLE 12.2
Model Parameters Considered in Different Papers

Reference	Holding Cost	Order Quantity	Shortage Cost	Purchase Cost	Deterioration Rate/Cost	Carbon Tax/ Emission Cost	Sensitivity Analysis
Al-Aomar and Alshraideh (2019)	✓			✓		✓	✓
Bazan et al. (2015)	✓					✓	
Bazan et al. (2017)	✓					✓	
Bouchery et al. (2012)	✓						✓
Bozorgi et al. (2014)	✓						✓
Castellano et al. (2019)	✓	✓				✓	✓
Chen and Hsieh (2013)	✓	✓					
Chen et al. (2019)							✓
Chibeles-Martins et al. (2016)							✓
Dey et al. (2021)	✓						✓
Fichtinger et al. (2015)	✓						✓
Franco et al. (2016)	✓						
Gautam et al. (2022)							
George and Pillai (2019)							
Glock and Kim (2015)	✓					✓	✓
Halat and Hafezalkotob (2019)	✓					✓	✓
Hua et al. (2011)	✓					✓	✓
Jauhari (2020)	✓					✓	
Jemai et al. (2020)	✓		✓		✓	✓	
Ji et al. (2012)	✓						

(*Continued*)

TABLE 12.2 *(Continued)*
Model Parameters Considered in Different Papers

Reference	Holding Cost	Order Quantity	Shortage Cost	Purchase Cost	Deterioration Rate/Cost	Carbon Tax/ Emission Cost	Sensitivity Analysis
Kang et al. (2018)	✓					✓	✓
Karimi et al. (2017)	✓					✓	✓
Khan et al. (2016)						✓	✓
Konur et al. (2016)	✓						
Konur and Schaefer (2015)	✓						
Kuo et al. (2014)							✓
Lee et al. (2018)	✓		✓	✓		✓	✓
Li et al. (2019)	✓	✓				✓	✓
Marklund and Berling (2017)	✓					✓	
Mohammadnazari and Ghannadpour (2020)	✓						✓
Niknamfar et al. (2018)	✓		✓				✓
Paam et al. (2019)	✓						✓
Palak et al. (2013)	✓				✓		✓
Paul et al. (2022)	✓				✓	✓	✓
Rahimi et al. (2017)	✓						✓
Rau et al. (2018)	✓						✓
Roozbeh Nia et al. (2015)	✓		✓			✓	✓
Sazvar et al. (2014)	✓		✓		✓	✓	✓
Shah et al. (2022a)	✓		✓	✓	✓	✓	
Shah et al. (2022b)							

Shah et al. (2021)	✓			✓		✓
Shen et al. (2019)	✓			✓	✓	✓
Singh et al. (2021)	✓				✓	✓
Soysal et al. (2015)	✓			✓		✓
Taleizadeh et al. (2018)	✓	✓	✓		✓	✓
Tang et al. (2016)	✓				✓	✓
Tiwari et al. (2018)	✓			✓	✓	✓
Urvashi et al. (2013)	✓		✓			✓
Yadav et al. (2021)	✓			✓	✓	✓
Zic and Zic (2020)	✓					✓

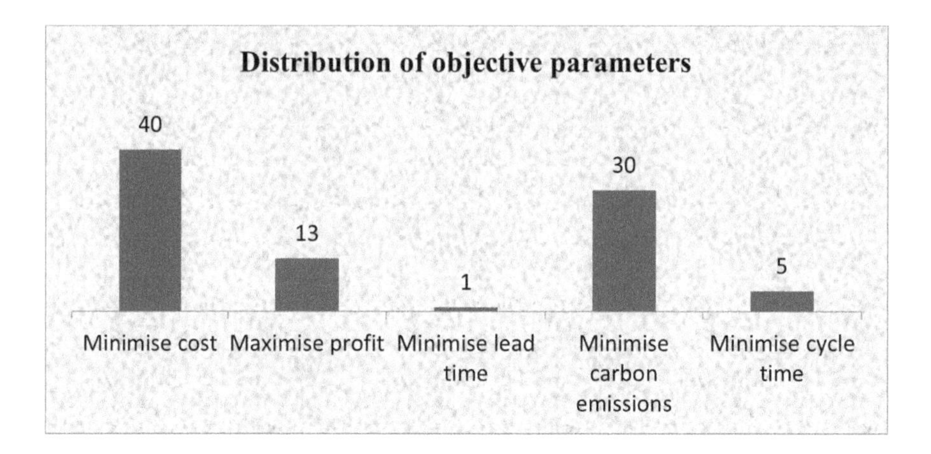

FIGURE 12.4 Number of publications per objective function.

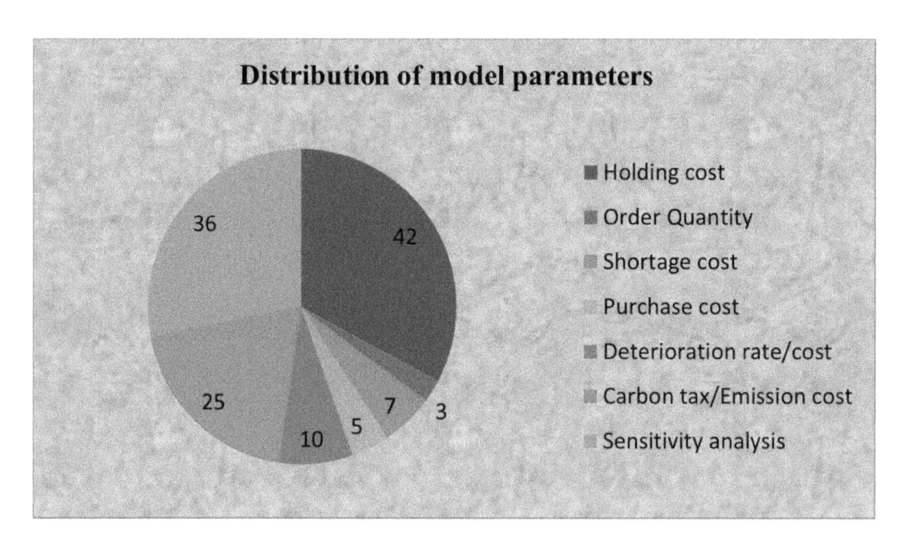

FIGURE 12.5 Number of publications per model parameter.

detailed parameters. *Holding cost* is an essential cost considered in the inventory model. Approximately 75% of articles deal with holding costs, and 44% included carbon emissions as a parameter. Sensitivity analysis is essential for studying relative changes in decision function to change model parameters. It is rare to fit a model with actual data in an inventory model due to the unavailability of data sources. Researchers can use sensitivity analysis to study the model's sensitivity by changing parameters. So many researchers (63%) have performed a sensitivity analysis in their studies.

12.4 CONCLUSIONS

The literature for sustainable supply chains based on green inventory and carbon emissions has already been completed up to 2022. This chapter highlighted the contribution of 57 research articles from different international journals and the reputed publication house proceedings. This work aims to combine all the scientific research articles linked to this emerging field. This work deals with a detailed study of the model parameters used, objective functions, journal, period of publication, and publication house. The detailed analysis of the chapter shows the trend in research. It helps find the gap and scope of research in this particular field.

REFERENCES

Ahi, P., Searcy, C. 2013. A comparative literature analysis of definitions for green and sustainable supply chain management. *J. Clean. Prod.* 52, 329–341.

Ajay, S. Y., Abid, M., Bansal, S., Tyagi, S. L., Kumar, T. 2020a. FIFO & LIFO in green supply chain inventory model of hazardous substance components industry with storage using simulated annealing. *Adv. Math. Sci. J.* 9, 5127–5132.

Ajay, S. Y., Kapil, K. B., Shivani, Seema, A., Vanaja, R. 2020b. FIFO in green supply chain inventory model of electrical components industry with distribution centres using particle swarm optimization. *Adv. Math. Sci. J.* 9, 5115–5120.

Ajay, S. Y., Kumar, A., Agarwal, P., Kumar, T., Vanaja, R. 2020c. LIFO in green supply chain inventory model of auto-components industry with warehouses using differential evolution. *Adv. Math. Sci. J.* 9, 5121–5126.

Al-Aomar, R., Alshraideh, H. 2019. A service-oriented material management model with green options. *J. Clean. Prod.* 236.

Bazan, E., Jaber, M. Y., el Saadany, A. M. A. 2015. Carbon emissions and energy effects on manufacturing-remanufacturing inventory models. *Comput. Ind. Eng.* 88, 307–316.

Bazan, E., Jaber, M. Y., Zanoni, S. 2017. Carbon emissions and energy effects on a two-level manufacturer-retailer closed-loop supply chain model with remanufacturing subject to different coordination mechanisms. *Int. J. Prod. Econ.* 183, 394–408.

Bouchery, Y., Ghaffari, A., Jemai, Z., Dallery, Y. 2012. Including sustainability criteria into inventory models. *Eur. J. Oper. Res.* 222, 229–240.

Bozorgi, A., Pazour, J., Nazzal, D. 2014. A new inventory model for cold items that considers costs and emissions. *Int. J. Prod. Econ.* 155, 114–125.

Castellano, D., Gallo, M., Grassi, A., Santillo, L. C. 2019. The effect of GHG emissions on production, inventory replenishment, and routing decisions in a single vendor-multiple buyers supply chain. *Int. J. Prod. Econ.* 218, 30–42.

Chen, H. C., Hsieh, Y. H. 2013. A newsboy model for apparel items under different back ordering situations and inflations of the green supply chain. *Adv. Mater. Res.* 2876–2881.

Chen, Y. K., Chiu, F. R., Chang, Y. C. 2019. Implementing green supply chain management for online pharmacies through a VADD inventory model. *Int. J. Environ. Res. Publ. Health* 16.

Chibeles-Martins, N., Pinto-Varela, T., Barbosa-Póvoa, A. P., Novais, A. Q. 2016. A multi-objective meta-heuristic approach for the design and planning of green supply chains – MBSA. *Expert Syst. Appl.* 47, 71–84.

Dey, B. K., Bhuniya, S., Sarkar, B. 2021. Involvement of controllable lead time and variable demand for a smart manufacturing system under a supply chain management. *Expert Syst. Appl.* 184.

Fichtinger, J., Ries, J. M., Grosse, E. H., Baker, P. 2015. Assessing the environmental impact of integrated inventory and warehouse management. *Int. J. Prod. Econ.* 170, 717–729.

Franco, C., López-Santana, E. R., Méndez-Giraldo, G. 2016. A column generation approach for solving a green bi-objective inventory routing problem. In: *Lecture Notes in Computer Science (Including Subseries Lecture Notes in Artificial Intelligence and Lecture Notes in Bioinformatics)*. Springer Verlag, pp. 101–112.

Gautam, P., Maheshwari, S., Jaggi, C. K. 2022. Sustainable production inventory model with greening degree and dual determinants of defective items. *J. Clean. Prod.* 367.

George, J., Pillai, V. M. 2019. A study of factors affecting supply chain performance. In: *Journal of Physics: Conference Series*. Institute of Physics Publishing.

Glock, C. H., Kim, T. 2015. Coordinating a supply chain with a heterogeneous vehicle fleet under greenhouse gas emissions. *Int. J. Logist. Manag.* 26(3), 494–516.

Halat, K., Hafezalkotob, A. 2019. Modeling carbon regulation policies in inventory decisions of a multi-stage green supply chain: a game theory approach. *Comput. Ind. Eng.* 128, 807–830.

Hua, G., Cheng, T. C. E., Wang, S. 2011. Managing carbon footprints in inventory management. *Int. J. Prod. Econ.* 132, 178–185.

Jauhari, W. A. 2020. Lot-sizing decisions in manufacturer-retailer inventory system under carbon emissions reduction. *IOP Confer. Ser. Mater. Sci. Eng.* 943(1).

Jemai, J., do Chung, B., Sarkar, B. 2020. Environmental effect for a complex green supply-chain management to control waste: a sustainable approach. *J. Clean. Prod.* 277.

Ji, S., Wang, W., Dong, Y. 2012. ILP model research of electronic manufacturing enterprise green supply chain inventory & transportation decision-making. *Adv. Mater. Res.* 255–259.

Kang, K., Pu, W., Ma, Y. 2018. A dynamic programming-based sustainable inventory-allocation planning problem with carbon emissions and defective item disposal under a fuzzy random environment. *Math. Probl. Eng.* 1–18.

Karimi, M., Niknamfar, A. H., Pasandideh, S. H. R. 2017. Two-stage single period inventory management for a manufacturing vendor under green-supplier supply chain. *Int. J. Syst. Assur. Eng.* 8(4), 704–718.

Khan, M., Hussain, M., Saber, H. M. 2016. Information sharing in a sustainable supply chain. *Int. J. Prod. Econ.* 181, 208–214.

Konur, D., Campbell, J. F., Monfared, S. A. 2016. Economic and environmental considerations in a stochastic inventory control model with order splitting under different delivery schedules among suppliers. *Omega* 71, 46–65.

Konur, D., Schaefer, B. 2015. Economic and environmental comparison of grouping strategies in coordinated multi-item inventory systems. *J. Oper. Res. Soc.* 67, 421–436.

Kovacs, G., Illes, B. 2019. Development of an optimization method and software for optimizing global supply chains for increased efficiency, competitiveness, and sustainability. *Sustainability* 11(6).

Kuo, T. C. 2011. The study of production and inventory policy of manufacturing/remanufacturing environment in a closed-loop supply chain. *Int. J. Sustain. Eng.* 4, 323–329.

Kuo, T. C., Chen, G. Y. H., Wang, M. L., Ho, M. W. 2014. Carbon footprint inventory route planning and selection of hot spot suppliers. *Int. J. Prod. Econ.* 150, 125–139.

Lee, A. H. I., Kang, H. Y., Ye, S. J., Wu, W. Y. 2018. An integrated approach for sustainable supply chain management with replenishment, transportation, and production decisions. *Sustainability* 10(11).

Li, L., Yang, Y., Qin, G. 2019. Optimization of integrated inventory routing problem for cold chain logistics considering carbon footprint and carbon regulations. *Sustainability* 11.

Marklund, J., Berling, P. 2017. Green inventory management. *Sustain. Supply Chains* 189–218.

Mohammadnazari, Z., Ghannadpour, S. F. 2020. Sustainable construction supply chain management with the spotlight of inventory optimization under uncertainty. *Environ. Dev. Sustain.* 23, 10937–10972.

Niknamfar, A. H., Niaki, S. A. A., Karimi, M. 2018. A series-parallel inventory-redundancy green allocation system using a max-min approach via the interior point method. *Assemb. Autom.* 38, 323–335.

Paam, P., Berretta, R., Heydar, M., García-Flores, R. 2019. The impact of inventory management on economic and environmental sustainability in the apple industry. *Comput. Electron. Agric.* 163.

Palak, G., Eksioglu, S. D., Geunes, J. 2013. *Models for Cost Efficient and Environmentally Friendly Inventory Replenishment Decisions for Perishable Products.* Industrial and Systems Engineering Research Conference, San Juan, Puerto Rico.

Paul, A., Pervin, M., Roy, S. K., Maculan, N., Weber, G. W. 2022. A green inventory model with the effect of carbon taxation. *Ann. Oper. Res.* 309, 233–248.

Rahimi, M., Baboli, A., Rekik, Y., 2017. Multi-objective inventory routing problem: a stochastic model to consider profit, service level and green criteria. *Transp. Res. E Logist. Transp. Rev.* 101, 59–83.

Rau, H., Budiman, S. D., Widyadana, G. A. 2018. Optimization of the multi-objective green cyclical inventory routing problem using discrete multi-swarm PSO method. *Transp. Res. E Logist. Transp. Rev.* 120, 51–75.

Roozbeh Nia, A., Hemmati Far, M., Niaki, S. T. A. 2015. A hybrid genetic and imperialist competitive algorithm for green vendor managed inventory of multi-item multi-constraint EOQ model under shortage. *Appl. Soft Comput. J.* 30, 353–364.

Sazvar, Z., Mirzapour Al-e-hashem, S. M. J., Baboli, A., Akbari Jokar, M. R. 2014. A bi-objective stochastic programming model for a centralized green supply chain with deteriorating products. *Int. J. Prod. Econ.* 150(1), 140–154.

Shah, N. H., Rabari, K., Patel, E. 2022b. Greening efforts and deteriorating inventory policies for price-sensitive stock-dependent demand. *Int. J. Syst. Sci. Oper. Logist.* 10(1), 1–7.

Shah, N. H., Shah, P. H., Patel, M. B. 2022a. Supply chain coordination with flexible payment policy under effect of green technology investments. *Yugoslav J. Oper. Res.* 29.

Shah, P. Shah, N. H., Patel, M. B. 2021. An inventory model with carbon emission dependent demand for a manufacturer-distributor supply chain. *Adalya J.* 10(6), 151–163.

Shen, Y., Shen, K., Yang, C. 2019. A production inventory model for deteriorating items with collaborative preservation technology investment under carbon tax. *Sustainability* 11(18), 5027.

Singh, S., Yadav, D., Sarkar, B., Sarkar, M. 2021. Impact of energy and carbon emission of a supply chain management with two-level trade-credit policy. *Energies* 14(6), 1569.

Soysal, M., Bloemhof-Ruwaard, J. M., Haijema, R., van der Vorst, J. G. A. J. 2015. Modeling an inventory routing problem for perishable products with environmental considerations and demand uncertainty. *Int. J. Prod. Econ.* 164, 118–133.

Taleizadeh, A. A., Hazarkhani, B., Moon, I. 2018. Joint pricing and inventory decisions with carbon emission considerations, partial backordering and planned discounts. *Ann. Oper. Res.* 290(1), 95–113.

Tang, J., Ji, S., Jiang, L. 2016. The design of a sustainable location-routing-inventory model considering consumer environmental behavior. *Sustainability* 8.

Tiwari, S., Daryanto, Y., Wee, H. M. 2018. Sustainable inventory management with deteriorating and imperfect quality items considering carbon emission. *J. Clean. Prod.* 192(1), 281–292.

Ugarte, G. M., Golden, J. S., Dooley, K. J. 2016. Lean versus green: the impact of lean logistics on greenhouse gas emissions in consumer goods supply chains. *J. Purch. Supply Manag.* 22, 98–109.

Urvashi, Singh, S. R., Singh, N. 2013. Green supply chain model with product manufacturing under volume flexible environment. *Procedia Technol.* 10(1), 216–226.

Yadav, D., Kumari, R., Kumar, N., Sarkar, B. 2021. Reduction of waste and carbon emission through the selection of items with cross-price elasticity of demand to form a sustainable supply chain with preservation technology. *J. Clean. Prod.* 297.

Zic, J., Zic, S. 2020. Multi-criteria decision making in supply chain management based on inventory levels, environmental impact and costs. *Adv. Prod. Eng. Manage.* 15, 151–163.

13 Biological System for Waste Management and Alternative Biofuel Production

Oindrila Gupta, Srishti Joshi, Shweta Shukla, and Satarupa Banerjee

13.1 INTRODUCTION

Since billions of tons of carbon pollutants are released into the atmosphere each year, the recently contaminated planet is rapidly coming to an end and posing a danger to climatic change. Studying scientific literature suggests that biofuel is a popular subject, gaining particular attention from the global scientific community, particularly from those working in underdeveloped nations. Because of this, biofuel is seen as an essential component of many future energy scenarios. The primary source of energy for many developing nations is biomass feedstock. Biofuels are a better option than petroleum-based fossil fuels to decrease carbon emissions into the atmosphere and the likelihood of future energy crises. Fossil fuels having a negative environmental impact tends to determine the biofuel feasibility and understanding.

Waste is also considered as "second-generation" pollution problem (Saldana-Duran et al., 2020) which, if solved, can eventually reduce other environmental problems. Waste is an unavoidable by-product of any human activity but can be processed to make it useful. Though there are many sources of waste, the principal sources of solid waste are municipal, agricultural, and industrial wastes. Municipal waste is generated mostly from households, schools, offices, and other institutions. Waste includes, but is not limited to, paper, glass, textile, food, electronics, etc. Municipal waste generates 14% of the total waste. But this percentage varies according to the rural or urban area. Another type of waste is produced in agriculture. which mostly consists of wastes from arable land and horticulture. These kinds of wastes not only comprise crop residues like dry leaves, wood, and silages but also agrochemical wastes, such as fertilizers, pesticide, and waste produced from livestock, including animal manure. A major part of agricultural waste can be utilized efficiently in biofuel production. Industries also produce a major percentage of total waste from their manufacturing, mining, and construction activities. Waste produced from industries can be non-hazardous or hazardous, like heavy metals, radioactive substances, and chemicals. Some of the non-hazardous waste after its processing can be used

DOI: 10.1201/9781003517108-13

in biofuel production, which includes waste from food processing, paper, and pulp (Casares et al., 2005).

Waste can be repurposed or reused for other different applications, which reduces pollution, greenhouse gas, and landfills. Out of the total waste produced from different areas, if this waste is treated properly, 70% of it can be reused for different applications. A very basic example is preparing compost out of it, which would increase the fertility of the soil rather than contaminating it. Mostly, non-biodegradable waste can be recycled or upcycled to produce other products, reducing the need for virgin materials. Bioenergy production has been gaining popularity recently and has the potential to replace current methods of energy production. It can be of different types – biomass, biogas, biofuel, and biochar (charcoal-like substance). It could not only help in waste management but also reduce pollution generated during conventional energy production and enhance infrastructural development (Gautam et al., 2012).

In most countries, municipal solid waste can be segregated and used for landfilling method. Again, some countries have utilized these wastes as raw materials for heat and power generation. To begin a waste management system, one must first monitor the waste production, followed by the process of waste collection and transportation. This step leads to waste treatment and disposal. The negative impact of an improper waste management system is on the environment since it leads to the production of many pollutants. One basic way to increase the usage of solid waste is to use it as a raw material for biofuel production. Biofuel production is a popular subject, gaining particular attention from the global scientific community, particularly from those working in underdeveloped nations. Because of this, biofuel is seen as an essential component of many future energy scenarios. The primary source of energy for many developing nations is biomass feedstock (Nanda & Berruti, 2021).

Biofuels include fuels that are solid, liquid, and gaseous. Certain biofuels are produced using biomass feedstocks, which may replace conventional fuel sources and reduce carbon emissions. Depending on the types of feedstocks used and changes in production techniques, the first generation of biofuels will be referred to as conventional biofuels, while the second and third generations may be referred to as contemporary biofuels. Conventional liquid biofuels include bioethanol made from starch and sugar yields as well as biodiesel made from oil yields. They can presently be produced cheaply. Modern biofuels are substantially more environmentally friendly and can reduce environmental damage by lowering carbon emissions than previous biofuels. Agricultural and forest lignocellulosic biomass as well as algal feedstock may be used to create advanced renewable energy sources. The liquid biofuels produced throughout time are biotechnologically or chemically catalyzed paraffin fuels or hydrocarbons manufactured by humans from plant carbohydrates. The use of lignocellulosic biomass as a raw material for the production of biofuel is gaining much importance.

Lignocellulosic materials such as agricultural, hardwood, and softwood residues are seen as potential sources for sugar fermentation to produce alcohol. These materials have cellulose and hemicellulose elements, which are long molecular chains of sugars. They are protected by the polymer that has the glue to hold all the materials together.

These lignocellulosic biomasses have the potential to form biofuels and could give optimum yield.

Various types of carbon sources, which include glucose, biomass-derived sugars, organic acids, and glycerol, can be used for lipid production. In most processes, glucose can be considered an ideal source for microorganisms. These microorganisms using glucose as the source can then be used for the production of biofuel.

Microalgae biomass is a raw material that contains a variety of biochemical components, including carbohydrates, lipids, and enzymes. These components may be converted into biofuels, such as biodiesel, bio-oil, bio-methane, and bio-crude oil. Agricultural residues, which include straws, hulls, seeds linter, and other similar by-products, have high saccharide and low polymer content. Dilute acid treatment along with vasoconstrictor steam treatment are the initial pre-treatment steps for the production of biofuel.

Depending upon the feedstock, biofuels can be categorized into four types: first-, second-, third-, and fourth-generation biofuels. First-generation biofuels are created from sugar, starch, and oil-based plants. Second-generation are generally non-food yields and are mainly produced from agricultural and woodland residues. Third-generation biofuels are produced from algae and can be produced on a large scale. The fourth generation of biofuels employs newly discovered and quickly developing synthetic cyanobacterial development (Alalwan et al., 2019).

13.2 TYPES OF WASTES

1. *Industrial waste.* A lot of waste is generated through industries annually. This has also become a cause of global concern, since treatment of all waste is not easy or possible. There are various categories of industrial waste according to their nature, composition, and origin:
   - *Hazardous waste.* Waste which is harmful for human health and environment, such as chemicals, radioactive materials, and heavy metals.
   - *Non-hazardous waste.* Includes paper, plastics, and food waste.
   - *Electronic waste.* Includes waste from discarded electronic devices, like computers, cell phones, and other machines which may contain lead, mercury, etc.
   - *Medical waste.* Materials generated from the healthcare industry, like syringes, protective equipment, laboratory specimen, expired drugs, which may cause infections.
   - *Textile waste.* Produced from the textile industry, which includes fabric scraps, threads, etc.
   - *Construction and demolition waste.* Materials which are generated from the construction industry, which include concrete, bricks, asphalt.
   - *Mining waste.* Includes waste rock, slag, etc.
   - *Nuclear waste.* Most hazardous waste, produced from nuclear power plants, nuclear radiations from research, etc. (Woodard, 2001)

   Among these wastes, only a few can be recycled and reused for useful applications, such as in construction, landscaping, energy generation, manufacturing and packaging, and recycled textile.

2. *Food waste and municipal solid waste.* Food waste generally begins from agricultural production due to pest and diseases, which can reach up to 500 million, followed by inefficient post-harvest and storage. A lot of waste is produced during the processing and distribution in the markets. A major proportion gets wasted due to non-consumption and end of shelf life of the packaged food product. It has been observed that over 1/3 of the food produced worldwide gets wasted, which is equivalent to 1.3 billion a year. As per the UN, by the year 2030, the goal is (1) to reduce half of the global food waste at the retail and consumer level and (2) to reduce food loss during the process of production and supply chains. Food production requires energy, pesticides, and a large area of land. It leads to the emission of a large amount of greenhouse gases. Hence, it has become a necessity that we focus on the ways to utilize food waste to produce different forms of energy in order to reuse the resources. It must be noted that food production causes a lot of wastage of water equivalent to 250 km^2 of fresh water. Thus, it has become a necessity to work on wastewater treatment for reuse of the resources being lost (McCarty et al., 2011). Municipal solid waste, which is commonly known as trash, consists of regular household items that are thrown post-usage. It mainly consists of food product packaging, clothing, plastic bottles, electronic appliances, furniture, etc.
3. *Agricultural waste.* Agricultural wastes include lignocellulosic biomass that can be used directly as a primary fuel for the burning and generation of heat and electricity. These wastes are further processed to produce different biofuels, such as bioethanol, biogas, bio-hydrogen, and biodiesel. The biomass that produces this biofuel can be categorized into several groups, which include whole plants, agricultural residues, and agro-wastes such as slide cattle manure and forest biomass. There are many advantages to using agricultural wastes as the source of biofuel, which include reduction of dependency on the woody biomass from the forest, which results in deforestation. Agricultural wastes are classified according to the residues obtained from production and processing of the products, which include crops, fruits, vegetables, meat and poultry, and dairy products.

At the field level, crop leftovers such as leaves, stovers, straws, and seedpods make up many of the waste residues produced by direct agricultural production. The projected yearly output of agricultural leftovers throughout the world is 2,802 MT (million tons). The most prevalent and affordable organic waste that is readily converted into various value-added products is agricultural residue, which is derived from crop wastes.

Three main crop wastes are utilized to produce bioethanol across the world: rice straw, wheat straw, and maize stover. These crops are available all year long, and only a very small amount is used to produce biofuel or fodder.

Agro-industry processing waste is included in the second category of agricultural waste. This includes waste products from food processing businesses, such as vegetable and fruit peels, fruit pomace left over after juice extraction, starch residue from businesses that make starch, sugarcane bagasse, molasses from businesses that make sugar, de-oiled seed cake from businesses that make edible oils, and chicken skin, egg, meat, and animal fat from businesses that process meat. Once the juice is extracted from sugarcane, one of the main agricultural by-products is bagasse: 180.73 MT (million tons) of sugarcane bagasse are readily available worldwide.

These wastes may be converted either biochemically or thermochemically to produce biofuel. In the biochemical approach, the biomass is broken down into intermediates (sugars, amino acids, or short-chain fatty acids) and then transformed into liquid or gaseous fuels, like biogas, bioethanol, bio-butanol, and biodiesel, using various microbes and enzymes. The thermochemical approach is straightforward and uses heat and chemicals either alone or in combination to produce syngas (a mixture of H2 and CO), bio-oil, biochar, and bio-coal.

13.3 STRATEGIES TO ESTIMATE THE AMOUNT OF FOOD WASTAGE ON A GLOBAL SCALE

As per the model suggested by the United Nations, the first step involves estimating the food waste being produced on an annual basis from the different sectors of the country: (1) search and collate the existing data; (2) filter data on scope and applicability to the needs of the current study; (3) adjust the data for consistency; (4) extrapolate for the countries which have insufficient data; (5) estimate the percentage.

Food waste can be classified as household, food service, and retail. The second step involves calculating the Food Waste Index for the country; here the waste is classified into edible and inedible parts. The proper means of using the same is identified to reduce wastage (Lahiri et al., 2023). The main processes which are considered are animal feed production, biomaterial processing, anaerobic digestion, compost or aerobic digestion, controlled combustion, land application, landfill, discard, thrown in the sewer. The third step involves identifying the supplementary indicators related to food wastage, such as including additional destinations which were not accounted for in the second step. After the calculation of the food loss index for the chosen country, certain measures need to be taken, such as (1) using the "target-measure-act" approach, which strengthens food security and cuts costs to households by integrating food wastage; (2) co-creating and adopting game-changing solutions through the guidelines of UN Food Systems Summit; and (3) participating in the regional food waste working groups. These groups provide the knowledge of designing the national strategies to food waste prevention.

13.4 IMPACT OF WASTE ON THE ENVIRONMENT

13.4.1 Industrial Waste

1. *Terrestrial ecosystem.* Organic waste from industries, untreated or landfilled, produce greenhouse gases which, on reaction with water, produces ammonium nitrate, nitric acid, and sulfuric acid, causing acid rain. These greenhouse gases also cause climate change by trapping heat in the Earth's atmosphere. It leads not only to habitat loss but also to rising sea levels and extreme weather conditions. Other contaminants, such as arsenic, mercury,

and benzene, found in the environment also lead to several health issues, like respiratory problems, increased heart rate, kidney damage, immune disorders, decrease in sperm quality, and high-exposure permanent brain damage, or even death. Commercial waste salts also affect soil fertility due to changes in salt concentration and pH. Wastewater from oil mill industry also damages soil quality. It may cause toxicity in plants and soil bacteria, increased mortality, and decreased seed germination. Also, this wastewater pollutes water bodies, damaging aquatic life and making water unfit for human use (Gupta & Verma, 2015).

2. *Aquatic ecosystem.* Pesticides and fertilizers from the agricultural industry are mostly soluble in water, thus entering water bodies or the food chain. Severe health problems, like Alzheimer's disease, Parkinson's disease, reproductive disorders, and cancer, are some side effects of water pollution. Heavily loaded contaminants from industries are discharged in water bodies without proper treatment. Toxic effects are reported on sea algae and aquatic animals which affect their reproductive health, alter sex ratio, decrease egg production, and lead to death (Gaur et al., 2020).

13.4.2 Municipal Solid Waste

It has been observed that in both urban and rural regions, the major focus is on the removal of garbage from human inhabitant regions and later throwing the same trash into open spaces; this leads to contamination through vectors such as flies, mosquitoes, and rodents. These vectors become a source of major infectious diseases, such as gastrointestinal, dermatological, and respiratory infections. The improper dumping of waste leads to the increase of mosquito-borne diseases, such as malaria. It has been predicted by scientists that the improper dumping of waste would lead to it being a major cause of cancer and congenital malformations (Ogundele et al., 2018). Burning waste is a major source of carbon dioxide production, thus becoming a concern for public health. Another major problem occurs when the same garbage is thrown into the river bodies; improper wastewater management leads to high content of metal and inorganic ions in water bodies, thereby leading to the death of aquatic flora and fauna. The potential increase in the summer droughts is a major cause of the decrease in water levels (Leu & Boussiba, 2014).

13.4.3 Agricultural Waste

The effects of these agricultural wastes on the environment can be seen as a rise in microbial resistance, which is typically brought on by contact with human and animal bacterial species, exposure to bacteria present in hospitals, farms, and home care settings, and finally, contact with bacteria present in sewage and biological wastes, which is concerning, in addition to toxicological effects related to direct contact with pharmaceuticals.

Assessing the ecotoxicological effects of pharmaceuticals is often done in a laboratorial setting using assays for acute toxicity to a particular substance. The fish *Danio rerio* and *Pimephales promelas*, as well as the crustacean *Daphnia magna*, are the most often utilized creatures for assessing the amounts that induce oxidative stress and/or death to the organism being examined. The effect concentration 50% (EC50), one of the most used techniques, establishes the concentration at which 50% of people experience an undesirable impact. As a result, a substance is toxic when its EC50 is between 1 and 10 g L1, hazardous when it is between 10 and 100 mg L1, and very toxic when it is less than 1 mg L1 (Barakat et al., 2008). Nevertheless, owing to the particular substances and species investigated, as well as the range of toxicological endpoints evaluated, data comparability in the literature is quite complicated. The EC50 value for a collection of compounds examined used *D. magna*, *V. fischeri*, *A. flos-aque*, *P. subcapitata*, *S. vacuolatus*, and *S. obliquus*.

Some of the impact of agricultural wastes can be seen as nitrate leaching into groundwater, leading to high concentrations of nitrate that can be dangerous to human health. Groundwater contamination, particularly from nitrogen, can damage human health.

Certain pesticides, including nitrogen, metals, and other substances, may leak into groundwater via polluted wells, posing risks to the environment.

Soil erosion, which causes rivers to be very turbid, silts up habitat on the bottom, etc. Disruption and alteration of the hydrologic cycle, sometimes accompanied by the disappearance of perennial streams, have a negative impact on public health.

13.5 CONFLICT OF FOOD VS. FUEL

The current conflict between food and fuel generated by the manufacture of bioethanol from grains is avoided by the green gold petroleum made from lignocellulosic wastes. According to Kim and Dale, agricultural leftovers and discarded crops can create 491 GL (gigaliter, billion liters) of bioethanol annually, which is around 16 times more than the actual global bioethanol output (Srivastava et al., 2020). Lignocellulosic biomass can produce 442 GL of bioethanol. The materials made of cellulosic fibers are abundant, inexpensive, and renewable. It contains grasses, sawdust, wood chips, agro-waste, and agricultural leftovers. For the last 20 years, several academics and researchers have been studying the synthesis of ethanol from lignocellulosic materials. Hence, the synthesis of bioethanol may be the key to the efficient use of agricultural waste and by-products. In terms of the amount of biomass available, rice straw, wheat straw, maize straw, cotton seed hair, seaweed, paper, pineapple leaf, banana stem, jatropha waste, poplar aspen, and sugarcane bagasse are the main agricultural residues. Three procedures, including pre-treatment, enzyme hydrolysis, and fermentation, are necessary for the generation of bioethanol from the cellulose material of agro-residues.

13.6 BIOFUEL PRODUCTION

With the increasing global population, the demand for industrial production is increasing at a very rapid pace. As discussed earlier, these industries produce a lot of waste. Biofuel is currently the only commercially operational process that utilizes these waste products. Not only does biofuel solve the waste management problem, but it also has the potential to fulfil global energy demand (Hafizan & Zainura, 2013).

Biofuels are the fuels derived from biomass of organic wastes, such as plants, animal waste, and other organic substances, which have the potential to produce energy. Fossil fuels, being non-renewable and unsustainable, are now being replaced with biofuels as a green energy source. The production process involves several steps, from feedstock preparation to fermentation or conversion to purification process. Industrial waste can efficiently produce several biofuels with the help of microorganisms in various stages of production. There are several types of biofuels that can be produced through industrial waste, such as ethanol, butanol, biodiesel, and biogas. Biomethanol produced through fermentation is easier to recover than bioethanol, as it does not form azeotropes. But bioethanol has received more attention than biomethanol due to its corrosive and toxic nature. Recently, biodiesel has gained a lot of popularity due to various reasons, like the cheap, sustainable, renewable, and less-toxic by-products produced. Bio-based butanol has the potential as next-generation biofuel. This may be because of its superior properties than ethanol, like low vapor pressure, high energy density, and less corrosivity.

However, it is still a challenge to produce bio-based butanol since end-product inhibition decreases its productivity and raises the product recovery cost (Guan et al., 2016). Biofuels are classified as primary and secondary biofuels. Primary biofuels are those which can be produced directly from organic products, whereas secondary biofuels are those fuels which require microorganisms or other living system to get converted from organic material to biofuel. Secondary biofuels can be further subdivided into three generations according to the type of biofuel they produce. The first-generation consist of the fermentation of starch or sugars producing bioethanol. Second-generation consist of the production of biofuels from plants like lignocellulose through conventional methods. Lastly, third-generation produced biodiesel of bio-based butanol using microorganisms (Rodionova et al., 2016). An ideal microbial strain should be able to produce high amount of product, be able to tolerate end product inhibition, withstand physiological conditions like high temperature and pH, be able to utilize many forms of sugars, have enzymes for waste degradation, and be able to synthesize one major product rapidly.

BASIC STRATEGY ON HOW TO DESIGN A MODEL TO USE FOOD WASTE FOR BIOFUEL PRODUCTION:

1. *Data gathering.* To identify a suitable raw material that can be feasible for biofuel production. Collecting relevant data regarding the same.
2. *Feature engineering.* With the help of proper data collected, one can create the layout of the biofuel production plan, which would be economically favorable.
3. *Feature selection.* Production depends upon several factors, such as the amount of food waste available, the potential equipment, the management system of utilization of waste, to biofuel production.
4. *Model development.* A potential model is designed once all the features and requirements have been met with.
5. *Testing the model.* Initially, a small-scale model is chosen to make sure that the aim of the project is being fulfilled and final product quality is as per requirement. This is followed by a large-scale production (Chowdhury et al., 2021).

13.7 MICROORGANISMS INVOLVED IN WASTE MANAGEMENT AND BIOFUEL PRODUCTION

There are various microorganisms that have the capability to produce biofuels:

- *Yeast.* Eukaryote which is mostly used to produce ethanol. Yeast is added to sugar and water to cause fermentation and produces ethanol and carbon dioxide.
- *Bacteria.* They are commonly used to produce biogas. Bacteria break down organic material in anaerobic digestion to produce biogas and carbon dioxide.
- *Algae.* Used to produce biodiesel and bio-oil, which can be purified to produce efficient biofuel.
- *Fungi.* It breaks down complex lignocellulosic biomass into simple sugars, which can then be fermented to produce biofuel (Silva et al., 2014).

Mostly, bacteria are used for the production process as they are cheaper and easy to grow. Two types of bacteria are used: obligate anaerobes and facultative anaerobes. Examples of facultative microorganisms are *Enterobacter cloacae*, *Bacillus coagulans*, *Citrobacter freundii*, *Bacillus amyloliqufaciens*, etc. Obligate anaerobes examples are *Clostridium butyricum*, *Clostridium acetobutyricum*, *Thermotoga neoplantia*, etc. (Rodionova et al., 2016). In the food industry, *Aspergillus tubingensis* is used to process palm waste from palm oil industry to produce biodiesel (Intasit et al., 2023). Pulp wastewater from paper and pulp industry, under dark fermentation with the help of *Thermoanaero bacterium* sp., produces biohydrogen. Also, cellulase enzyme obtained from *Trichoderma reesei* hydrolyzes paper and pulp effluent to produce biofuel (Gaur et al., 2020). Microbes are involved in municipal waste management and biofuel production: the predominant fatty acids that are produced from microbes mainly consist within the range of C14–C18. One major source for biodiesel production is vegetable oil or animal fats (Kargbo, 2010). Another successful process to produce biofuel is to use microalgal-bacterial co-cultivation (Leong et al., 2019). Microbe *M. parvicella* has been popularly used for the production of biofuel (Muller et al., 2014). Microalgae can be used as a raw material for producing biofuel from wastewater sludge (Srimongkol et al., 2022). Municipal solid waste is considered to be a major source of cellulose, and the strain *Trichoderma viride* has been a promising source of the biodegradation of organic waste. This fungus shows high cellulolytic activity (Gautam et al., 2012). Studies recommend the use of *Clarias gariepinus* for treating the toxicants present in the leachate obtained from landfill of the municipal solid waste (Oshode et al., 2008). The fly ash obtained from waste incinerations can be removed using *Acidithiobacillus thiooxidans*. They have the ability to dissolve heavy metal contents by means of the sulfuric acid produced in their cells (Sarkodie et al., 2021).

13.8 MECHANISM/METHODS FOR BIOFUEL PRODUCTION

To produce biofuel, various methods are involved, like treatment of waste as the primary step. The second step involves the actual conversion of waste into biofuel through various methods, but here we will be focusing on the use of a biological system for waste conversion to biofuels. Lastly, the purification or recovery of biofuels produced to obtain high-quality fuel.

The major methods that need to be followed to convert food waste into biofuel can be classified into two groups, namely, biochemical technology and thermochemical technology. Biochemical technology consists of: (1) *Anaerobic digestion.* Procedure followed to produce methane gas and carbon dioxide. It consists of three major steps, namely, enzymatic hydrolysis and fermentation, followed by acidogenesis and, lastly, methanogenesis. (2) *Photobiological hydrogen production.* These are mostly observed in the biofuels produced from microalgae. (3) *Transesterification technique.* A widely known method to obtain biodiesel from waste cooking oil and animal fat. Thermochemical technology consists of mainly three methods, namely, (1) gasification, a method useful to produce biofuel from food waste containing high sulfur content; (2) pyrolysis, a method of decomposition of biomass via thermal decomposition in the absence of oxygen; and (3) liquefaction, where the biofuel is produced at low-temperature condition and elevated pressure in the presence of hydrogen.

13.9 TREATMENT PROCESS TO USE INDUSTRIAL WASTE FOR BIOFUEL PRODUCTION

The conventional method of treatment of industrial waste is costly and less effective as it is unable to remove certain hazardous compounds, like heavy metals, inorganic materials, and even radioactive materials. These substances, being nonbiodegradable cannot be used for biofuel production. Chemically treating these substances has limitations of handling and environmental damage. Also, hydrothermal treatment will lead to more energy input than output. Biological approaches are used efficiently these days to treat waste before converting it into biofuels. This method can help in tackling various limitations faced by conventional methods. Prior treatment produces not only cleaner by-products but also higher-quality fuel. A lot of waste has high potassium content which, when not filtered out, makes the machinery inefficient in which the biofuel is used. Potassium forms crust or slag on the surface of tubes of the combustion chamber, which has low melting temperature. This causes hindrance in heat transfer and decreases the efficiency of biofuel (Intasit et al., 2020). In paper and pulp industry, the presence of recalcitrant pollutants such as AOX in bleaching effluents and lignin derivatives present in liquor also needs to be removed before further processing. Many bacteria are employed for the treatment of paper and pulp mill effluent, such as *Bacillus subtilis*, *Micrococcus luteus*, and *Rhodococcus* sp., and fungi, like *Phanerochaete crysosporium* (Kamali et al., 2019).

Water plays a very crucial role in industries, with a lot of applications. But wastewater excreted from industries does not get properly treated and contains a high proportion of heavy metals, such as mercury and arsenic, and inorganic materials, like phosphorus, nitrogen, and other toxics. This water can be used for biofuel production since it is not fit for human consumption. The removal of toxic substances is necessary before its further processing. Activated sludge method can be employed to treat organic waste substances. It is a culture containing microorganisms like bacteria, yeast, and protozoa which feed on organic substances to be removed (He et al., 2017).

Many bacteria have the capability to degrade hydrocarbons from organic substances into carbon dioxide, water, etc. To remove radioactive substances and heavy metals from wastewater, the biosorption method is gaining popularity as it is efficient, is cost-effective, has no nutritional requirement, and requires no special disposal method. Algae biomass, especially microalgae, has the capability to absorb heavy metals and radioactive substances since they are rich in functional groups, like carboxylic, sulfonic, and hydroxyl group, in its polysaccharide. Marine red algae *Pterocladia capillacea* is one of the many examples which can remove chromium from wastewater (Rajasulochana & Preethy, 2016). Other examples of algae that can remove heavy metals are *Chlamydomonas reinhardtii*, *Scenedesmus obliquus*, *Chlorella pyrenoidosa*, and *Chlorella vulgaris*. Plants are also a candidate for waste treatment, like *Canna indica* and *Typha angustifolia* (Kamali et al., 2019).

13.10 EXTRACTION/PURIFICATION PROCESS OF THE BIOFUEL

After the production of biofuel in the microsystem, it is essential to extract fuel from it. For this purpose, the cell, especially the cell wall, of the microorganisms needs to be disrupted. Various techniques have been used to achieve the extraction process, like ultrasound, pulsed electric field, and hydrolysis. To extract lipids from oleaginous microbes, methods like solvent extraction, supercritical extraction, and other forms of extractions are used for biodiesel production. Although oleaginous microbes are rich in lipid content, they also contain high amounts of proteins and carbohydrates. These extra substances may produce nitrogen dioxide or cause catalyst poising during combustion. So the proper purification of biofuel is an essential step in biofuel synthesis. In addition to this, the purification process itself can cost 90% of the total production cost.

New microsystems have been explored to overcome this problem. Microalgae are one such example. The pyrolysis of microalgae in the presence of a zeolite produces aromatic hydrocarbons from denitrification or deoxygenation of algae biomass. These aromatic compounds are themselves of great importance as they enhance the octane rating of the fuel (Zhang et al., 2021).

After the extraction process is done, lipids should be converted to simple alkyl esters, that is, biodiesel. This process is known as transesterification. The process reduces the viscosity of biodiesel by its reaction with alcohols like ethanol. Though chemical catalysis methods are being used for transesterification, enzymatic catalysis methods are also prevalent. Microorganisms like *Candida antarctica* are being used for this process (Chintagunta et al., 2021).

13.10.1 SOURCES OF INDUSTRIAL WASTE USED FOR BIOFUEL PRODUCTION

Among all the waste produced by various industries, lignocellulose is found in the majority of them. Food, paper and pulp, textile, oil, agricultural industry, etc. utilize plant-based products in some way or another. Plant is the major source of lignocellulose, which is made up of hemicellulose and cellulose. Many non-edible and oil-producing plants have been discovered, like *Jatropha curcus*, *Nicotina tabacum*, *Simmondsia chinensis*, *Ricinus communis*, etc. (Voloshin et al., 2016).

Microorganisms have various enzymes that contain hemicelluloses or cellulases that can degrade lignocellulose into simple sugars. Cellulolytic enzymes are of three types: endogluconase, exogluconase, and beta-glucosidase (Adegboye et al., 2021). Endogluconase can cut cellulose at random sites, producing oligosaccharides of different length. On the other hand, exogluconase cuts at the end of cellulose, producing glucose or cellobiose. Enzyme beta-glucosidase has the capability to hydrolyze complex cellulose into simple sugars. All these enzymes work together to convert lignocellulose into simpler sugar, a method called saccharification, which can be utilized by microbes in the fermentation process to produce biofuel. Hemicellulose is degraded by enzymes xylanase, beta-mannanases, beta-xylosidases, etc (Malode et al., 2021). Some of the microbes containing these enzymes are *Aspergillus niger*, *Trichoderma longibrachiatum*, and *Ustilago maydis* (Chukwuma et al., 2021).

To produce biodiesel, oleaginous microorganisms are used since they have higher lipid accumulation capability from waste substances. This lipid is then transesterified to produce biodiesel. Saccharification is done prior since most of the oleaginous microbes cannot utilize polysaccharides from the wastes. Some examples of such microbes are *Rhodosporidium toruloids*, *Mortierella isabelline*, *Chlorella* sp., etc. The growth kinetics curve of these bacteria starts with the growth phase under balanced nutrition condition, followed by the oleaginous phase under nitrogen-limiting conditions, and later reserved lipid turnover phase that occurs after the depletion of carbon source in the media. Glycolysis and pentose phosphate pathways are involved in the biomass growth, and lipid triacylglycerol is synthesized only after the depletion of nitrogen source. There are various factors that influence the production of oil during the oleaginous phase, such as nitrogen content, mineral depletion, and balanced production of reactive oxygen species.

The biosynthetic pathway that occurs inside an oleaginous microorganism has the capability of producing fatty acids from acetyl-CoA and NADH. An excess amount of carbon source and limiting nitrogen upregulate AMP deaminase, which decreases the AMP concentration. Through a cascade of events, citrate synthesis is upregulated. Citrate gets converted to acetyl-CoA, which is a precursor of fatty acids. This biosynthesis process *in vitro* can be carried out through three different processes: separate hydrolysis and lipid production in a two-step process (SHLP); simultaneous saccharification and lipid production (SSLP), which requires only one reactor; and consolidated bioprocessing (CBP), where enzyme production, saccharification, and lipid production take place in a single step (Chintagunta et al., 2021).

Bioalcohols like biomethanol, bioethanol, and biobutanol are being produced as a substitute for fossil fuels and also efficiently manage wastes produced from industries. The most preferred microorganism for ethanol production is *Saccharomyces cerevisiae* since it has high tolerance for ethanol concentration and inhibitors produced during the fermentation process. Co-culturing *Clostridium phytofermentans* (cellulolytic) and *Saccharomyces cerevisiae* (ethanol-producing) is also done to produce bioethanol from cellulose. For bio-based butanol production, co-culturing of *Clostridium acetobutylicum* and *Bacillus subtilis* is done. *Bacillus subtilis* utilizes all the oxygen present in the medium and creates anaerobic condition, which is required for growth of *C. acetobutylicum.* Two types of approaches are used – dark fermentation and light fermentation. Dark fermentation occurs in the absence of light and is done by facultative anaerobes or anaerobes, whereas light fermentation occurs in the presence of light (Malode et al., 2021). Fermentation occurs either through solid-state fermentation (SoSF) or submerged fermentation (SmF) at room temperature around 30–37°C. Due to low operational cost, simplicity, higher production cost, and easier product recovery, SoSF has been used widely for biofuel production.

Various factors also play a key role in the quality and quantity of biofuel produced through the microbial system. For SoSF, moisture content is the major determining factor as it affects the mass transfer of nutrients between the solid and liquid phases. Hence, lower moisture content is preferred. Also, since lipid content increases with decrease in nitrogen source, lower nutritional supplementation is required. It was seen that by increasing the volume of anaerobic digestion of effluents, enzyme activity of cellulase and xylanase increased. Biodiesel quality–determining parameters are cetane number (CN), iodine value (IV), cold filter plugging point (CFPP), and oxidative stability (OS). CN value determines the combustion property of biodiesel quantitatively, which tells about the time required before ignition (it is largely dependent on the chain length and degree of unsaturated fatty acids present), whereas OS value tells about the shelf life of biodiesel over storage period. IV values measure the number of unsaturated fatty acids present in the biodiesel. CFP value is the filterability of biodiesel at low temperature. This is completely dependent on the number of saturated fatty acids present in the biodiesel (Intasit et al., 2020).

13.11 EXTRACTION/PURIFICATION OF THE BIOFUEL

After the production of biofuel in the microsystem, it is essential to extract fuel from it. For this purpose, the cell, especially the cell wall, of the microorganisms needs to be disrupted. Various techniques have been used to achieve the extraction process, like ultrasound, pulsed electric field, and hydrolysis. To extract lipids from oleaginous microbes, methods like solvent extraction, supercritical extraction, and other forms of extractions are used for biodiesel production. Although oleaginous microbes are rich in lipid content, they also contain high amounts of proteins and carbohydrates. These extra substances may produce nitrogen dioxide or cause catalyst poising during combustion. So the proper purification of biofuel is an essential step in biofuel synthesis. In addition to this, the purification process itself can cost 90% of the total production cost (Patel et al., 2023).

New microsystems have been explored to overcome this problem. Microalgae are one such example. The pyrolysis of microalgae in the presence of a zeolite produces aromatic hydrocarbons from the denitrification or deoxygenation of the algae biomass. These aromatic compounds are themselves of great importance as they enhance the octane rating of the fuel.

After the extraction process is done, lipids should be converted to simple alkyl esters, that is, biodiesel. This process is known as transesterification. The process reduces the viscosity of biodiesel by its reaction with alcohols like ethanol. Though chemical catalysis methods are being used for transesterification, enzymatic catalysis methods are also prevalent. Microorganisms like *Candida antarctica* are being used for this process (Chintagunta et al., 2021).

13.12 TREATMENT PROCESS TO USE MUNICIPAL SOLID WASTE FOR BIOFUEL PRODUCTION

Scientists have used these basic principles and processes and designed innovative methods. As mentioned here, in the method using microalgal–bacterial co-cultivation, the sludge is treated initially using the freshwater algal species *Chlorella vulgaris* in a sequencing batch reactor. The bioreactor is continuously aerated and illuminated with white, fluorescent light to increase algal yield. Under suitable pH and temperature, one can obtain the microbial biomass concentration which can be harvested for biomass and lipid yield. It has been observed that this mixed microbial wastewater produces efficient yield in a very short span of time. With an increase in the ratio of microalgae–microbes, there was an increase in biomass. In biological wastewater production, organisms, namely, *Candidatus* sp. and *M. parvivicella*, can be cultured, and using the Fischer–Tropsch process, biofuels can be obtained. The popular bacterial species which majorly accumulate triacylglycerols in their cytoplasm are *Mycobacterium*, *Rhodococcus*, *Nocardia*, or *Microthrix*. When these microorganisms are cultured in an activated sludge–based biological wastewater treatment, specific triacylglycerols can be obtained. Another approach to obtain biofuel is to use enzymatic transesterification by the usage of extracellular and intracellular lipases. Bio-oils can be produced using waste sludge by means of pyrolysis. The calorific value is up to 36 MJ.kg. The bio-oil obtained from biological wastewater treatment consists of hydrocarbons ranging from C6 to C20.

13.13 TREATMENT PROCESS TO USE AGRICULTURAL PRODUCTION FOR BIOFUEL PRODUCTION

Pre-treatment. The most crucial and first step in the process of separating free cellulose from agro-residues is pre-treatment. The second step of enzyme hydrolysis, carried out by effective microorganisms with the capacity to release cellulose enzyme, is equally crucial (Ullah et al., 2018). The hydrolysis of cellulose into glucose is a function of this enzyme. Cellulase enzyme production is possible in several microbial species, including *Clostridium*, *Cellulomonas*, *Thermonospora*, *Bacillus*, *Bacteriodes*, *Ruminococcus*, *Erwinia*, *Acetovibrio*, *Microbispora*, and *Streptomyces*. Many fungi have also been shown to produce cellulase, including *Trichoderma*, *Penicillium*, *Fusarium*, *Phanerochaete*, *Humicola*, and *Schizophillum* sp. After this, bacteria such as *Saccharomyces cerevisiae*, *Escherichia coli*, *Zymomonas mobilis*, *Pachysolen tannophilus*, *Candida shehatae*, *Pichia stipitis*, *Candida brassicae*, *Mucor indicus*, etc. are needed for the fermentation process to convert glucose to ethanol (Jin et al., 2015). These procedures are necessary for the sustained manufacture of bioethanol from cellulosic agro-residues. The primary goal of this analysis is to provide a succinct summary of the methods that are already in use and easily accessible for producing bioethanol from key agro-residue cellulosic materials. The manufacture of bioethanol from agricultural leftovers is an environmentally sound, socially acceptable technique that also lowers greenhouse gas emissions.

Chemical treatment. Chemical technologies that are commonly used include sodium hydroxide, perchloric acid, peracetic acid, acid hydrolysis using sulfuric and formic acids, ammonia freeze explosion, and organic solvents such as n-propylamine, ethylenediamine, n-butylamine, etc (Gupta & Verma, 2015). Yet the main barrier to chemical pre-treatment is how the use of such chemicals impacts the overall economics of cellulosic biomass bioconversion.

The chemical pre-treatment has demonstrated that among chemical treatments, the dilute sulfuric acid–based pre-treatment is the most popular by means of enzymatic hydrolysis using biomasses such as cashew apple pulp and coffee pulp in India, which contain about 23–27% fermentable sugars on a dry weight basis. This method has previously and extensively been used in the paper industry for lignin demolition in cellulosic materials to produce high-quality paper products.

Biological treatment. Different microorganisms like brown-rot, white-rot, and soft-rot fungi may be used in the breakdown of the lignocellulosic complex to liberate cellulose in the biological pre-treatment process. In order to form amorphous cellulose, this pre-treatment also aids in the breakdown of lignin and hemicelluloses (Adeniyi et al., 2018). White-rot fungi appear to be the most efficient microorganism. White- and soft-rots target both cellulose and lignin, while brown-rot aids in attacks on cellulose. Because there is less mechanical support, the rate of hydrolysis is often relatively low during biological pre-treatment, making it safe and energy-efficient. Low hydrolysis rates and low yields make it difficult to adapt because it requires no chemicals (Cruz-Casas et al., 2021). By utilizing a variety of microorganisms, this biological pre-treatment is an economical and environmentally responsible method of releasing the sugars from the lignocellulosic matrix of waste sugarcane.

Enzyme hydrolysis. The crucial stage in the production of bioethanol is enzyme hydrolysis, which converts complex carbs into simple monomers. Compared to acid hydrolysis, it takes less energy and creates a softer atmosphere. In this process, cellulose enzyme is the most important enzyme, which is naturally occurring in cellulolytic microbes, for example, *Clostridium*, *Cellulomonas*, *Thermonospora*, *Bacillus*, *Bacteriodes*, *Ruminococcus*, *Erwinia*, *Acetovibrio*, *Microbispora*, *Streptomyces*, and other fungi, such as *Trichoderma*, *Penicillium*, *Fusarium*, *Phanerochaete*, *Humicola*, and *Schizophillum* sp. (Ambat et al., 2018) The cellulose can be converted into glucose or galactose monomer by these enzymes. Similarly, pH 4–5 and a temperature of 50°C have been listed as the ideal conditions for xylanase. In contrast to acid or alkaline hydrolysis, enzymatic hydrolysis is therefore advantageous due to its low toxicity, low utility cost, and low corrosion. Moreover, enzymatic hydrolysis does not produce any inhibitory by-products. The cellulase enzyme has a very narrow substrate range. The linkages of cellulose and hemicellulose are respectively broken by the enzymes cellulase and hemicellulase. Hemicellulose and cellulose both include diverse sugar units, including mannan, xylan, galactan, and arabinan. There are three different kinds of cellulase enzymes, including endo- and exoglucanase and glucosidases. Endoglucanase (endo 1,4-D glucanhydrolase, or E.C. 3.2.1.4) attacks the low-crystallinity areas of the cellulose fiber, exoglucanase (1,4-D glucan cellobiohydrolase, or E.C. 3.2.1.91) removes the cellobiose units from the free chain ends, and finally, glucosidase (E.C. 3.2). Hemicellulolytic enzymes are more complicated and are a mixture of at least eight enzymes, such as endo-1,4-β-D-xylanases, exo-1,4-β-D xylocuronidases, α-L-arabino-furanosidases, endo-1,4-β-D mannanases, β-mannosidases, acetyl xylan esterases, α-glucoronidases, and α-galactosidases. Hemicellulose produces a number of pentoses and hexoses, while cellulose is degraded to glucose.

13.14 ADVANTAGES OF BIOFUEL PRODUCTION

With a carbon-neutral technique, biomass is transformed into energy. There are now automobiles that run on a blend of gasoline and up to 85% ethanol, which is mostly utilized as fuel additives to reduce emissions. Economic, environmental, and social sustainability are the three basic pillars of bioenergy. Bioethanol reduces carbon dioxide emissions, air pollution, global warming, and other negative environmental effects. Fossil fuel combustion causes air pollution by releasing a variety of GHGs into the atmosphere (the main GHGs are CO_2, N_2O, CH4, SF6, and chlorofluorocarbons). Both the world's increasing energy demand and the GHG emissions from fossil fuels are decreased by biofuel, especially ethanol. In a similar vein, Galbe and Zacchi said that bioethanol, the most widely used renewable resource in the world, reduces humankind's reliance on fossil fuels for transportation. The principal biofuels include bioethanol, biodiesel, biogas, biomethanol, biosyngas (CO H_2), bio-oil, biochar, biohydrogen, Fischer–Tropsch liquids petroleum, and vegetable oil. Bioethanol and biodiesel are liquid transportation fuels that are utilized as an additive source. When combined as an additive, biodiesel can replace diesel, and bioethanol can replace gasoline, in order to cut down on greenhouse gas emissions.

13.15 APPLICATIONS OF BIOFUEL PRODUCTION USING WASTE MATERIALS

India has the largest population in the world. With the increase in population, energy demands are also increasing, which is fed by non-renewable sources of energy, like fossil fuels. India, by 2030, will require 1,516 billion tons of oil to meet its energy demands. The conventional sources used are not only depleting but also have harmful impact on our environment. Recently, biofuels have gained popularity because of their renewable source and less-harmful impact on the environment. But one of the major limitations of using biofuel is the availability of biomass. India, being an agricultural country, produces 350 billion tons of agricultural waste per year. This waste can solve the biomass availability problem very efficiently. Also, wastewater produced from industries can be treated and used for biofuel production.

The Indian government has released a statement on the need to switch to biofuels from conventional methods. Since then, various policies and proposals have been made by the government, like five-year plan to achieve 20% biofuel blending with petroleum and usage of waste oilseeds for biofuel synthesis (Joshi et al., 2017).

13.16 LIMITATIONS OF BIOFUEL PRODUCTION

1. *Sludge collection.* During the collection of primary and secondary sludge, the effect of temperature, the concentration of acid catalyst, and the methanol-to-sludge ratio causes an increase in the cost of production.
2. *Production challenges.* Difficult to obtain the optimal ratio of 2.2:1 of methanol to oil at the temp of 43°C. To reach the appropriate acid value tends to remain a challenge for the producers.

3. *Maintaining product quality.* The primary sludge which is obtained from wastewater treatment is very low-quality; hence, it fails to meet the requirement of the biofuel industry.
4. *Pharmaceutical chemicals.* The fatty acids which are lipid regulators in the cosmetic and pharmaceutical industries are also found in the wastewater, which can lead to a hindrance in biofuel production.
5. *Economics of biodiesel production.* Pre-treatment cost, enzymes, production cost, and purification make the synthesis of biofuel a more expensive process than conventional biofuel cost.
6. *Environmental effects.* The combustion of biofuels might produce carbon dioxide, which contributes to greenhouse gases.
7. *Stability.* Biodiesel is made up of fatty acid esters, which make it highly susceptible to auto-oxidation.
8. *Compatibility with the machine.* Engines of machines are made according to the conventional oil used. So sometimes the biofuel might get leached inside the machine's engine, which might degrade the engine's durability. (Joshi et al., 2017).

REFERENCES

Adegboye, M., Ojuederie, O., Talia, P., Babolola, O. (2021). Bioprospecting of microbial strains for biofuel production: metabolic engineering, applications and challenges. *Biotechnology for Biofuels*, 14, 5.

Adeniyi, O. M., Azimov, U., Burluka, A. (2018). Algae biofuel: current status and future applications. *Renewable and Sustainable Energy Reviews*, 90, 316–335.

Alalwan, H., Alminshid, A., Aljaafari, H. (2019). Promising evolution of biofuel generation. *Renewable Energy Focus*, 28, 127–139.

Ambat, I., Srivastava, V., Sillanpää, M. (2018). Recent advancement in biodiesel production methodologies using various feedstock: a review. *Renewable and Sustainable Energy Reviews*, 90, 356–369.

Barakat, A. R., Schreiber, M. N., Flaschar, J., Georgieff, M., Schraag, S. (2008). The effective concentration 50 (EC50) for propofol with 70% xenon versus 70% nitrous oxide. *Anesthesia & Analgesia*, 106(3), 823–829.

Casares, M., Ulierte, N., Mataran, A., Ramos, A., Zamorano, M. (2005). Solid industrial waste and their management in Asegra. *Waste Management*, 25, 1075–1082.

Chintagunta, A., Zuccaro, G., Kumar, M., Kumar, S., Garlapati, V. (2021). Biodiesel production from lignocellulosic biomass using oleaginous microbes: prospects for integrated biofuel production. *Frontiers in Microbiology*, 12(658284), 23.

Chowdhury, H., Chowdhury, T., Barua, P., et al. (2021). Biofuel production from food waste biomass and application of machine learning for process management. *Advanced Technology for the Conversion of Waste into Fuels and Chemicals*, 96–117.

Chukwuma, O., Rafatulla, M., Tajarudin, H., Ismail, N. (2021). A review on bacterial contribution to lignocellulose breakdown into useful bio-products. *International Journal of Environmental Research and Public Health*, 18(11), 6001–6007.

Cruz-Casas, D. E., Aguilar, C. N., Ascacio-Valdés, J. A., Rodríguez-Herrera, R., Chávez-González, M. L., Flores-Gallegos, A. C. (2021). Enzymatic hydrolysis and microbial fermentation: The most favorable biotechnological methods for the release of bioactive peptides. *Food Chemistry: Molecular Sciences*, 3, 100047.

Gaur, V., Sharma, P., Sirohi, R., Awasthi, M., Dussap, C. (2020). Assessing the impact of industrial waste on environment and mitigation strategies: a comprehensive review. *Journal of Hazardous Materials*, 1–62.

Gautam, S. P., Bundela, P. S., Pandey, A. K., Jamaluddin, Awasthi, M. K., & Sarsaiya, S. (2012). Diversity of cellulolytic microbes and the biodegradation of municipal solid waste by a potential strain. *International Journal of Microbiology*, 2012(1), 325907.

Guan, W., Shi, S., Tu, M., Lee, Y. (2016). Acetone-butanol-ethanol production from Kraft paper mill sludge by simultaneous saccharification and fermentation. *Bioresource Technology*, 200, 713–721.

Gupta, A., Verma, J. P. (2015). Sustainable bio-ethanol production from agro-residues: a review. *Renewable and Sustainable Energy Reviews*, 41, 550–567.

Hafisan, C., Zainura, N. (2013). Biofuels: advantages and disadvantages based on life cycle assessment perspective. *Journal of Environmental Research and Development*, 7(4), 1444–1449.

He, L., Du, P., Chen, Y., Lu, H., Cheng, X. (2017). Advances in microbial fuel cells for wastewater treatment. *Renewable and Sustainable Energy Reviews*, 71, 388–403.

Jin, M., Slininger, P. J., Dien, B. S., et al. (2015). Microbial lipid-based lignocellulosic biorefinery: feasibility and challenges. *Trends in Biotechnology*, 33(1), 43–54.

Joshi, G., Pandey, J., Rana, S., Rawat, D. (2017). Challenges and opportunities for the applications of biofuel. *Renewable and Sustainable Energy Reviews*, 79, 850–866.

Kamali, M., Borazjani, S., Khodaparast, Z., Khajal, M., Jahanshahi, A. (2019). Additive and additive-free technologies for pulp and paper mill effluents: advances, challenges and opportunities. *Water Resource and Industry*, 21, 100–109.

Kargbo, M. (2010). Biodiesel production from municipal sewage sludges. *Energy Fuels*, 24, 2791–2794.

Lahiri, A., Daniel, S., Kanthapazham, R., Vanaraj, R., Thambidurai, A., Peter, L. S. (2023). A critical review on food waste management for the production of materials and biofuel. *Journal of Hazardous Materials Advances*, 10, 100266.

Leong, H. W., Zaine, S., Ho, Y., et al. (2019). Impact of various microalgal-bacterial populations on municipal wastewater bioremediation and its energy feasibility for lipid-based biofuel production. *Journal of Environmental Management*, 249.

Leu, S., Boussiba, S. (2014). Advances in the production of high-value products by microalgae. *Industrial Biotechnology*, 10(3), 169–183.

Malode, S. J., Prabhu, K. K., Mascarenhas, R. J., Shetti, N. P., Aminabhavi, T. M. (2021). Recent advances and viability in biofuel production. *Energy Conversion and Management: X*, 10, 100070.

McCarty, P., Bae, J., Kim, J. (2011). Domestic wastewater treatment as a net energy producer–can this be achieved. *Environment Science Technology*, 7100–7106.

Muller, E., Sheik, A., Wilmes, P. (2014). Lipid-based biofuel production from wastewater. *Current Opinion in Biotechnology*, 30, 9–16.

Nanda, S., Berruti, F. (2021). Municipal solid waste management and landfilling technologies: a review. *Environmental Chemistry Letters*, 19, 1433–1456.

Ogundele, O., Rapheal, O., Abiodun, A. (2018). Effects of municipal waste disposal methods on community health in Ibadan – Nigeria. *Polytechnica*, 61–72.

Oshode, A., Bakare, A., Adeogun, A., et al. (2008). Ecotoxicological assessment using Clarias Gariepinus and microbial characterization of leachate from municipal solid waste landfill. *International Journal of Environment Residues*, 391–400.

Patel, A., Dong, C., Chen, C., Pandey, A., Singhania, R. (2023). Production, purification and application of microbial enzymes. *Biotechnology of Microbial Enzymes*, 2, 25–57.

Rajasulochana, P., Preethy, V. (2016). Comparison of efficiency of various techniques in treatment of waste and sewage water-a comparative review. *Resource Efficient Technologies*, 10.

Rodionova, M., Poudyal, R., Tiwari, I., Voloshin, R., Zharmukhamedov, S. (2016). Biofuel production: challenges and opportunities. *International Journal of Hydrogen Energy*, 1–12.

Saldaña-Durán, C. E., Bernache-Pérez, G., Ojeda-Benitez, S., Cruz-Sotelo, S. E. (2020). Environmental pollution of E-waste: generation, collection, legislation, and recycling practices in Mexico. In *Handbook of Electronic Waste Management* (pp. 421–442). Butterworth-Heinemann.

Sarkodie, E. K., Juang, L., Li, K., Yang, J., Guo, Z., Shi, J., . . . Liu, X. (2022). A review on the bioleaching of toxic metal(loid)s from contaminated soil: insight into the mechanism of action and the role of influencing factors. *Frontiers in Microbiology,* 13, 1049277.

Silva, T., Gouveia, L., Reis, A. (2014). Integrated microbial processes for biofuels and high value-added products. *Applied Microbiology and Biotechnology*, 98, 1043–1053.

Srimongkol, P., Sangtanoo, P., Songserm, P., Watsuntorn, W., Karnchanatat, A. (2022). Microalgae-based wastewater treatment for developing economic and environmental sustainability: Current status and future prospects. *Frontiers in Bioengineering and Biotechnology*, 10, 904046.

Srivastava, R. K., Shetti, N. P., Reddy, K. R., et al. (2020). Biofuels, biodiesel and biohydrogen production using bioprocesses: A review. *Environmental Chemistry Letters*, 18, 1049–1072.

Ullah, K., Sharma, V. K., Ahmad, M., et al. (2018). The insight views of advanced technologies and its application in bio-origin fuel synthesis from lignocellulose biomasses waste, a review. *Renewable and Sustainable Energy Reviews*, 82, 3992–4008.

Voloshin, R., Rodionova, M., Zharmukhamedov, S., Veziroglu, T., Allakhverdiev, S. (2016). Review: biofuel production form plant and algal biomass. *International Journal of Hydrogen Energy*, 41(39), 17257–17273.

Woodard, F. (2001). *Industrial Waste Treatment Handbook* (vol. 1, pp. 61–119). Butterworth Heinemann.

Zhang, L., Loh, K., Kuroki, A., Dai, Y., Tong, Y. (2021). Microbial biofuel production from industrial organic wastes by oleaginous microorganism: current status and prospects. *Journal of Hazardous Materials*, 402, 123–143.

14 Syngas Production from Lignocellulosic Biomass and Its Applications

*Diptimayee Padhi, Deepika Devadarshini, Sunanda Mishra, Deviprasad Samantaray, Mahendra Kumar Mohanty, and Pradip Kumar Jena**

14.1 INTRODUCTION

In nature, a wide array of biomass is abundantly available that can be processed for the generation of bioenergy. The biomass is derived from four major sources, viz., (1) energy crops, including woody vitality crops, herbaceous vitality crops, modern harvests, agrarian yields, and amphibian harvests; (2) agricultural trash and residues, like crop waste and creature squander; (3) forestry waste and buildups, including factory wood wastage, logging deposits, trees, and bush encroachments; and (4) mechanical and municipal squanders, such as municipal solid waste (MSW), sewage slime, and industry wastage (Meng et al., 2006). Further, four thermo-processes, namely, pyrolysis, gasification, burning, and ignition, are used for the transformation of biomass to syngas. Additionally, five different organic methods, including direct and indirect bio-photolysis, natural water–gas shift reaction, photo-fermentation, and dark fermentation, are also used for syngas generation. *Ignition* is the process of direct consumption of biomass in the air into heat, mechanical electricity, or power, utilizing stoves, heaters, boilers, or steam turbines individually. However, the burning practice is unsuitable for the generation of bioenergy because of low energy productivity (10–30%) and releases of pollutants. In contrast, the usage of biofuels helps minimize the burning of fossil fuel and CO_2 production. These biofuels are derived from renewable resources, like plants, or some organic waste. In general, plant growth is mediated by CO_2, in return releasing oxygen to the environment. Biofuels produced from plant waste or lignocellulosic biomass have enormous potential to decrease the level of CO_2 as well as usage of non-renewable fossil fuel (Osamu and Carl 1989). Currently, academia, in collaboration with industry, is shaping a substitute for the production of biofuel and other value-added green chemicals to strengthen bio-economy, especially in rural sectors (Stevens and Verhe, 2004).

In the 20th century, fossil feedstocks were exploited for the production of petroleum, natural gas, and coal. During the 21st century, focus has been given on liquid as well as gaseous fuel. Hydrogen plants are essential for refineries and small

 DOI: 10.1201/9781003517108-14

units producing hydrogen for fuel cells. The hydrocarbon mixtures should have an overall hydrogen and carbon ratio of around 02, which is primarily necessary for turning fossil fuels into fuels for transportation (Rostrup-Nielsen, 1994). This calls for hydrotreating and hydrocracking heavy crude to add hydrogen, with the option to also remove metals, nitrogen, and sulfur. A significant portion of the hydrogen required is provided by catalytic reforming units, but due to restrictions on the amount of aromatics in diesel and gasoline, its availability may be limited. As a result, there is a growing demand for hydrogen production facilities that can run on a variety of hydrocarbon streams, including natural gas, naphtha, and off-gases (Rostrup-Nielsen, 2004). Carbonaceous materials are heated up during the gasification process to produce gaseous fuel, typically with low to medium heating value. Complete combustion is not included in this definition because the resulting flue gas has no remaining thermal value as a result of the fuel's complete combustion. It comprises hydrogenation, fuel-rich combustion, and partial oxidation of fuel. The oxidant, also known as the gasifying agent, in the partial oxidation process might be air/oxygen, steam, carbon dioxide, or a combination of two or more of these gasifying agents (Ahmed and Gupta, 2009). To achieve the required chemical composition of syngas and process efficiency, the gasifying agent is chosen, and its ratio to carbonaceous feedstock is changed correspondingly (Ahmed and Gupta, 2009; Furimsky, 2006).

First-generation (1G) biofuel such as biodiesel, vegetable oil, biogas, bioalcohol, and syngas are derived from feedstock like starch, sugar, animal fats, and vegetables through conventional techniques. Second-generation biofuels are mainly produced from lignocellulosic feedstock, which primarily includes agricultural residues, etc. (Gomez et al., 2008). Plant biomass contains a huge number of lignocellulosic materials which comprise inexpensive non-food material present abundantly that are utilized during biofuel production. At present, fuel production is expensive, because of some technical issues which should be planned out for a superior and influential process for the generation of biofuels (Eisberg, 2006). One of the most plentiful and neglected biological resources, plant biomass is portrayed as a viable supply of material for fuels and raw materials that can be ignited to produce both electricity and heat. A few parts of biofuel production can be fulfilled by agricultural by-products; however, these give a lesser amount of biofuel than others. The process of biofuel production from different crop feedstocks has generated a scope and dedication in the respective research field (Gomez et al., 2008). There are two major categories of biofuels, namely, primary biofuel, which are an unprocessed, natural source utilized for the production of electricity and heating (firewood, plants, forest products, animal dung, and other materials are examples of primary biofuels), and secondary biofuels, which are made from biomass that has been treated and further categorized into two types: first- and second-generation biofuels (as shown in Figure 14.1). Many research reports are available in the public domain pertaining to the biorefinery concept, including wheat straw biorefinery (Deswarte et al., 2008), corn biorefinery (Haung et al., 2008), forest residue–based biorefinery (Mabee et al., 2005), etc. Further, data related to bioethanol, biomass, chemicals derived from glycerol, and green diesel are also available. However, adequate information is not available on 1G and 2G biofuels and several other green chemicals derived from non-food crops.

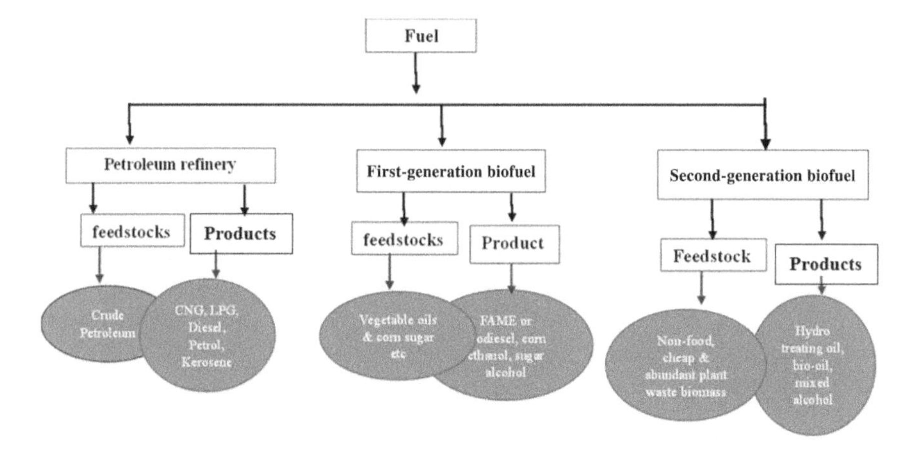

FIGURE 14.1 Comparative account on production of fossil fuel vs. biofuels of different generation.

14.2 RENEWABLE FEEDSTOCK USED FOR SYNGAS PRODUCTION

Biomass generated from trees, agro-forest, grass, aquatic plants, and crops serves as the most productive feedstock for the chemical industry. Agricultural wastes and forest residues consist of lignocellulose that can be utilized for biofuel production either by physical, biochemical, thermochemical, or chemical processes, as shown in Figure 14.2. These feedstocks are composed of primary metabolites, like carbohydrate (simple sugar, cellulose, hemicellulose, starch, etc.) and lignin, which altogether comprise lignocellulose that can be converted into biofuels. The biochemicals found in gums, rubber, waxes, resins, steroids, terpenes, tannin, terpenoids, triglycerides, plant acids, alkaloids, and other irregularly occurring substances are the precursor of secondary metabolites (Clark et al., 2007). It is also used for the preparation of food flavors, feeds, pharmaceuticals, cosmeceuticals, nutraceuticals, etc. via an integrated process which is initiated from secondary metabolites. Non-edible feedstocks like mustard oilseeds, Indian beech, Jatropha, canola green seeds, and microalgae are used in the production of 2G biofuel. Aquatic biomass can also be used to produce bioethanol and biodiesel.

14.2.1 Different Biomass for Syngas Production

Lignocellulosic material is mainly alienated into three constituents, as shown in Figure 14.3: cellulose (30–50%), hemicelluloses (15–35%), and lignin (10–20%) (Pettersen et al., 1984; Badger, 2000; Mielenz, 2001; Girio et al., 2010). About 70% of the biomass is made up of cellulose and hemicellulose, which are tightly bound to the lignin component by covalent and hydrogenic connections to create a structure that is exceedingly durable and resistant to any treatment (Mielenz, 2001; Knauf and Moniruzzaman, 2004).

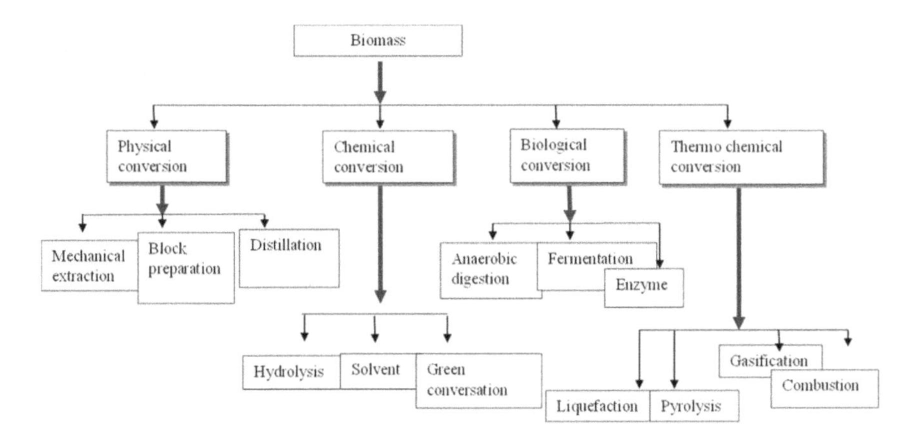

FIGURE 14.2 Different routes for biomass conversion.

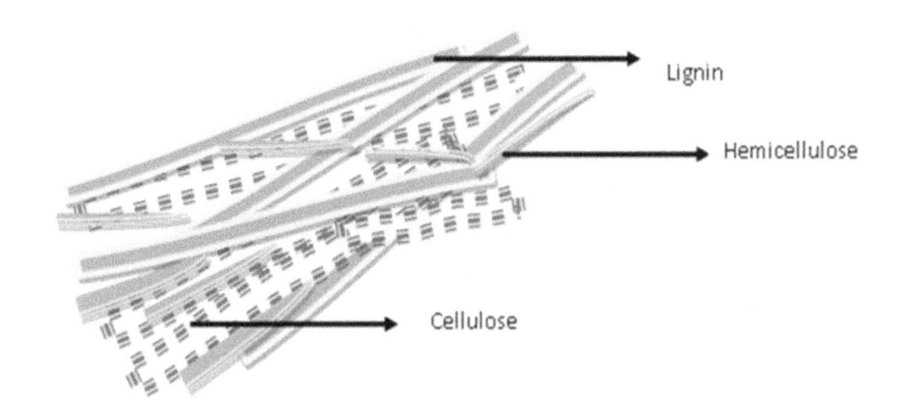

FIGURE 14.3 Various components of lignocellulosic biomass.

14.2.1.1 Biomass from Agricultural Residues

Agricultural wastes like cornstalks, maize stover, wheat straws, and rice straw, as well as sugarcane bagasse, are most of the wide range of crop residues. Globally, around 1,400 MT (million tons) per year of agriculture biomass are harvested annually with residues (USDEBP, 2009; RFA Outlook, 2010; Perlack et al., 2005), mainly originating from maize stalks, wheat, and rice straws. Compared to woody biomass, crop leftovers include a greater amount of hemicellulosic material (approximately 25–35%) (Demirbas, 2005). Besides being an Earth benevolent procedure, agricultural residues help stay away from dependence on forest woody biomass and, in this way, diminish deforestation (non-sustainable cutting plants) (Knauf and Moniruzzaman, 2004; Kim and Dale, 2004).

14.2.1.2 Biomass from Forest Residues

There are two sorts of woody materials, categorized as softwoods and hardwoods. Conifers and gymnosperm trees are the sources of softwoods, which have lower densities and grow more quickly (Perlack et al., 2005; Hoadley, 2000). Evergreen tree species like cypress, spruce, pine, fir, cedar, hemlock, and redwood are most of the gymnosperm tree family (Boone et al., 1988). All angiosperm trees are classified as hardwoods, and most of them are deciduous (Markwardt and Wilson, 1935), including aspen, cottonwood, willow, poplar, and oak. Hardwood trees make up over 40% of all trees in the United States (AHEC, 2002). Populus contains 35 species, the majority of which are large (cottonwood) fast-growing species that are useful for producing bioethanol (Kennedy, 1985; Zhu and Pan, 2010). Compared to agricultural residues, woody feedstock has a higher lignin content and a lower ash level (close to zero). Due to its distinctive properties, primarily its high density and low ash content, woody biomass is more favorable for greater bioethanol conversion if recalcitrance is stunned. Since it contains lesser pentoses than agricultural biomass, it is also particularly appealing for transportation that is affordable (Zhu and Pan, 2010). Forestry wastes like sawdust from slashes, sawmills, branches from dead trees, and wood chips have also been used as bioethanol feedstocks (Perlack et al., 2005). The Indian Institute of Science (IISc) has created an electronic chartbook for the Ministry of New and Renewable Energy (MNRE) that offers a perspective on the nation's biomass resources, with a particular focus on their potential for power generation. The Biomass Atlas is a graphical map book that shows the demographic and land use details for many Indian states at the state, region, and taluka levels. Table 14.1 shows the assessed biomass asset and corresponding force potential for the categories of agricultural, forest, and wasteland residues.

14.3 EFFECT OF OPERATION PARAMETERS ON SYNGAS COMPOSITION

The type of gasification agent used has significant influence on the syngas. A higher S/W (steam-to-biomass) ratio means half the weight is heavier, improving the response of water vapor moving to one side, making H_2 more than CO. Improved endothermic gasification reactions result in higher gasification temperatures for H_2 and CO production. Considering the low tar concentration and ambient conditions, the bed temperature must be higher than necessary. When it comes to biomass gasification using pure gas, this is not easy. Also, the high alkali content in the biomass should not exceed the maximum allowable limit, because the alkali content affects the temperature of the gasification; ash also causes slight agglomeration problem (Corella et al., 2008). A long stay of gas in the bed will cause the tar to deteriorate, but this will require a larger bed if the gas leaving the bed must be kept constant. Regardless of the temperature of the gasifier bed, tar production can be reduced, and power output can be improved by increasing B/W. The presence of nitrogenous species in the biomass controls the amount of NH_3 in the syngas (Zhang et al., 2007).

TABLE 14.1
List of Agro- and Forest Residue (State-Wise) in India

State	Agro-Residue Biomass Generation (kT/yr)	Forest and Wasteland Residue Biomass Generation (kT/yr)
Andhra Pradesh	24,871.7	3,601.0
Arunachal Pradesh	400.4	8,313.1
Assam	11,443.6	3,674.0
Bihar	25,756.9	1,248.3
Chhattisgarh	11,272.8	13,592.3
Goa	668.5	180.7
Gujarat	29,001.0	12,196.3
Haryana	29,034.7	393.3
Himachal Pradesh	2,896.9	3,054.6
Jammu and Kashmir	1,591.3	11,461.7
Jharkhand	3,644.9	4,876.6
Karnataka	34,167.3	10,001.3
Kerala	11,644.3	2,122.1
Madhya Pradesh	33,344.8	18,398.2
Maharashtra	47,624.8	18,407.1
Manipur	909.4	1,264.0
Meghalaya	61.1	1,705.9
Mizoram	511.1	1,590.9
Nagaland	492.2	843.8
Odisha	20,069.5	9,370.2
Punjab	50,847.6	398.5
Rajasthan	29,851.3	9,541.6
Sikkim	149.5	531.5
Tamil Nadu	22,507.6	4,652.4
Telangana	19,021.5	1,550.7
Tripura	40.9	1,035.5
Uttar Pradesh	60,322.2	5,478.4
Uttarakhand	2,903.2	4,559.2
West Bengal	35,989.9	1,430.7
Total	**511,040.9**	**155,473.9**

Source: Ministry of New and Renewable Energy Biomass Portal Based on Survey Report (2013).

14.4 SYNTHESIS OF SYNGAS

Syngas, Fischer–Tropsch liquid, methanol, ammonia, etc. are the main products produced from the gasification of biomass. Syngas has a promising future, and it is produced from renewable resources, such as lignocellulosic biomass. The H_2/CO ratio is the main factor affecting process performance. The characteristics of the raw material and the type of gasifier used, which are important factors that affect the production of syngas, are explained in what follows.

14.4.1 Gasification

Gasification is a thermochemical process in which solid biomass is converted into low-heating-value fuel having a calorific value between 1,000 and 1,200 kcal/Nm^3 (kilocalorie per normal cubic meter). The gasification of solid fuels involves the creation of flammable gas, which could be used as a source of energy. In recent times, this process has also been used to modify the liquid hydrocarbon fractions into gases or chemicals which are also known as syngas or producer gas. It is mainly composed of hydrogen (H_2), carbon monoxide (CO), nitrogen (N_2), carbon dioxide (CO_2), methane (CH_4), water vapor (H_2O), and few hydrocarbons in extremely limited quantities, and carbon particles, ash, and tar are examples of pollutants. The standard calorific value of the syngas obtained from various biomasses is approximately 4–10 MJ/m^3, with carbon conversion efficiency of 50–70%. In the gasification process, thermochemical conversion of the solid fuel includes two essential parts, namely, devolatilization and oxidation, where it follows four steps: drying, pyrolysis, oxidation, and reduction, as stated in the following (Demirba, 2002):

- *Drying.* Moisture present in different lignocellulosic biomass ranges from 5 to 35%. At temperatures over 373 K, water is expelled and changed into steam. Then deterioration happens in fuel, which predicts that there ought not be an aggravation to the composition of the biomass during the drying procedure.
- *Pyrolysis.* It is the thermal decay of biomass that happens without oxygen, where the unstable parts of a strong carbonaceous feedstock are disintegrated into different products. Then products are wedged at the working conditions and concoction composition of biomass fuels.
- *Oxidation.* In the oxidation zone, the air is acquainted with oxygen, water fumes, and dormant gases, which are non-responsive, like nitrogen and argon. The heterogeneous response occurs between oxygen noticeable all around and strong carbonized fuel that structures carbon monoxide, which happens at a temperature 975 to 1,275 K. The plus sign indicates the release of heat energy, and the minus sign represents the supply of heat energy, individually:

$$C + O_2 \rightarrow CO_2 + 393.8 \text{ MJ / k mol}$$

The hydrogen present in the fuel reacts with the oxygen present in the air, resulting in the formation of steam.

$$H_2 + 1/2\,O_2 \rightarrow H_2O + 242 \text{ MJ/k mol.}$$

- *Reduction.* Presuming a procedure in gasification using lignocellulosic biomass as crude material, thermal decomposition of lignocellulose components is the essential step with the arrival of some output products, for example, char and volatiles. In the reduction zone, a few synthetic reactions show up at high temperatures without oxygen. The primary gasification reactions in the reduction process are (Demirba, 2002; Maschio et al., 1994) given here:

Boudouard's reactions keep equilibrium between carbon's reactivity and its various gaseous phases of CO and CO_2. It is the reaction of Boudouard, constraining step of the reduction process.

$$CO_2 + C \rightarrow 2CO - 172.6 \text{ MJ / k mol}$$

Steam reaction is given as: $C + H_2O\ CO + H_2 - 131.4$ MJ/k mol

Water-shift reaction is represented as: $CO_2 + H_2\ CO + H_2O + 41.2$ MJ/k mol

Methanation reaction: $C + 2H_2\ CH_4 + 75$ MJ/k mol

In the primary reaction, it is observed that during the reduction process, heat is required. So the temperature of the gas drops at this stage. If the gasification process continues, at the end, after the entire gasification process, all the carbon will be reduced to carbon monoxide, and some other will disintegrate. The remaining unborn carbons will leave as char. In the gasification process, various reactions may occur. For example:

$$C + CO_2 \rightarrow 2CO \text{ and } CH_4 + H_2O \rightarrow CO + 3H_2$$

14.5 TYPES OF GASIFIERS USED FOR SYNGAS GENERATION

The development of biomass gasifiers has taken place in a number of ways. A fixed-bed updraft system, a bubbling fluidized bed system, a fixed-bed downdraft system, and a circulating fluidized bed system can all be classified under biomass gasifiers, which are the four major categories. It is dependent on how the biomass is sustained in the reactor vessel, how a stream of biomass and oxidant is carried, and the way heat is supplied in the reactor that determines differentiation (Rampling, 1993). Figure 14.4 depicts the four most common types of gasifiers used in syngas generation.

14.5.1 Fluidized Bed (FB) Gasifiers

Fixed-bed gasifiers are generally preferred over fluidized bed gasifiers because of the consistent temperature dispersion that is achieved in the gasification zone. In order to maintain a constant temperature, a fine-textured bed material is used, the bed material is fluidized, and biomass feed and combustion gases are snugly blended. Two fundamental kinds of fluidized bed gasifier are used:

- Circulating fluidized bed
- Bubbling fluidized bed

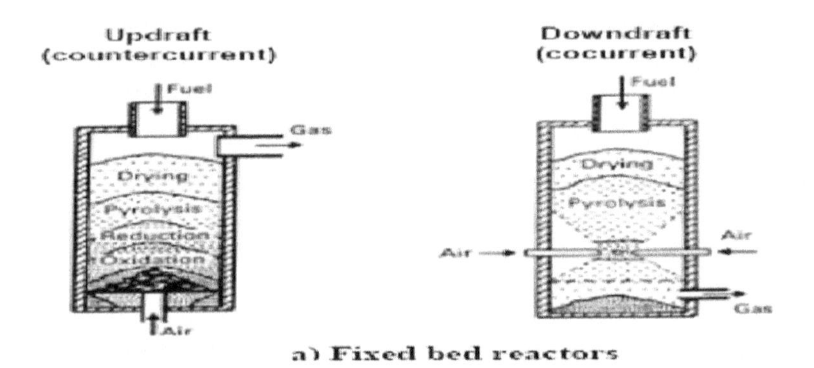

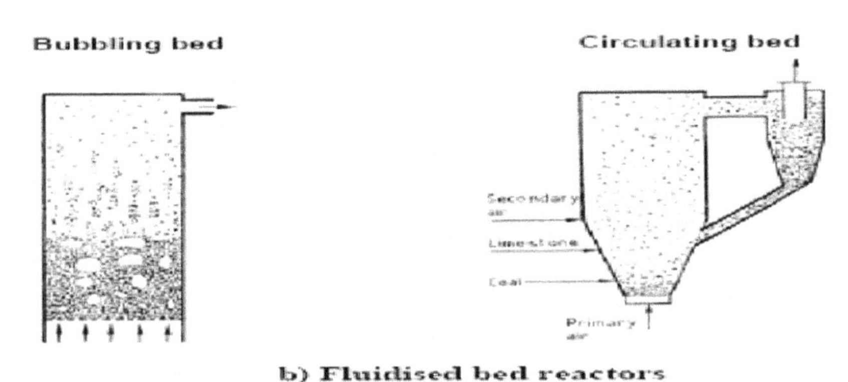

FIGURE 14.4 Types of gasifier.

14.5.2 Fixed-Bed Gasifier

Gasifiers with a fixed-bed mesh help the solid fuel and maintain the reaction zone's stationary nature. Light and medium forces (under 1 MW) are becoming increasingly acceptable for applications involving them because they are relatively easy to send and use. However, keeping the reaction zone at a constant temperature and mixing the right ingredients can be very challenging. Consequently, the pay varies based on the final compositions of the gas fuel. The two primary kinds of fixed-bed gasifiers are:

- Counter-current (i.e., updraft)
- Co-current (i.e., downdraft)

Table 14.2 shows the syngas production (CO/H_2 composition/volume of syngas) of different feedstock, mainly agricultural and forest wastes, when undergoing gasification in different types of gasifiers.

TABLE 14.2
Comparison between Syngas Production from Different Agricultural Waste and Forest Waste

SL No.	Feedstock	Syngas Amount	Types of Gasifiers	References
1	Cedarwood	249 m^3	Updraft	(Ding et al., 2018) (Umeki et al., 2012)
2	Cedarwood	261 m^3	Updraft Fixed	(Zhang and Zheng, 2016)
3	Soya bean stalk and Chinese redwood	CO: 20–25% H_2: 1–1.5%	Downdraft	(Gai and Dong, 2012)
4	Corn straw	CO: 19.81% H_2: 13.51%	Downdraft	(Olgun et al., 2011)
5	Wood chips	CO: 17.63% H_2: 14.95%	Downdraft	(Sheth and Babu, 2009)
6	Rice husk	CO: 14.9% H_2: 13.6%	Downdraft	(Galindo et al., 2014)
7	Rice husk pellet	CO: 20.2% H_2: 18.6%	Downdraft	(Martínez et al., 2011)
8	Eucalyptus	CO: 19.2% H_2: 16.78%	Downdraft	(Bhattacharya et al., 2001)
9	Eucalyptus	CO: 19.04% H_2: 16.78%	Downdraft	Guangul et al., 2012)
10	Charcoal and coconut shell	CO: 20–25% H_2: 13–15%	Downdraft	(Nisamaneenate et al., 2015)
11	Oil palm fronds	CO: 25–30% H_2: 13–15%	Downdraft	(Vyas and Singh, 2007)
12	Peanut shell	CO: 47.2% H_2: 49.3%	Downdraft	(Zainal et al., 2002)
13	Jatropha seeds	CO: 19.259% H_2: 10.619%	Downdraft	(Zainal et al., 2002)
14	Rice husk	CO: 9.50% H_2: 1.98%	Downdraft	(Bhoi et al., 2006)
15	Sawdust	CO: 11.69% H_2: 2.17%	Downdraft	(Bhoi et al., 2006)
16	Babul wood	2.75 m^3/Kg	Downdraft	(Bhoi et al., 2006)
17	Ground nutshell	2.76 m^3/Kg	Downdraft	(Ueki et al., 2011
18	Cashew nutshell	2.76 m^3/Kg	Updraft	(Erlich et al., 2011)
19	Black pine pellets Wood pellet	CO: 27.30% H_2: 5%	Downdraft	(Erlich et al., 2011)
20	Bagasse	CO: 25.7 ± 1.7% H_2: 11.9 ± 1.1%	Downdraft Fixed-bed	(Erlich et al., 2011)
21	Empty fruit branch Wood sawdust pellet and sunflower seed pellet	CO: 23.7 ± 1.2% H_2: 9.9 ± 0.6%	Downdraft	(Simone et al., 2012)

(Continued)

TABLE 14.2 *(Continued)*
Comparison between Syngas Production from Different Agricultural Waste and Forest Waste

SL No.	Feedstock	Syngas Amount	Types of Gasifiers	References
22	Peach wood	CO: 17.4 ± 1.5% H_2: 12.9 ± 0.3%	Downdraft	(Machin et al., 2015)
23	Olive wood	CO: 20.6% H_2: 17.6%	Downdraft	(Machin et al., 2015)
24	Pinewood	CO: 17.7% H_2: 15.0%	Downdraft	(Machin et al., 2015)
25	Wood pellet	CO: 17.4% H_2: 13.21%	Downdraft Fixed-bed	(Montuori et al., 2015)
26	Birchwood	CO: 20.6% H_2: 17.6%	Downdraft	Sarker and Nielsen, 2015)
27	Oak	CO: 27.47% H_2: 7.13%	Downdraft	
28	Poplar	CO: 19.4 ± 3.8% H_2: 11.9 ± 2.3%	Downdraft	
29	Willow Pinewood block	CO: 21.7 ± 5.1% H_2: 9.0 ± 2.3%		
30	Camphor wood chips	CO: 19.7 ± 2.6% H_2: 15.3 ± 2.8%		

14.6 APPLICATIONS OF SYNGAS

Syngas is used for different purposes worldwide. Figure 14.5 shows some major use of syngas in the present day.

14.6.1 Integrated Biomass Gasification Power Plant (IBGPP)

Syngas serves as an energy source for internal combustion (IC) motors, which are connected to a synchronous alternator that generates power. Assuring complete utilization of the waste power entering the gasification plant for end use, the thermal energy created as ignition gases is used for the age of thermal energy, either as boiling steam, water, thermal oil, organic Rankine cycles (ORC), high-temperature water, and so on.

14.6.2 Thermal Energy Generation

Direct burning of syngas in heat-generating equipment, such as steam boilers, concrete ovens, dryers, and so on, to provide thermal energy. This thermal energy can be utilized in a scope of parts, for example, the mechanical segment, in synthetic compounds, cement, nourishment, and so on, just as in tertiary sector, for workplaces, lodgings, private (district heating and cooling), or agriculture (greenhouses).

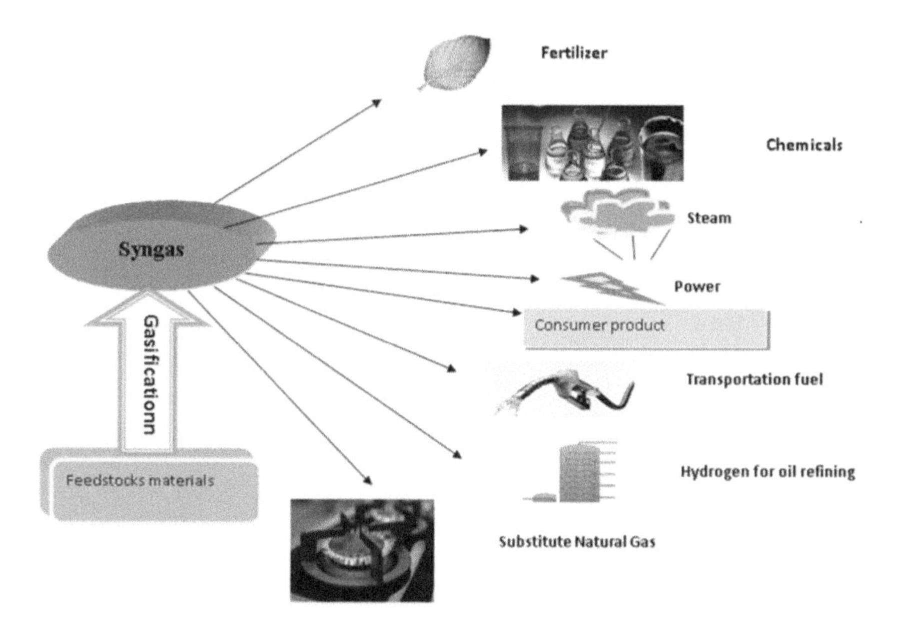

FIGURE 14.5 Uses of syngas.

14.6.3 Synthesis of Chemicals

Syngas is useful for the production of bioethanol, bio-SNG (synthetic natural gas), biochemical (DME, ethylene glycol, methanol), and so on. These are utilized for the production of transportation fuels, chemicals and plastics, fuel infusion into pipelines, etc. The synthesis of some important chemicals from syngas through chemical and biochemical route are discussed next.

14.6.3.1 Methanol

Methanol can be created from fossil or renewable resources and utilized as a fuel for transportation either directly or after being further processed into hydrocarbons. It is created from syngas by hydrogenating carbon oxides over an appropriate catalyst made of zinc oxide, copper oxide, or chromium oxide:

$$CO + 2H_2 \leftrightarrow CH_3OH$$

$$\mathbf{CO_2 + 3H_2 \leftrightarrow CH_3OH + H_2O}$$

Methanol is usually produced in a two-phase system, with the catalyst forming the product phase and the reactants and products forming the gas phase. Germany's BASF took the lead in methanol production from syngas in 1992. The process used a weak zinc oxide/chromium oxide catalyst and required a high pressure of 300 to 1,000 bar and a temperature of about 400°C.

Syngas is used for the production of methanol by methanotrophic microbes. These anaerobic microbes are distributed in diverse ecological niches, including muck, swamps, rice paddies, meadow soils, rivers, oceans, streams, ponds, sediments, sewage sludge, and deciduous woodlands. Several strains of methanotrophs are able to survive in unfavorable conditions, like the acidic peat wetlands of northern Europe and Siberia, the alkaline lakes of Central Asia and Kenya, the hot springs of Hungary, and the soda lakes of Antarctica. They have methane monooxygenase enzyme that transforms syngas into methanol. Further, the obtained methanol is again converted to formaldehyde through methanol dehydrogenase enzyme, which is used to generate a variety of multi-carbon compounds.

14.6.3.2 Ethanol

The following catalysts are mainly used in CO hydrogenation to create ethanol: (1) Cu-based catalysts, (2) Rh-based catalysts, (3) modified Fischer–Tropsch catalysts, and (4) Mo-based catalysts. The most promising catalysts have been those based on rhodium, but their high price and scarcity prevent their application as industrial catalysts. Copper-based catalysts, which are considerably less expensive, are a desirable choice. Here, a copper catalyst causes the direct synthesis of ethanol from syngas. Ethanol synthesis seems to be favored by a low H_2/CO ratio, temperatures in the range of 280–310°C, and pressures in the range of 55–70 bar.

On the other hand, acetogens like *Clostridium carboxidivorans*, *Clostridium autoethanogenum*, *Clostridium ljungdahlii*, *Clostridium ragsdalei*, and *Butyribacterium methylotrophicum* have the ability to utilize syngas for the production of ethanol via the Wood–Ljungdhal (WL) pathway. Here, a unicarbon compound like CO or CO_2 is converted to ethanol through the intermediate acetyl-CoA. They have aldehyde-ferredoxin oxidoreductase (AOR) enzyme, which can reduce acetaldehyde to ethanol.

14.6.3.3 Acetic Acid

The conventional method for converting syngas to acetic acid involves converting syngas to methanol over a copper-zinc-aluminum-oxide catalyst, and then carbonylating the resultant methanol in the presence of a homogeneous rhodium catalyst. In the first process, methanol is directly prepared from syngas by using Cu/ZnO catalyst. Methanol synthesis takes place at a pressure of approximately 5 MPa and a temperature of 200 to 250°C. This reaction is also equilibrium-limited, and cool feed gas is added between catalyst zones to allow higher conversion. Then the methanol is treated with a rhodium catalyst, which forms acetic acid.

Syngas is also used for the production of acetic acid by acetogens through WL pathway. Here, unicarbon compounds like CO or CO_2 are converted to acetyl-coenzyme A. Then, the acetyl-coenzyme reacts with phosphotransacetylase to form 2-acetylphosphate, which is further transformed to acetic acid by acetate kinase. In this process, the CO_2 molecule is reduced to enzyme-bound CO in the carbonyl branch of the WL pathway, while CO_2 or CO molecule is reduced to formate in the methyl branch, which is activated by binding to the coenzyme tetrahydrofolate (THF). Then converted to methyl-THF. Transfer of the methyl group to the iron-sulfur-corrinoid protein liberates THF. These two arms combine in the CO dehydrogenase/acetyl-CoA synthase complex, where CoA, carbonyl, and methyl groups combine to form acetyl-CoA.

14.6.4 BioSNG

Synthetic natural gas, often known as "Substitute Natural Gas" (SNG), is natural gas manufactured from coal or biomass. The typical catalyst for methanation is nickel, and the main reaction is $CO + 3H_2 \rightarrow CH_4 + H_2O$.

Ni-based catalysts are also useful for the hydration of higher hydrocarbons like olefins and the water–gas shift. A water–gas shift reactor is typically used to achieve the required H_2:CO ratio of 3, which is required for methanation. There is no need to modify the H_2:CO ratio externally, because the water–gas shift can occur in some reactor types, such as fluidized beds, concurrently with methanation. Although methanation can occur at atmospheric pressure, higher pressure is preferred by thermodynamics. Gas treatment is crucial before the methanation because catalysts made up of nickel are susceptible to sulfur toxicity. Depending on the operation temperature, the produced gas from almost all gasifiers contains some methane. The production of BioSNG is greatly facilitated by the synthesis gas with high methane content. Since methane is not converted during the methanation process, therefore, indirect gasifiers that contain syngas with a methane level of at least 10 vol% are beneficial for methanation.

14.7 CONCLUSION

Bioethanol production originates from the most essential renewable resources, namely, lignocellulosic biomass, which is economically profitable and environmentally friendly. Bearing in mind the evolvement and development of second-generation biofuel, rice straw emerges as an encouraging and influential candidate due to its rich abundance and well-formed composition. The motivation is that biofuels are starting to build resilience, especially in poor countries with ecological concerns. Since the transition to non-renewable energy sources occurs independently of these concerns, the best way to gain a competitive advantage is through the negative impact of bioenergy production from food. The influence of various parameters on the final combination of syngas was analyzed based on the data.

The effects of biomass type, reactor type, oxidant and reactor operating conditions on the syngas termination map were evaluated. In addition, the properties and process error of the biomass and the design of the gasifier are also important aspects of the gasification process. The appropriate design and alteration of gasifier diminish limitations, thus enhance the efficiency of the gasifiers. Although a huge amount of biomass (agro- and forest waste) is produced in a year, the lion's share is being used to burn it rather than exploit its potentiality in different fields. So the right technique, potential strain, and proper processing technique may be developed for the optimum use of syngas for biofuel production, along with some effective chemicals through both chemical and biochemical route, production of thermal energy, and hydrogen for oil refining.

ACKNOWLEDGMENTS

This work was supported by BPCL-OUAT biofuel project.

REFERENCES

Ahmed I, Gupta AK. Syngas yield during pyrolysis and steam gasification of paper. *Appl Energy* (2009), 86:1813–1821.

American Hardwood Export Council (AHEC) (2002), http://www.ahec.org/hardwoods/species.html.

Badger PC. *Trends in new crops and new uses*, Jannick J, Whipsekey A, editors. Alexandria, VA: ASHS Press (2000), pp. 17–21.

Bhattacharya SC, Shwe Hla S, Pham HL. A study on a multi-stage hybrid gasifier engine system. *Biomass Bioenergy* (2001), 21(6):445–460.

Bhoi PR, Singh RN, Sharma M, Patel SR. Performance evaluation of open core gasifier on multi-fuels. *Biomass Bioenergy* (2006), 30(6):575–589.

Boone RS, Kozlik CJ, Bois CJPJ, Wengert PPJEM. *Dry kiln schedules for commercial woods temperate and tropical.* General Technical Reports FPL_GTR_57. Madison, WI: U.S. Department of Agriculture, Forest Service, Forest Products Laboratory (1988).

Clark JH. Green chemistry for the second generation biorefinery-sustainable chemical manufacturing based on biomass. *J Chem Technol Biotechnol* (2007), 82:603–609.

Corella J, Toledo J, Molina G. Biomass gasification with pure steam in fluidized bed: 12 variables that affect the effectiveness of the biomass gasifier. *Int J Oil Gas Coal Technol* (2008), 1:194–207.

Demirba A. Hydrogen production from biomass by the gasification process. *J Energy Sources* (2002), 24:59–68.

Demirbas A. Bioethanol from cellulosic materials: a renewable motor fuel from biomass. *Energy Source* (2005), 27:327–337.

Deswarte FEI, Clark JH, Hardy JJE, Rose PM. The fractionation of valuable wax products from wheat straw using CO_2. *Green Chem* (2008), 8:39–42.

Ding L, Yoshikawa K, Ismail TM, Abd El-Salam M. Assessment of the carbonized woody briquette gasification in an updraft fixed bed gasifier using the Euler-Euler model. *Appl Energy* (2018), 220:70–86.

Eisberg N. Harvesting energy. *Chem Indus* (2006), 17:24–25.

Erlich C, Fransson TH. Downdraft gasification of pellets made of wood, palm-oil residues respective bagasse: experimental study. *Appl Energy* (2011), 88(3):899–908.

Furimsky E. Gasification of sand coke: review. *Fuel Process Technol* (2006), 56:262–290.

Gai C, Dong Y. Experimental study on non-woody biomass gasification in a downdraft gasifier. *Int J Hydrog Energy* (2012), 37(6):4935–4944.

Galindo AL, Lora ES, Andrade RV, Giraldo SY, Jaén RL, Cobas VM. Biomass gasification in a downdraft gasifier with a two-stage air supply: effect of operating conditions on gas quality. *Biomass Bioenergy* (2014), 61:236–244.

Girio FM, Fonseca C, Carvalheiro F, Duarte LC, Marques S, Bogel-Lukasic R. Hemicellulose. *Bioresour Technol* (2010), 101:4775–4800.

Gomez LD, Clare GS, McQueen-Mason J. Sustainable liquid biofuels from biomass: the writings on the walls. *New Phytol* (2008), 178:473–485.

Guangul FM, Sulaiman SA, Ramli A. Gasifier selection, design and gasification of oil palm fronds with preheated and unheated gasifying air. *Bioresour Technol* (2012), 126:224–232.

Haung HJ, Ramaswamy S, Tschirner UW, Ramarao BV. A review of separation technologies in current and future biorefineries. *Sep Purif Technol* (2008), 62:1–21.

Hoadley RB. *Understanding wood: a craftsman's guide to wood technology*, 2nd ed. Newtown, CT: Taunton Press (2000). https://forumofregulators.gov.in/Data/Meetings/Minutes/TC/24.pdf

Kennedy JHE. *Cottonwood, an American wood.* Washington, DC: U.S. Department of Agriculture, Forest Service (1985).

Kim S, Dale BE. Global potential bioethanol production from wasted crops and crop residues. *Biomass Bioenergy* (2004), 26:361–375.

Knauf M, Moniruzzaman M. Lignocellulosic biomass processing. *Persp Int Sugar J* (2004), 106:147–150.

Mabee WE, Gregg DJ, Saddler JN. Assessing the emerging biorefinery sector in Canada. *Appl Biochem Biotechnol* (2005), 124:765–778.

Machin EB, Pedroso DT, Proenza N, Silveira JL, Conti L, Braga LB, Machin AB. Tar reduction in downdraft biomass gasifier using a primary method. *Renew Energy* (2015), 78:478–483.

Markwardt LJ, Wilson TRC. *Strength and related properties of woods grown in the United States*. Technical Bulletins 479. Washington, DC: U.S. Department of Agriculture, Forest Service, U.S. Government Printing Office (1935).

Martínez JD, Silva Lora EE, Andrade RV, Jaén RL. Experimental study on biomass gasification in a double air stage downdraft reactor. *Biomass Bioenergy* (2011), 35(8):3465–3480.

Maschio G, Lucchesi A, Stoppato G. Production of syngas from biomass. *J Bioresour Technol* (1994), 48:119–126.

Meng N, Leung DYC, Leungand MKH, Sumathy K. An overview of hydrogen production from biomass. *Fuel Process Technol* (2006), 87:461–472.

Mielenz JR. Ethanol production from biomass: technology and commercialization status. *Curr Opin Microbiol* (2001), 4:324–335.

Montuori L, Vargas-Salgado C, Alcázar-Ortega M. Impact of the throat sizing on the operating parameters in an experimental fixed bed gasifier: analysis, evaluation and testing. *Renew Energy* (2015), 83:615–625.

Nisamaneenate J, Atong D, Sornkade P, Sricharoenchaikul V. Fuel gas production from peanut shell waste using a modular downdraft gasifier with the thermal integrated unit. *Renew Energy* (2015), 79:45–50.

Olgun H, Ozdogan S, Yinesor G. Results with a bench scale downdraft biomass gasifier for agricultural and forestry residues. *Biomass Bioenergy* (2011), 35(1):572–580.

Osamu K, Carl HW. *Biomass handbook*. New York: Gordon Breach Science Publisher (1989), pp. 23–46.

Perlack RD, Wright L, Turhollow LA, Graham RL, Stokes B, Erbach DC. *Biomass as feedstock for a bioenergy and bioproducts industry: the technical feasibility of a billion-ton annual supply*. Oak Ridge National Laboratory Report ORNL/TM-2005/66. Oak Ridge, TN: US Department of Energy (2005).

Pettersen RC. The chemical composition of wood. In Rowell RM, editor. *The chemistry of solid wood*. Advances in Chemistry Series, vol. 207. Washington, DC: American Chemical Society (1984), pp. 115–126.

Rampling TWA. Fundamental research on the thermal treatment of wastes and biomass: thermal treatment characteristics of biomass, vol. 208 ETSU B/T1. UK: Harwell Laboratory, Energy Technology Support Unit, United Kingdom (1993).

RFA. *Ethanol industry outlook: climate of opportunity* (2010), http://www.ethanolrfa.org/page/-/objects/pdf/outlook/RFAoutlook2010_fin.pdf?nocdn¼1.

Rostrup-Nielsen JR. Catalysis and large-scale conversion of natural gas. *Catal Today* (1994), 21(2–3):257–267.

Rostrup-Nielsen JR. Steam reforming of hydrocarbons: A historical perspective. *Stud Surface Sci Catal* (2004), 147:121–126.

Sarker S, Nielsen HK. Assessing the gasification potential of five woodchips species by employing a lab-scale fixed-bed downdraft reactor. *Energy Convers Manag* (2015), 103:801–813.

Sheth PN, Babu BV. Experimental studies on producer gas generation from wood waste in a downdraft biomass gasifier. *Bioresour Technol* (2009), 100(12):3127–3133.

Simone M, Barontini F, Nicolella C, Tognotti L. Gasification of pelletized biomass in a pilot scale downdraft gasifier. *Bioresour Technol* (2012), 116:403–412.

Stevens CV, Verhe R. *Renewable bioresources scope and modification for non-food application*. Hoboken, NJ: John Wiley and Sons Ltd. (2004).

Ueki Y, Torigoe T, Ono H, Yoshiie R. Gasification characteristics of woody biomass in the packed bed reactor. *Proc Combust Inst* (2011), 33(2):1795–1800.

Umeki K, Namioka T, Yoshikawa K. Analysis of an updraft biomass gasifier with high temperature steam using a numerical model. *Appl Energy* (2012), 90(1):38–45.

Vyas DK, Singh RN. Feasibility study of Jatropha seed husk as an open core gasifier feedstock. *Renew Energy* (2007), 32(3):512–517.

Zainal ZA, Rifau A, Quadir GA, Seetharamu KN. Experimental investigation of a downdraft biomass gasifier. *Biomass Bioenergy* (2002), 23:283–289.

Zhang W, Söderlind U, Göransson K. *Coal gasification – for LKAB pelletizing furnace*. Department of Engineering, Physics and Mathematics, Mid Sweden University, Sweden (2007).

Zhang Y, Zheng Y. Co-gasification of coal and biomass in a fixed bed reactor with separate and mixed bed configurations. *Fuel* (2016), 183:132–138.

Zhu JY, Pan HJ. Woody biomass treatment for cellulosic ethanol production: technology and energy consumption evaluation. *Bioresour Technol* (2010), 101:4992–5002.

Index

A

AAO (*Aquilaria agallocha* oil), 180
accumulate, 93, 94, 95, 96, 97, 98, 99
acetic acid, 254
acidogenic fermentation (AF), 197, 202, 204
adaptogenic, 34, 37, 38
agar-bit (biologically agarwood-inducing technique), 176, 177, 195
agar-wit (whole-tree agarwood-inducing technique), 176, 177
agarwood policy, 190
agricultural residues, 243, 245
agricultural wastes, 90, 92, 93, 103
agro-waste (AW), 196, 197, 198, 199, 200, 201, 202, 203, 204
air velocity, 6, 16
algae, 54, 55
alkaloids, 3
alternative, 89, 93, 94
ammonia, 48, 51, 53, 54, 55, 56, 58, 59, 159, 162
Ansys Fluent, 2, 16
antibacterial, 35
antibacterial treatment, 157
anticancer, 33, 37
antidepressant, 37
antifungal, 35
anti-inflammatory, 34, 38, 42, 45, 46, 180
antimicrobial resistance, 161
antioxidant, 33, 36, 41
antiparasitic, 36
antiproliferative, 34, 36, 43
anti-stress, 34, 37, 40
antiviral, 35
aquifer, 114, 115, 118, 119, 120, 122, 123, 124, 125, 126
artificial induction, 170, 173, 175, 176, 191
ashwagandha, 35, 36, 37, 38, 44
Aspen Plus, 2, 4, 5, 6, 8, 12, 15, 16
Aspergillus flavus, 37
Aspergillus niger, 37
Ayurveda, 33
Aβ plaques, 37

B

backwash, 19, 20, 22, 24, 27, 29, 31
backwash–backflush, 22, 27, 29, 31
bacteriotherapy, 161
BCD (burning chisel drilling), 175
biochar, 150, 151, 152, 153, 159, 160, 162, 163
biocompatible, 103
bioconversion/biotransformation, 48, 58
biodegradable, 89, 103, 104
biodiesel, 157, 158, 159
bioethanol, 197, 203, 204, 243, 244, 246, 255
biofertilizer, 48, 53, 55, 56, 57, 58, 59
biofilm, 158
biofuel, 48, 53, 54, 242, 243, 244, 255
bio-gas, 150, 159, 160
biological inducer, 177, 190, 191
biomass conversion, 245
bio-oil, 150, 151, 152, 153, 155, 156, 157, 158, 161, 162, 163
bioplastics, 104
biopolymers, 90, 91, 96, 97, 103, 104, 197, 203
bioremediation, 54, 133, 134, 135, 138, 139, 141, 143, 145, 146, 147, 148
BioSNG, 255
biosynthesized nanoparticles, 136
BOD19, 22, 23
bottom zone height, 1, 6, 7, 8, 9, 10, 11, 12, 13, 14, 15, 16
bubble, 1, 2, 6, 8, 10, 11, 12, 13, 14

C

CA Kit (Cultivated agarwood kit), 176, 177
Candida albicans, 37
carbon, 151, 159, 160, 163
carbon emissions, 208, 209, 210, 211
carbon source, 90, 92, 93, 94, 95, 96, 97, 99, 100, 101, 102, 103
carbon tax, 208, 209, 211
cardiovascular, 33, 36, 38
cash crop, 179
catalyst, 153, 160, 161, 162
cell dry weight (CDW), 90, 93, 94, 95, 96, 97, 98, 99
cheap, 90, 98, 99
chemical fertilizer, 48, 50, 53, 56, 57, 59
chemical inducer, 175, 177
chromone, 188, 189, 190, 193, 194
CITES (Convention on International Trade in Endangered Species of Wild Fauna and Flora), 171, 172, 191, 192
cleaning19, 20, 24
climate change, 208, 209
coastal and near-shore ecosystems, 112, 113
computer nitrogen load modeling, 113, 114

condenser, 156
conduction, 155, 158
conventional, 175, 176, 191
conventional septic system, 113, 116, 118, 119, 123
critically endangered, 171
cytokines, 35, 42

D

dairy wastewater, 48, 50, 51, 52, 53, 54, 55, 56, 57, 58, 59
denitrification, 113, 115, 117, 119, 120, 122, 123, 124, 125
dental caries, 157, 158, 162, 167
DGEIS (draft generic environmental impact statement), 114, 129
diabetic, 36, 41
disposal, 89, 93, 94, 99, 104
dolomite, 160, 161
drinking water aquifer contamination, 114

E

EAA (ethyl acetate extract of *Aquilaria agallocha*), 181, 194
eco-friendly, 48, 50, 59
economic, 90, 91, 104
economic importance, 177
effluent, 48, 49, 50, 51, 52, 53, 54, 55, 56
emulsification, 159
endophytic bacteria, 186, 187, 188
endophytic fungi, 169, 181, 182, 183, 184, 185, 187, 193, 194, 195
essential ecosystem services, 113
esterification, 159

F

fast bed, 153
FCC (fluid catalytic cracking), 160, 161
feed solution, 20, 22
feedstock, 90, 92, 93, 94, 96, 97, 100, 101, 102, 103, 104, 150, 151, 157, 162
fermentation, 90, 93, 94, 95
FGEIS (final generic environmental impact statement), 112, 114, 118, 120, 121, 125, 130
fish kills, 114
flow rate, 19, 20, 22, 24, 29
fluidizations, 1, 2, 4, 5, 6, 7, 12, 15, 16
fluidization velocity, 1, 2, 4, 5, 6, 7, 12, 15, 16
fluidized bed, 1, 2, 3, 5, 7, 8, 12, 15
fluorescent carbon nanoparticles, 140
FORDA (Forestry Research and Development Agency), 171, 177
fouling, 19, 20, 22, 23, 24, 29
FPP (farnesyl pyrophosphate), 189
FPS (farnesyl pyrophosphate synthase), 189
FTPEC (flindersia-type 2-(2-phenylethyl) chromone), 189
Fusarium moniliformis, 37
Fusarium oxysporium, 37

G

gasification, 158, 162, 163, 242, 243, 244, 246
gas–solid, 1
glacial moraine, 120, 121, 124, 125
glacial outwash, 120, 121, 123, 124, 125
gray water, 18, 20, 22, 23, 29
green currency, 191
greenhouse gases (GHGs), 196, 199
green inventory, 208, 209, 211
green nanotechnology, 134, 146
green-sensitive demand, 211
groundwater, 113, 114, 115, 117, 119, 122, 123

H

harmful algae blooms, 114
heavy metals, 152, 160, 161, 163
herbal medicine, 33
high-quality, 170, 174, 175
holding cost, 218
hopper system, 156
hydrocarbon, 150, 158, 161
hydro-deoxygenation, 159
hydrodynamics, 1, 2
hydrogen, 152, 159, 161
hyporheic zone, 115, 116, 122, 123, 124, 125

I

immunomodulation, 167
immunomodulatory, 38
incineration, 149, 152, 153, 163
inert gas, 153, 156
intercropping, 179
IUCN (International Union for Conservation of Nature), 171

K

kinetics, 151, 163

L

lactobacillus, 174
leaching lines, 115, 117
leaching pools, 117, 118, 119, 123, 126
lignocellulosic biomass (LCB), 196, 201, 248, 253, 255
liquid biofertilizer, 48, 53

M

MAP (mevalonic acid pathway), 189
MAPK (mitogen-activated protein kinase), 190, 194
MeJA (methyl jasmonic acid), 190
MEP (methylerythritol phosphate pathway), 189
metabolic constituents, 188
metal/metal oxide, 136, 137
methanol, 247, 253, 254
microorganisms, 89, 90, 91, 92, 94, 100, 101, 102, 103, 134, 136, 138, 145
minimum fluidization, 2, 4, 5, 6, 7, 12, 15, 16
most expensive wood, 179
municipal solid waste, 149, 151, 153, 155, 157, 159, 160, 161, 163

N

nanocatalysts, 138, 148
nanoparticles, 37, 134, 137, 139, 140, 143, 146, 148
natural induction, 174, 176, 186
neuroprotective, 34, 38, 45
neuropsychiatric, 36
NFG (nitrogen focus group), 112, 114, 118, 119, 120, 121, 125, 126
nitrogen, 152, 159, 162, 163
nitrogen attenuation, 114, 115, 117, 121
nitrogen loading, 114, 115, 117, 121, 123, 125, 126
non-conventional, 175, 176, 191
non-toxic, 103
NTFP (non-timber forest product), 170

O

oleoresin, 173, 175, 181, 190, 192
OMT (O-methyltransferase), 189
oral biofilms, 157
oral health, 157, 158, 174
organic matter, 157, 162
overall NRE, 115, 116, 117, 123, 124, 125, 126
oxygen, 150, 151, 152, 153, 156, 160, 161, 163

P

P450 (cytochrome P450–dependent mono-oxygenase), 189
PAMP (pathogen-associated molecular pattern), 190
Parkinson's, 38
particle size, 2, 5
pathogenic microbes, 173, 174
peat, 122
PECs (2-phenylethyl chromones), 189
permeate, 19, 20, 22, 23, 24
pharmaceutical, 179, 192
phytochemicals, 33, 34
PKs (polyketide synthases), 189
plastics, 89, 90, 91, 96, 104
pollution, 89, 92, 94, 97, 104, 133, 145, 146, 148
polyhydroxyalkanoates (PHAs), 89, 90, 91, 92, 93, 94, 95, 96, 97, 98, 99, 100, 101, 102, 103, 104
polyhydroxybutyrate (PHB), 90, 93, 94, 97, 203
polypropylene, 158, 161
pore size, 19, 22
potential, 92, 93, 94, 96, 97, 99, 104
pressure, 1, 2, 4, 5, 6, 7, 12, 15
primary treatment, 116
probiotics, 157, 158, 161, 162, 167, 174
production, 90, 91, 92, 93, 94, 95, 96, 97, 98, 99, 100, 101, 102, 103, 104
propellant, 158, 159
PTP (partial trunk pruning), 175
PVC (poly vinyl chloride), 158, 161
pyrolysis, 150, 151, 152, 158, 161, 242, 248
pyrolyzer, 152

R

reactive oxygen species (ROS), 37, 43
reactor, 156, 157, 158, 161, 162, 163
recovery, 19, 20, 24, 27, 31
reject, 22
renewable, 90, 92, 103, 104
renewable plastic, 158
reuse, 19, 20, 29
riparian zone, 115, 116, 122, 124, 125, 126
RO membrane, 19, 20, 22, 24, 27

S

scavenger plant, 118, 119
SCS (selective cutting system), 170
secondary treatment, 115, 117
self-defense mechanism, 188
sensitivity analysis, 218
septic plume, 119, 124
sesquiterpene, 177, 188, 189, 190, 193, 195
SesTP (sesquiterpene synthase), 189
shellfish, 114
single cell protein (SCP), 197, 203, 204
socioeconomic, 174, 179
solid particles, 1, 2, 5
solid volume fraction, 1, 6, 8
steam reforming, 158, 161
steroidal lactones, 34, 35
Streptococcus, 174
sugar, 96, 98, 101, 102
supply chain management, 208, 211
surfactant, 19, 20, 22, 24, 27, 29, 31

sustainable, 91, 94, 97, 103, 104
sustainable development, 39
syngas, 151, 152, 161, 162, 163, 242, 246, 247, 255
syngas fermentation (SF), 197, 202, 203, 204

T

TDH, 2, 4, 5, 6, 7, 15
TDS, 19, 20, 22, 27, 29
terminal velocity, 2
TFs (transcription factors), 190
thermochemical conversion, 149, 150, 157
tooth demineralization, 161
total height, 1, 8, 15
treated, 19, 20, 24
Trichophyton rubrum, 37
turbidity, 19, 20, 22, 24, 27, 29

U

UF membrane, 19, 20, 23, 24, 25
urbanization, 113

V

vadose, 115, 116, 118, 119, 124, 125, 126
value added product, 48, 55
visceral leishmaniasis, 36
vitamin D, 167
volatile, 150, 151, 160, 162
volatile fatty acids (VFAs), 197, 204
volume fraction, 1, 6, 8, 9, 12, 13, 14, 15

W

waste management, 150, 160, 163
wastes, 89, 90, 91, 92, 93, 94, 95, 96, 99, 100, 101, 102, 103, 104
wastewater nutrient cycling, 113
wastewater treatment, 133, 134, 139, 143, 147
water scarcity, 49, 50
Wistar albino, 36, 38, 40, 41
withaferins, 34, 37
Withania somnifera, 33, 34, 35, 36, 37, 38, 39, 40, 41, 42, 43, 44, 45
withanolides, 33, 37
Wood-Ljungdahl (WL) pathway, 203